Ministry of Agriculture, Fisheries and Food

Reference Book 381

An Introduction to farm business management

London: Her Majesty's Stationery Office

© Crown Copyright 1980
First published 1980

ISBN 0 11 241177 0

Foreword

This book is intended as an introduction to the analysis and planning techniques available to help farmers and growers solve their business management problems. It may also be of use to teachers, students and others interested in the subject.

The material and examples shown have been prepared by the farm management advisory officers of the Agricultural Development and Advisory Service. Whilst the financial data used in the examples rapidly become outdated the underlying principles remain fairly static.

The principal contributors are T. W. D. Theophilus, Regional Farm Management Advisory Officer, Eastern Region; J. H. Clift, Farm Management Advisory Officer, Northern Region; P. D. Mills, Farm Management Advisory Officer, National Agricultural Centre, Stoneleigh; K. Butterworth, Regional Farm Management Advisory Officer, Northern Region and R. W. Doel, Farm Management Advisory Officer, Eastern Region.

R. J. Dancey
Senior Farm Management Advisory Officer
Agricultural Development and Advisory Service

Ministry of Agriculture, Fisheries and Food
May 1980

Contents

Page

5   Introduction

7   1. Application of economics to Farm Management

12 2. Records and Accounts for Farm Management
    Farm business records; The balance sheet; Physical Records; Accounts for management purposes; The annual trading account

36 3. Examination of the farm business
    Adjusting the accounts for inter farm comparison; Definition of input and output terms; Calculation of performance measures; Sources and allocation of funds; Gross margin analysis; Records and preparation of gross margin data; Interpreting the gross margin summary

59 4. Planning the farm business
    Identification of the problem; Planning the solution; Partial budgets; Whole farm budgets; Marginal capital; Labour planning; Example of whole farm planning; Cash flow budgets; Programme planning; Linear programming

90 5. Financial control

104 Appendix 1. Farm Financial Analysis
    (Calculation of Annual Cash Flow, Balance Sheet) MA 9

109 Appendix 2. Information required for Farm Management Analysis
    (Farm Trading Account etc.) MA 1

119 Appendix 3. Farm Management Report MA 2

123 Appendix 4. Information for Farm Management Analysis
    (Enterprise Financial Information etc.) MA 4

135 Appendix 5. Multi Stage Cash Flow Projection MA 7

Introduction

In recent years, economic trends have emphasised the need to make the most efficient use of the resources in a farm business. At the same time, widespread experience has been gained on individual farms in various methods of analysing accounts and budgeting for changes in farm systems. It is now widely appreciated that both technical efficiency and good business organisation are needed to obtain satisfactory results.

Four basic requirements for the successful management of a farm business are:
- an understanding of the principles on which the successful operation of a farm business depends;
- a knowledge of the existing position of the farm business derived from the analysis of farm accounts and records;
- the ability to budget long and short-term plans based on a knowledge of what is technically and financially possible on the individual farm;
- institution of a system of record keeping to ensure accurate control of the business (budgetary control).

This book An Introduction to Farm Business Management gives up-to-date information under these four headings to assist advisers and farmers to practise farm management successfully.

In the last few years there have been several developments in the field of farm management, and the sections relating to management methods have been expanded to cover current practice. The method of accounts analysis has become well established since it was described in earlier editions, but it has been critised as being too complicated, concerned over much with the past history of the farm and inter farm comparisons of management efficiency, and as providing too little information when it came to budgeting improvements. The gross margin approach to farm business analysis and planning has certain advantages over the accounts analysis approach, particularly in that the emphasis is placed on the contribution of different enterprises to farm income rather than on comparisons of the farm as a whole with other financial results achieved by similar type and size of farms. Most advisers now regard gross margin analysis as being complementary to the whole farm comparison method of accounts analysis.

Farm business analysis and planning now frequently combine gross margin and accounts analysis techniques. Gross margin data are an essential requirement for the more advanced types of planning, such as programme planning and linear programming.

Care must be taken to ensure that, whatever method is used on the farm, policy decisions are always based on the best available information. Sometimes too simple an approach can only result in a superficial assessment of a farm business, with all the difficulties that inevitably follow. As a general rule, those farmers wishing to apply the methods of analysis and budgeting described will be well advised to get in touch with their agricultural advisory officer who can supplement much of the information in this booklet from his own knowledge of local and regional farming conditions.

1 The application of economics to farm management

There are a number of general economic principles which are directly relevant to the successful operation of a farm business. First and foremost, it is generally assumed that farm businesses are planned for the maximum profit consistent with good husbandry. However, when objectives other than profit maximisation are considered, for example, the satisfaction of family interests, leisure activities or livestock showing and judging, the costs (ie. reduction in potential profit) should be taken into account. Economic principles can be used to indicate the best allocation of resources for attaining the chosen objectives.

Within limits, the organisation of a farm business is decided by the natural and physical conditions of the farm. The Fens in the eastern counties of England have a natural advantage over other parts of the country for intensive arable cropping. Likewise, some lowland areas in the west of England have climatic advantages that favour livestock husbandry, particularly dairy farming. Thin soils, altitude and rigorous winter conditions in some parts of Wales or the north of England impose restrictions on the scope of farming operations and limit agricultural activities to some form of livestock rearing.

In spite of many natural limitations, there is still plenty of choice open to a farmer in the organisation of his farm. The fenland farmers, for instance, may choose between wheat, potatoes and sugar beet or some more intensive crop, such as carrots or celery. The west of England farmer may decide between dairy farming or beef cattle and sheep. The upland farmer has a more limited choice, but he can choose between cattle and sheep, or intensify his business by the introduction of supplementary enterprises. The final choice to be made by a farmer depends upon his objectives, and takes account of his personal preferences, skills and expectations of the returns he can get from the land, labour and capital available to him.

His decision on an appropriate system will be influenced by certain economic criteria:

- how much of each commodity to produce;
- resources to be used in producing each commodity;
- selection and combination of enterprises.

In considering these criteria the farmer is influenced by many factors, such as the maintenance of soil fertility or a personal aptitude for cow keeping rather than egg production.

How much of each commodity to produce

The answer to this question is concerned essentially with the well-known law of diminishing returns. This law is applicable to all types of crop and livestock production. Basically the problem is to decide how much of a particular input or resource should be used. For this to be determined it is necessary to estimate the effects of successive increments of this input upon the total output of the product. In dairying, for example, this would be the relevant problem in deciding how much concentrate to feed; with pig fattening it would be to choose the optimum slaughter weight, having regard to quantities of feed fed; and in crop production to decide on the most economic levels of fertiliser application. The particular point is determined by consideration of both physical input-output relationships and the relative price levels. If, for the sake of argument, the price of milk is 12p a litre and the cost of concentrate feed is £110 tonne then the optimum level of milk output will be different from the situation obtained should the respective prices be 10p litre and £140 tonne. The optimum level of output will be obtained by producing up to the point at which the value of the additional product is just sufficient to cover the cost of the last unit of input of resources. This assumes that the variable input is available in unrestricted quantities.

Resources to be used in producing each commodity

Other things being equal, the cheapest combination of resources should be used to produce a given level of output. Thus, in deciding between a number of alternative inputs of practices, the main consideration is that the least-cost combination of inputs should be chosen. An example of this problem would be the method of harvesting arable crops. A beet grower, for instance, could decide on a comparative cost basis whether it would be cheaper to hire a contractor's beet harvester or own his own harvester; the corn grower could likewise choose between buying a combine or employing a contractor. Also for two animal feedingstuffs which substitute for each other at a constant rate, the least-cost situation occurs when the one with the lower cost per unit of output is used exclusively.

These examples are relatively straight forward. In other cases the choice of a least-cost combination of inputs is more complex particularly when a combination of inputs, rather than two direct substitutes, is involved. For example, a dairy cow may be fed various combinations of hay and concentrates or a pig can be fattened on various proportions of grain and protein supplement. When there is a diminishing rate of substitution between various inputs the least-cost combination—which involves both physical relationships and relative prices—rarely leads to the exclusive use of one or other of the ingredients. In these circumstances, the cheapest combination of resources to produce a given level of output is found when the cost of any resource replaced is just greater than the cost of the resource introduced.

Selection and combination of enterprises

The farmer has first to consider any comparative advantage which he may enjoy for particular lines of production. To exploit such comparative advantage, he should concentrate on those enterprises where the relative net returns are greatest, having regard to yields and prevailing cost and price conditions. For most farmers the availability of land, labour or capital is limited and having selected the enterprises for which he is best suited, the farmer needs to consider allocating his scarce resources among the various enterprises to produce the greatest profit from the farm as a whole. This is determined by the principle of equi-marginal returns: profits are greatest when the last £1 invested in an enterprise earns the same as if invested in any of the other selected enterprises; then, if he allocates his resources in any other way, his profits will fall. Inter-relationships between the various farm enterprises, in terms of their use of resources and possible benefits to one another also have to be considered. These are referred to as the complementary and supplementary aspects.

Fixed and Variable Costs

For the purpose of calculating gross margins (output less variable costs) variable costs are defined as those costs which can both be readily allocated to a specific enterprise and vary in approximately direct proportion to changes in the scale of that enterprise, that is, number of hectares or head of livestock. The main variable costs are seed, fertilisers, sprays, concentrate feedingstuffs and much of the casual labour and contract work. Fixed costs are those costs which cannot readily be allocated to a specific enterprise and/or will not vary in direct proportion to small changes in the scale of the individual enterprises on the farm. Fixed cost items include regular labour, machinery depreciation, rent and rates and general overheads. Fuel and repairs are normally treated as fixed costs mainly because of the difficulty of recording and allocating them but also because they are not usually large enough items to affect the level of gross margin very significantly and their total cost on any farm is little affected when small changes occur in the farm plan.

In farm management work, however, there is a broader interpretation of fixed and variable costs other than that assumed when calculating and using gross margins. In this wider and extremely important context, the distinction between fixed and variable costs depends on the particular situation being investigated and here the length of period under consideration is of paramount importance. This distinction is basic to partial budgeting. In making a decision which affects only a short period of time, for example whether a crop of cabbages is worth harvesting, nearly all costs can be regarded as fixed so far as this decision is concerned, including all the growing costs that have already been incurred; this is because they are inescapable and cannot affect the decision; in this situation what has to be decided is whether the returns are sufficient to cover the extra (ie. variable) costs that will be incurred by harvesting and marketing the crop. At the other extreme, however, an arable farmer who is thinking of starting a dairy herd needs to consider every cost associated with that enterprise, including the annual depreciation of extra buildings and equipment, additional regular labour, and so on; during the time he is reaching a decision these are variable costs to him, in that these will be extra costs that he does not have to incur but which he will have to bear if he decides to start the enterprise; he needs to consider whether the extra returns will be sufficient to cover these costs and leave him a worthwhile return on his investment.

Of the costs referred to as Fixed Costs in Gross Margin analysis, some can and will alter given a large enough change in the farm plan. However, they will not do so gradually but by large amounts at a time such as an extra man may need to be employed or a larger machine may have to purchased. Such resources and other fixed resources such as land, are sometimes referred to as 'flow' resources. Similarly Variable Costs can be turned on and off at will and are sometimes referred to as 'Stock' resources.

Risk and uncertainty

Farmers, like all other businessmen, do not have perfect knowledge about the future and have to formulate their plans under conditions of risk and uncertainty. In decision making no one can say that a decision is right or wrong at the time it is made; only time can prove this. What the businessman is trying to do is to make a good decision as opposed to a bad one, in the light of all the facts available to him at the time.

Budgeted plans or forecasts of the future aim to give a good indication of the likely outcome of a project or of the business as a whole, but the decision maker does not know how often he will obtain this result or how often he will be above or below the budgeted outcome.

In a situation of certainty the farmer would have perfect knowledge about the relevant facts and could forecast with certainty. In practice such certainty rarely occurs, and so he must try to assess the degree of risk and uncertainty there is in his plans.

The distinction between risk and uncertainty is one of degree. Risk is defined as the situation characterised by the fact that a farmer can insure against it such as fire, animal diseases or death, where, through past experience, it is possible to account the probabilities of these happenings occurring. Insurance companies, through their actuarial departments, are concerned with establishing the probability of the situation arising by reference to their considerable amount of experience in past histories. Uncertainty is characterised by the fact that little certain knowledge exists and it would not be possible, from previous experience, to set any mathematical probability about the likely occurrence of the outcome. In framing it may be possible from previous records to arrive at a probability that 3 years out of 10 will be bad, although of course it is uncertain when these bad ones will occur. The future is uncertain too in the situation where a farmer is applying new technology in the early days, or when he moves into a new enterprise which has not previously been undertaken in his area; also regarding prices and costs, which could rise or fall.

The formulation of plans to meet these various types of risk and uncertainty constitutes perhaps the principal function of the farmer as a manager. The differing attitudes of individuals to the acceptance of risk largely amount for varying management decisions under a similar environment. Risks involved in variable crop yields, for instance, may be countered by diversification, but the exact choice of enterprises is largely influenced by the individual's own predictions. Several stages can be distinguished in this decision making process, namely the formulation of predictions, the making of plans and putting them into effect and bearing the likely consequences of the action. An estimation of the latter stage has great significance, and to deal with it the farmer should obtain as much background information as possible about the possibilities, and likely consequences, of a given procedure and take precautions accordingly. Such precautions against risk and uncertainty may include the diversification of crop and livestock enterprises having regard to the available resources and the use of insurance against the various personal and farm contingencies.

2 Records and accounts for farm management

The reason for keeping records and analysing financial accounts is to provide information for tax purposes and to help to improve the returns from the farm or to reduce costs in order to make it a more efficient unit. Records and accounts kept properly can be used to:

- assess tax liability;
- make a critical analysis of the farm business as it existed in the immediate past, which should reveal any weaknesses in organisation and possibly indicate ways of improvement;
- provide information which can be used to assess the probable effect of any changes in organisation which are proposed, or which might be indicated by the analysis of the accounts;
- give information which will show whether plans are being operated correctly;
- provide information on the farmer's track record.

The Trading and Profit and Loss Account

A Trading and Profit and Loss Account records the financial transactions and the resulting Farm Profit or Loss for the accounting period, normally one year. It includes an opening valuation of livestock, crops, cultivations and stocks of purchased materials on the farm at the beginning of the period; the Costs and Trading Revenue for the same period; and a closing valuation of livestock, crops, cultivations and stocks of purchased materials on the farm at the end of the period.

Information required

The basic documents required by an accountant preparing a Trading Profit and Loss Account are:

- Bank statements;
- Paying in slips (receipts) and cheque book stubs (expenses);
- Payments advice, credit and sales notes (receipts) and statements, invoices and delivery notes (expenses);
- The receipts and expenses Analysis sheets.

The bank statement is the evidence that money has passed in and out of the business. Only those transactions which have taken place through the Bank should be recorded in the Analysis sheets. Maximum use should be made of the

Bank since it is the hub of the whole recording system. All cash receipts should be banked and as many payments as possible made by cheque, with adequate detail recorded on the cheque stubs and 'paying in' slips. Details of other transactions should be kept in a petty cash book or cash diary.

The accountant checks whether the merchant's statements, on which receipts and payments were made, were correct, and must see each invoice mentioned on the statement. Invoices should, therefore, be allocated a number and filed with the statement to which they belong and all delivery notes should be affixed to their invoices, as recommended in STL 47 Office Organisation— Business Records. Statements can be filed either in order of payment, month by month; or in separate files for each supplier.

Finally, the accountant will examine the Analysis sheets of Receipts and Expenses to check:

- If all items are entered;
- Whether the entries are in the right columns;
- Whether the totals have been entered correctly.

Farm business records

An adequate system of farm recording is required, such as that provided in The NFU Farm Business Records Book.

It is intended as a simple form of recording which will provide the basic information required for farm management purposes, while also providing the financial records needed by an accountant in the preparation of the Trading and Profit and Loss Account.

The NFU Farm Business Records Book is divided into: Section 1 Physical Records; Section 2 Financial Records for the production of a Trading and Profit and Loss Account or an MAI[1] for farm management purposes; Section 3 Financial Records needed for the preparation of Gross Margin accounts. The farmer should select the sections and forms which are appropriate to his requirements.

The items required by Section 2 Financial Records are discussed first, ie cash analysis of expenses and receipts.

Completing the Cash Analysis Sheets

The Cash Analysis section of a Farm Business Records Book is divided into sheets for 'Expenses' and 'Receipts'. Each entry, whether it is an expense or a receipt, is recorded twice; once as a total in the 'Amount' column and again, on same line, itemised in one or more of the remaining columns to the right under the appropriate heading(s). Thus payments of £708·00 and £285·00 for fertilisers and seed respectively and a combined payment of £350·00 for seed potatoes and tyres would be entered as in table 1.

---

[1] See Appendix 2.

Table 1 Example of payments analysis

Date	Detail	Amount	Cheque No.	Contra	Petty Cash	Livestock Purchases	Fertiliser	Seeds	Machinery Repairs	VAT Charged on Inputs
		£		£	£	£	£	£	£	£
24.9.80	J Bloggs 6 tonnes fertiliser	708·00	12 345	—	—	—	615·65	—	—	92·35
29.9.80	W Smith 1½ tonnes seed Wheat	285·00	12 346	—	—	—	—	285·00	—	—
30.9.80	Oxshire farmers T tyres & 1 tonne seed potatoes	350·00	12 347	—	—	—	—	120·00	200·00	30·00

The 'VAT Charged on Inputs' column in the example does not attempt to record the information in a manner suitable for completing a VAT return. It performs the simple function of providing a column to which VAT items can be assigned in order that the sum of all items appearing on one line will reconcile with the figure in the 'Amount column, thus providing a useful built-in check on accuracy.

Each double page of the section of The NFU Farm Business Records Book for 'receipts' usually contains 20 columns; similarly there are 20 columns on the 'expenses' page. The column headings have been left blank because needs vary from farm to farm. Those which are required can be selected from the lists below. The abbreviations refer to the coding column described later. Example codes are given but the code chosen is entirely a personal matter.

Recommended headings for Analysis Sheets

Expenses

    Contra
    Petty Cash (this may be included with Private Drawings with PC in the coding column)
    Livestock Purchases
    Feed
    Seeds, Plants (P) and Bulbs (Bu)
    Fertilisers and Lime (F), Sprays (S)
    Sundries (expendable items) including Baler twine (BT) Sack hire (SH)
    Veterinary expenses and Medicines (V)
    Wages and National Insurance, Regular Labour (R), Casual Labour (C)
    Machinery, Tractor and Vehicle Repairs
    Fuel and Oil (FO), Electricity (E) and Coal (C), Glasshouse Fuel (GF)
    Contract and Hire (CH), Transport (T)
    General Insurance (G), Vehicle Tax and Insurance (VTI)
    Rent, Rates (Ra), Water (W)
    Property Repairs (to Buildings, Fences, etc)
    Office Expenses and Professional Charges (O), Other

Ownership Expenses (OE)
Capital (Machinery, Buildings, etc)
Private Drawings
Bank Charges and Interest (B), Mortgage (M), Hire Purchase Payments (HP)
Deductible VAT (This is VAT paid by the business on purchases and reclaimable from Customs and Excise)
VAT paid to Customs and Excise

When entering the headings in the book, they may have to be suitably abbreviated.

Grouped items listed above can be split up into more than one column. The principle is that the column headings for the cash anaylsis book are chosen to suit a particular business and the enterprises it includes. It it unlikely that horticultural businesses require columns for livestock and feed and these columns can, therefore, be used for packing materials and bought-in produce; casual labour may also be recorded in a separate column.

On livestock producing farms, the farmer may wish to split up livestock purchases on the following lines:

Date	Detail	Amount	Cheque No.	Livestock Purchases Cattle	Livestock Purchases Sheep
		£		£	£
20.8	Borderway Sales 50 Lambs	800·00	12 345		L 800·00
3.9	Borderway Sales 20 Draft Ewes	580·00	23 456		E 580·00
20.3	Hotton Auction 4 Cows	1 884·00	34 567	C 1 884·00	
27.3	Hotton Auction 2 Heifers	980·00	34 578	H 980·00	

An alternative method is to use code numbers. For convenience, these may be selected from the relevant item code numbers in the summary form. If the records are to be summarised on the ADAS MA1 form (see Appendix 2), the following codes can be used:

Type of Livestock	MA1 Item Number
Dairy Cows	53
Dairy Heifers	55
Ewes	59
Store Lambs	62

At the end of the recording year, the coding column can be readily examined to identify all items coded '53' to summarise the expenditure on dairy cows.

It may be appropriate on a livestock farm to have a column heading 'veterinary expenses and medicines, AI fees and livestock sundries'. Within this column entries for different types of items can be distinguished by a system of coding if required;

Veterinary, medicines, AI fees and livestock sundries

Coding

Commission	C
Vet. charges and medicines	V
MMB Capital levy and co-responsibility levy	M
Dips	DIP
AI fees	AI
Detergents and Dairy Sundries	DS
Livestock Transport	T

Value Added Tax should be shown in separate columns and not included in the price of the item, except where it is not reclaimable (eg. cars).

Receipts
    Contra
    Milk
    Cattle, including guarantee payments
    Sheep, including guarantee payments and Wool (W)
    Pigs
    Poultry and Eggs
    Cereals
    (Put wheat, barley and oats separately if they are the main crops and there are enough spare columns; otherwise use (W) for wheat, (B) for barley (O) for oats in the coding column—see below)
    Potatoes
    Other crops—in separate columns as far as the number of columns allow
    Sundries, including Rents received (R), Wayleaves and Levies (W), Scrap sales (S), Miscellaneous receipts (M), Contract work (C), Keep let (K), Hay and Straw (H)
    Subsidies and grants (other than capital grants) including Hill Sheep (HS), (HS), Hill Cow (HC), FHDS (F)
    Capital Grants
    Capital Sales (including machinery)
    VAT refund received from Customs and Excise
    VAT charged on outputs (eg. wool and second hand machinery)
    Private—Receipts from non-farming activities (eg. share interest, loans, gifts)

The headings will be determined by the main sale products of the farm. Within each column entries for different types of items may again be distinguished by a system of coding.

Where VAT analysis is completed in the cash analysis book, an extra column is required on the expense analysis sheets for taxable inputs (ie. standard rated and zero rated purchases) and two extra columns are required on the receipt analysis sheet for (a) taxable outputs and (b) exempt outputs.

Further points

So far as the total number of columns allows, use separate columns for different crops and types of livestock and for feed purchased for specific types of livestock.

Include in the 'detail' column sufficient information to enable all items to be identified with certainty.

Take great care to ensure that each item is entered on the correct line and in the correct column.

The number of animals in a livestock transaction should be included in the detail column.

Include in the detail column quantities of produce sold.

If there are queries concerning any points, consult the accountant.

Coding

When there are insufficient columns (or when there are likely to be so few entries as not to warrant a separate column) particular items of costs or receipts may be identified by code using the letters shown in brackets in the above lists. These should be entered in the column before the £ columns on the analysis sheets.

Example
Power and Fuel
£
E   203·12
D   110·00
E = Electricity   D = Diesel

A key to the coding should be included on the inside cover of the Account Book.

Hire Purchase

Hire purchase payments on livestock should be included under 'Livestock Purchases', and other hire purchase payments under 'Capital expenditure'. In each case, the code HP should be inserted in the coding column.

Discounts

Some suppliers offer a discount, usually a percentage of the total purchase price. The more common method of recording of discounts is to reduce the cost of all items included in the statement by the percentage stated, so giving an accurate figure for the cost of the item to the business.

Checking entries and totals in analysis columns

To check that analysis sheets have been completed accurately, all columns are added up and totals put into the spaces at the bottom of each completed page. The totals of the 'AMOUNT' and 'CONTRA' columns added together must equal the totals of the columns to the right. When correct, the totals are carried over to the top of the next double page.

Contra Accounts

A contra transaction occurs when an account is settled by setting off one item against another, and only partly involving a cash payment. Difficulties in recording such transactions arise because no money passes; therefore, as far as possible the practice should be avoided. If this is not possible, full details must be recorded separately in the Expenses and Receipts analysis sheet and the invoice carefully preserved because the details of the transaction will not appear in the bank statement. A contra transaction involving the purchase of £1 200 of fertiliser, paid for by the sale of £800 of barley and cheque for £400 would be recorded as follows:

Expenses analysis

Date	Detail	Amount	Cheque No.	Contra	Fertiliser
11 Oct.	J. Smith 10 tonnes fertiliser	400	123 456	800	1 200

Receipts analysis

Date	Detail	Amount	Cheque No.	Contra	Barley
11 Oct.	J. Smith 8 tonnes barley	—	—	800	800

Cash Transactions (including Petty Cash)

The importance of banking all cash income and paying by cheque wherever possible has already been stressed. In good business practice cash payments should be kept to an absolute minimum, but some cash will inevitably be needed for minor day to day business expenses. The level at which these normally run will be known from experience. Petty cash can either be drawn from the bank as a separate item or as part of a sum which also covers wages and possibly private drawings.

Cash received for all sales at the door, etc. should be accurately recorded with details on the receipts side of the petty cash section of the book. When cash is paid into the bank it is recorded in the receipts analysis section as a payment for petty cash. At the end of the year all receipts must also be analysed and allocated to the appropriate columns in the cash analysis section.

Wages

Always draw wages from the bank even though it may not be possible to draw exactly the right amount on every occasion because of differences in the amount of overtime and casual rates to be paid. It is then perfectly acceptable to use the 'petty cash' float to 'make up' any small excess that is needed in addition to the original cheque or to credit to petty cash any amount drawn in excess of the amount of wages paid, provided this is properly recorded in the petty cash book and on the cash analysis sheets.

Household and Private Drawings

The money needed each week for private expenses should be drawn from the bank and not from petty cash. This can be drawn either by separate cheque or as part of the total amount drawn by a single cheque to cover also wages and petty cash.

Payment on Account

Payment on account should be avoided where possible. If the whole amount owing cannot be paid at one time, the statement for which part payment is being made must be identified with the cheque concerned. The number and amount of the cheque should be written on the statement, and the statement number on the cheque counterfoil. The balance outstanding is carried forward and the process repeated until the full amount has been paid.

The Use of the Bank Statement

The bank statement will include, together with all payments and receipts recorded in the analysis book, details of payments made directly by the bank, eg. standing orders, direct debits and bank charges. These must be entered in the expenses analysis section of the records book under the appropriate headings. Similarly, sundry payments made direct into the account, eg. by credit transfer, must be recorded in the receipts analysis section.

A bank reconciliation statement should be prepared at regular intervals, say monthly or quarterly. All entries in the bank statement should be checked against the entries in the cash analysis book, any necessary corrections made and any 'missing' items entered in the cash analysis sheets as explained above. The 'amount' columns for both receipts and expenses should be totalled for the period covered by the reconciliation.

Any cheques sent out and entered in the account book as paid, which have not been presented at the bank, have to be added at the end to obtain the final reconciliation with the actual bank balance. These unpresented cheques automatically cancel themselves out when presented so they do not affect the reconciliation in the following period.

Debtors and Creditors

The annual profit and loss account and any management account must include only those expenses and receipts which relate to the 12 month period covered. On the last day of the financial year there must be a number of transactions which are incomplete but which belong to the trading period that has just ended. For example, purchases made on account during the past month still have to be paid for, and payment will be due for produce sold during that month. In the same way, at the beginning of the financial year, a number of payments will have been made and recorded in the cash analysis book for goods bought and delivered in the previous accounting period. Payment will also have been received and recorded for produce sold during the previous period.

Expenses and receipts for the year can therefore be defined as follows:

Expenses for the year:
Amounts recorded in expense analysis
section of cash analysis book
during the year

    Plus

Amounts owing by the farm at the
end of the year   (Closing Creditors)

    Minus

Amounts paid during year for
items bought in previous year   (Opening Creditors)

and

Receipts for the year
Amounts recorded in receipts
analysis section of cash
analysis book during the year

    Plus

Amounts owed to farm at the
end of the year   (Closing Debtors)

    Minus

Amounts received during the year
for items sold in previous year   (Opening Debtors)

The closing debtors and creditors for the year 1980/81 automatically become the opening debtors and creditors for the year 1981/82. Thus, after the first year of starting to keep accounts, the list of debtors and creditors has only to be compiled once a year.

Once this list has been completed, all the items must be totalled under the same headings as those used in the cash analysis book, and the totals carried to the last page of the book at the end of the financial year.

Valuations

The cash analysis records, after adjustment for debtors and creditors, will provide details of the cash flow through the business, but they are insufficient to enable a full profit and loss account to be prepared because materials, produce, livestock, etc., are continually being held within the business. A valuation of these at the end of the financial year is therefore necessary to make allowances for changes in stocks on hand. This annual valuation automatically forms the opening valuation for the next year. The basis, which may vary according to the purpose for which the valuation is made, should be clearly stated and also be consistent throughout the accounting period.

Saleable crops in store are valued at estimated net market value, ie. market value less costs still to be incurred, such as marketing and storage costs. Market value and costs still to be incurred may be those either at the date of valuation or at the expected date of sale. Saleable crops ready for harvesting but still in the ground should preferably be valued in the same way as 'Saleable crops in store' less the estimated costs of harvesting. Alternatively they may be treated as 'growing crops'.

Growing crops (and cultivations) are valued at estimated accrued cost up to the date of valuation. This may be either at variable costs or at estimated total costs. For most purposes, variable costs are preferable. Residual manurial values need to be taken into account only on change of tenancy. Fodder stocks (home grown) may be valued at accrued variable costs or estimated net market value. If net market value is used, as in presenting whole business data, stocks of non-saleable crops, such as silage, should be valued in relation to hay value adjusted according to quality. Stocks of purchased materials (including fodder) are valued at cost net of any discounts and subsidies.

Livestock, whether for breeding, production or sale, are valued in their present condition at current net market value, including as appropriate any premia or subsidies. Fluctuations in market value which are expected to be temporary should be ignored.

Vehicles, machinery and implements are usually valued at historic cost (net of grants), less accumulated depreciation to date of valuation; depreciation is usually calculated on the reducing balance method. Buildings and fixed equipment are usually valued at historic cost less accumulated depreciation to date of valuation; depreciation is usually calculated on a straight line basis.

Farm accounting in a period of inflation

In a period of inflation the above approaches suffer from a number of shortcomings. In particular, historic cost profit does not equate to 'operating profit' and includes holding gains when prices are rising. Price rises also result in historic cost depreciation not providing adequately for the replacement of fixed assets in cases where such provision is considered desirable.

At the time of going to print, certain interim measures for specific use in farm accounting have been implemented. For example, in the Ministry's Farm Management Survey fixed assets (such as buildings and machinery) are revalued by applying appropriate indices to historic cost data. Depreciation is calculated on the revised figures. Additionally, in the presentation of these accounts, gains or losses resulting from changes in the unit (per head) values of breeding livestock (ie. stock appreciation) are excluded from Enterprise Outputs of the relevant individual livestock enterprises. Net Farm Income is shown including and excluding such gains or losses.

The Farm Trading and Profit and Loss Account

One of the principal reasons for keeping financial records of the farm business is to summarise the transactions which have taken place, and show the profit or loss on the year's working.

Table 2 An example of a typical farm trading and profit and loss account

	£	£		£	£
Opening valuation			Revenue (comprising		
Livestock	35 132		sales, subsidies, grants		
Harvested Crops	2 550		and other income)		73 386
Growing Crops and			Closing valuation		
Tillages	900		Livestock	35 989	
Goods in Store	4 540		Harvested Crops	3 150	
		43 122	Growing Crops and		
Costs (comprising			Cultivations	900	
Livestock purchases,			Goods in Store	5 060	
bought feed, seeds and					45 099
fertilisers, paid					
labour and other farm					
costs, inc. depreciation)	52 537				
Profit for the year	22 826				
		118 485			118 485

Note: Full details are required of all items included in the valuation to enable the accounts to be fully analysed.

Balance sheet

The annual accounts for a farm business will consist of a Balance Sheet and a Trading Profit and Loss Account.

The Balance Sheet is a statement of the assets and liabilities of a business at a specific point in time—usually the last day of the accounting year.

In a Balance Sheet prepared in conjunction with a Trading and Profit and Loss Account, valuations in each should be consistent. Land is normally entered at historic cost or simply omitted. However, for management purposes it may be desirable to prepare an additional Balance Sheet where all Assets are revalued (where necessary) on the basis of current agricultural market values

The schematic layout on page 23 shows how the funds of the business (liabilities of those who have provided funds in the business) have been deployed in the form of assets.

An example of a Balance Sheet is given in Table 3.

Table 3 Balance Sheet as at............

LIABILITIES	£	ASSETS	£
Creditors	2 870	Cash in Hand	120
Bank Overdraft	6 500	Debtors	3 100
Loans	12 394	Cash in Bank	2 880
Tax Owed	8 419	Valuation (Livestock, crops,	
		cultivations and stores)	45 099
Mortgage		Machinery	9 300
Owner Equity (Net Worth)	282 530	Land and Buildings	252 214
	312 713		312 713

The Balance Sheet

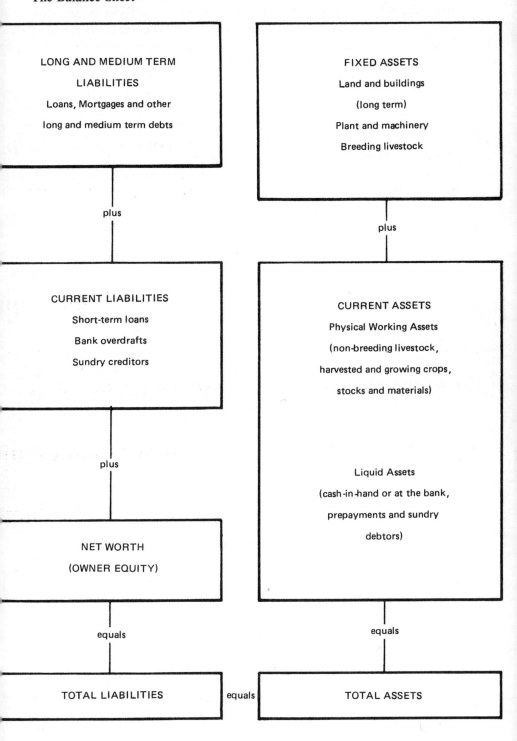

Owner Equity is referred to as 'Capital' by some accountants and an abbreviated Capital Account is often included as part of the Balance Sheet. Table 4 sets out the same Balance Sheet as in Table 3 but incorporating a Capital Account.

Table 4 Balance Sheet as at ....

LIABILITIES	£	£	ASSETS	£
Creditors		2 870	Cash in Hand	120
Bank Overdraft		6 500	Debtors	3 100
Loans		12 394	Cash in Bank	2 880
Tax Owed		8 419	Valuation (Livestock, crops,	
Mortgage			cultivations and stores)	45 099
Capital Account			Machinery	9 300
Capital (Net Worth at			Land and Buildings	252 214
beginning of year)	272 684			
Add: Farm Profit	21 278			
	293 962			
Less: Living Expenses	5 000			
Income Tax Paid	6 432			
Capital (Net Worth at				
end of year)		282 530		
		312 713		312 713

The precise form of a Balance Sheet varies but most broadly follow the pattern already illustrated. It is becoming more common for accountants to present the Balance Sheet in a columnar form as in Table 5.

The main difference in the two layouts is that from Total Assets (the sum of Fixed and Current Assets) Current Liabilities are deducted to obtain Net Assets. The final section then shows how these Net Assets less any Long Term Liabilities must equal Owner Equity (Net Worth).

The columnar form of Balance Sheet is used by the ADAS in their Analysis Form MA9 (a copy of which is included at Appendix 1).

Definitions of Terms Used in Balance Sheets

The following terms are associated with the use of the farm Balance Sheet for management purposes.

Liabilities are the total value of claims on the Assets of a business by the various suppliers of funds to it.

Total Liabilities comprise Long and Medium Term Liabilities.

Current Liabilities and Net Worth

Long Term and Medium Term Liabilities are loans, mortgages and other debts not liable to early recall under normal circumstances. Examples include Agricultural Mortgage Corporation mortgages, bank loans, and private and family loans (whether bearing interest or not). Current Liabilities are claims on the

Table 5 Example of a columnar balance sheet

	£
FIXED ASSETS (at current values, including improvements)	
Land (at current value)	250 000
Buildings and Fixtures	2 214
Machinery	9 300
Breeding Livestock	20 800
TOTAL FIXED ASSETS	282 314
CURRENT ASSETS	
PHYSICAL WORKING ASSETS	
Non-breeding livestock	15 189
Harvested Crops	3 150
Growing Crops and Tillages	900
Stores and Materials	5 060
Sub-Total	24 299
LIQUID ASSETS	
Debtors and Prepayments	3 100
Bank Balance and Deposits	2 880
Cash in Hand	120
Sub-Total	6 100
TOTAL CURRENT ASSETS	30 399
TOTAL ASSETS	312 713
CURRENT LIABILITIES	
Creditors (including HP outstanding)	2 870
Taxed owed	8 419
Bank overdraft	6 500
TOTAL CURRENT LIABILITIES	17 789
NET ASSETS	294 924
LONG TERM LIABILITIES	
Bank Loans	—
Private Loans	—
Other Long Term Loans	12 394
TOTAL LONG TERM LIABILITIES	12 394
OWNER EQUITY (NET WORTH)	282 530

business which may have to be met within a short period of time, usually not longer than a year. Examples include sundry creditors, bank overdrafts and short-term loans. Current Liabilities also include Tax owed.

Net Worth (or Owner Equity) is the Balance Sheet value of Assets available to the owner of the business after all other claims against these Assets have been met. The change in Net Worth between successive Balance Sheets is commonly presented in a capital account which incorporates adjustments for Profit or Loss in the relevant trading period, personal funds introduced (excluding loans) and withdrawals for personal expenditure, taxes paid and off-farm investments, etc. (see Capital Account in Table 4). The change in Net Worth will include the appreciation of business assets, which may be identified separately.

Assets are 'anything of value in the possession of the business or claims on anything of value in the possession of another'. In valuing the assets of a business, several methods can be used such as historic cost; written down value; current market value; cost of production (see 'valuations', pages 20–21).

Total Assets are usually considered under two headings, fixed and current.

Fixed Assets are durable assets representing relatively long-term investments that are used for more than one production cycle. Examples are breeding livestock, plant and machinery (medium-term), land and buildings (long-term).

Current Assets are usually sub-divided into two parts:
- Physical Working Assets comprising temporary assets normally intended for conversion into cash within a short space of time (generally a year or less). Examples are livestock (other than breeding stock), harvested and growing crops, stocks of livestock produce and items of deadstock, such as seed, feedingstuffs and fertilisers.
- Liquid Assets comprising the value of cash either in hand or at the bank, pre-payments and 'near cash' assets such as sundry debtors.

Net Assets are Total Assets less Current Liabilities. The sum of long-term liabilities and owner equity represents the long-term finance of the business and equals Net Assets. (See Table 5).

It may appear confusing to have the farmer's capital balance (owner's equity or net worth) shown as a liability: but it may help if the farmer and the business are looked upon as two separate entities. The capital balance will then be seen as the amount the business owes the farmer.

If a true picture of the progress of a business is to be obtained it is important not to treat a single Balance Sheet in isolation, but to examine a sequence of consecutive balance sheets, so that the trends in the movement of the component figures can be seen. This may be facilitated by the calculation of a series of balance sheet ratios.

Balance Sheet Ratios

In the examination of a Balance Sheet for a business, attention should be focused on its ability to meet its liabilities in the short term; on its medium and long-term stability and its capacity for growth. The method used in such an examination depends upon the use of Balance Sheet ratios. These ratios are not static and they depend a lot upon when the Balance Sheet was struck. The likely changes of ratio constituents and the inter-relationship of separate ratios should be considered when interpreting them. Thus there is no correct figure for any ratio. It is the trend established from a succession of ratios that is most useful in understanding their significance and the trend within a business is much more valuable than comparison with other businesses.

There are many ratios which can be used in Balance Sheet analysis; ADAS has chosen 5 ratios in the MA9 form (see Appendix 1).

These are:
- Owner Equity to Total Assets;
- Fixed Assets to Total Assets;
- Loan Capital to Owner Equity, known as 'Gearing';
- Current Assets to Current Liabilities, known as the Current Ratio;
- Liquid Assets to Current Liability, called the Liquidity Ratio.

These have all been calculated as percentages to avoid the inadvertant reversal of the ratios.

OWNER EQUITY TO TOTAL ASSETS indicates the proportion of total assets which would remain to the owner if the business were liquidated and all outside claims against it met. It is important in establishing the credit worthiness of the business, that is to say its ability to borrow further funds for expansion.

FIXED ASSETS TO TOTAL ASSETS measures the prominence of fixed assets in the composition of total assets. Assuming that they are being adequately utilised, the more prominent the fixed assets feature in the business the better, since it indicates a preponderance of earning assets. However, as with all ratios in isolation it must be remembered that the maintenance of a high proportion of fixed assets at the expense of liquidity is undesirable; it could result in 'over trading' if a high volume of production was undertaken.

LOAN CAPITAL TO OWNER EQUITY—the Gearing Ratio. A business which derives a large proportion of its long-term funds from fixed interest debt capital is highly geared; one which has little fixed interest borrowing in relation to owner's Equity is low geared; where all the funds are provided by the owner it has no gearing. The level of gearing of a business which can safely be undertaken depends upon whether profits are regular or cyclical since interest on loan capital is a first charge on any profits earned by the business.

CURRENT ASSETS TO CURRENT LIABILITIES—the Current Ratio. By expressing the relationship which exists between current assets of the business and its current liabilities, the stability of the business in the shorter term can be assessed. It measures the availability of funds to meet creditors who may present their claims within the next year. Its drawback is that it treats all current assets as being equally liquid. It should always be examined together with the Liquidity Ratio.

LIQUID ASSETS TO CURRENT LIABILITIES—the Liquidity Ratio which is the same as the Current Ratio with 'stocks' (Physical Working Assets) removed from current assets—ie. both partly finished livestock and harvested and growing crops. It indicates the cover afforded to short-term creditors in the form of current assets that can be turned quickly into cash, and is perhaps the best single measure of the liquidity of a business. It should, however, like all ratios be used with caution, for if the 'stocks' vary seasonally, so will its liquid assets. Thus on most farms the period of striking the Balance Sheet will materially affect the result, eg. a Balance Sheet for a cereal farm with large stocks in September would show a completely different ratio to that in March when stocks have been sold.

The method of valuing assets, as has already been stated, materially affects these ratios. In constructing a Balance Sheet for analysis by using ratios it is advisable to value assets at market value. Such a procedure is recommended in the ADAS form MA9.[1]

Physical records

The farm accounts provide a general picture of what has happened on the farm, together with limited information for specific enterprises. They need to be supplemented by additional records, which indicates exactly how the farmer has used certain resources of the farm. These records, which can be very simple, generally cover details of cropping and stocking and concentrated feed consumption, both purchased and home-grown. The importance of keeping adequate feed records, which enable the concentrates used to be allocated accurately to the various livestock enterprises, cannot be too strongly emphasised.

To be of any value, these records must be accurate and not based on estimates of any kind. Many farmers refuse to contemplate simple feed recording because they contend that they know what is happening, having given detailed orders to the staff concerned. However, a small daily error in feeding concentrates can mount up over a period to a substantial quantity, and it is well worth while keeping records to prevent this from occurring. An extra justification for recording concentrate use is the fact that feed conversion ratios are the main factors that determine the profitability of pigs and poultry and, to a lesser extent, of milk production.

There is no single way of recording the information required, but careful thought should be given to the layout of the forms to be used for feed recording.

The NFU Farm Business Records Book provides specific pages for feed recording where the use of home-grown grain and home mixing is practised. Where all the feed for each class of stock is bought as balanced concentrates these can be entered up under separate columns for each livestock enterprise in the Expense Analysis sheets.

Where home-grown grain is fed and home mixing practised, each livestock enterprise should have its own Feed Record Sheet which provides columns for home-grown concentrates, bought grain, other bought concentrates, and bought bulky feed (excluding hay and roots). Recording should be coupled with the regular use of a Barn Feed Record pad. The record slips details are transferred at regular intervals to the appropriate Feed Record Sheet for the class of stock. The Feed Reconciliation Sheet in the Record Book summarises the allocation of feed to each class of stock as shown on the Feed Record

---

[1] For further information the reader should consult The Farm Balance Sheet, its construction and interpretation, by G D D Davies and W J Dunford, Agricultural Economics Unit, University of Exeter report No. 167, September 1967.

Sheets, and reconciles these quantities with opening stocks, home-grown grain production, purchases, sales, grain used for seed, and closing stocks, both for grains and for bought concentrates other than grains.

The section on feed recording is a vital part of a farm recording system where livestock enterprises are involved. No useful analyses of farm performance nor data valuable for planning livestock enterprises are possible unless the proper recording of feed has been undertaken. In addition to this basic necessity for feed recording to produce accounts for management purposes, many farmers wish to keep records to control the performance of individual enterprises. Of course the type of Farm Business Record described above will (if feeds have been currently allocated to appropriate classes of stock) enable some measure of control to be exercised, for example to work out feed cost per litre of milk produced. However, more detailed enterprise records may be required if a serious attempt is to be made at control measures. This applies particularly where concentrate feed represents a major part of costs. There are numerous enterprise check schemes designed and offered by advisory, marketing and commercial organisations as a means of helping farmers to control specific enterprises. They not only provide a regular, eg. monthly, check on performance but can also incorporate a target against which the farmer can measure actual achievement.

The main function of these types of schemes is to provide a comparison between actual and planned performance, so that management can take appropriate steps to recognise deviations from the plan and to take counter-measures sures before the profit margin is eroded. There is little point in the exercise unless this comparison of actual and planned performance can be made.

The use of electronic data storage and computational techniques is obviously increasing the scope for Farm Enterprise Records as the availability of such equipment becomes ever more widespread to the individual farmer. Developments in computing are at such a rapid rate that it would be inappropriate to make more than a passing reference in this book. It is worth summarising here some of the more basic considerations which underline the need and suitability of enterprise records, leaving the details of the storage, retrieval and manipulation of the data to other more technically detailed publications.

Dairy Cows

Because milk producers receive a monthly statement of sales from the Milk Marketing Board production records can be very simple. An estimate of milk used for household purposes, by farm staff and for calf feeding and at the same time the number of cows in the herd, both in milk and dry, should be recorded each month. If concentrate use is also recorded monthly, it will be possible to calculate the production being obtained, per cow per day, from grassland in the summer, and from roughages in the winter. This will not only provide an opportunity for checking the feeding system followed, but will also provide the input-output data required for budgeting changes in the farm programme.

Pigs

The profitability of pig production depends largely upon the feed conversion rate achieved by the fattening pigs, the breeding performance of the sows and gilts, and the length of the fattening period. Records kept for pigs must therefore cover these points adequately. As far as possible breeding and fattening should be treated as two distinct enterprises, particularly insofar as feed used is concerned. Bulky feeds fed should be recorded in addition to the weight of concentrates used.

Poultry

As with pigs, it is desirable to separate the rearing unit from the laying flock. If laying birds are housed in several separate units, then feed used, egg production, mortality and numbers should be recorded for each of them.

Sheep

Unlike the three livestock enterprises considered so far, concentrate use for sheep is usually limited and feed recording should not be necessary. For management purposes the main records required are lambing percentages and details of the replacement rate of ewes. This information can be obtained by keeping simple breeding records for the ewe flock.

Grassland

For management purposes the area of grassland used by grazing livestock and for conservation purposes together with the quantities of seeds and fertilisers used should be recorded.

Arable Crops

For both arable cash crops and arable forage crops, record the area of crop grown, yields per hectare and quantities of seeds and fertilisers used.

Some farmers may have a particular need to keep other records. Advice on these and also on the interpretation of the results can be obtained from ADAS.

Livestock Reconciliation

A Livestock Reconciliation sheet drawn up under the same column headings for the various classes of livestock as the Monthly Livestock Record provides a valuable means of checking the physical records over the accounting year, from the 'Opening Number', through the record of purchases, transfers in, sales, deaths, and transfers out, arriving at the 'Closing Number'.

Example: Dairy Cows

Opening Number	80
+Purchases during year	4
+Transfers in during year	20
Sub-Total (1)	104
Sales during year	21
+Deaths during year	3
+Transfers Out during year	0
Sub-Total (2)	24
Closing Number (1-2)	80

Accounts for management purposes

Production on a farm can be described briefly as follows. Raw materials such as seeds, fertilisers, machinery and labour, are used in conjunction with land and livestock and fixed equipment to produce output. In order to analyse a farm business it is necessary, therefore, to have information about the land and livestock used, the raw materials and services applied to them, and the outputs which are achieved from them.

Financial accounts can supply much of the information required in terms of money cost and value and the farm's Balance Sheet gives information about the capital employed; but accounts must be supplemented by particulars of crop areas, yields and the numbers of livestock kept, hence the importance of detailed physical records.

When this information has been assembled, a number of measures of performance can be calculated as a means of assessing progress on a farm by comparing one year's results with those of previous years. If the effect of seasonal and price differences can be eliminated, this method will help to show if any changes made on the farm have been successful, but it can do little to reveal any important weaknesses in the organisation of a farm. To do this some standards of comparison are required.

Agricultural economists in England, Wales and Scotland collect financial accounts from a large number of farms and use them to provide standards which are derived from the results of a group of farms similar in size and type. The results achieved by a particular farm can be compared with those standards.

This method of comparative analysis will only be valid if like is compared with like. Therefore, the measures for an individual farm must be calculated in the same way and from the same sort of information as the standards of the group. The most satisfactory method of ensuring that the appropriate information is available is to make certain that the financial accounts provide it.

There are however, important limitations to the effectiveness of analysis based on financial accounts. Most farmers have more than one line of production, and, whilst it may be possible to calculate from the accounts the output of the individual enterprises, it is not possible to determine all the inputs used by individual enterprises for two reasons. Firstly, the information in the accounts does not necessarily indicate the use to which all the materials, services, etc. were put; for example, purchase of 5 tonnes of compound fertiliser gives no indication of the crop or crops to which it was applied. Secondly, the production from one enterprise on a farm may be used by another enterprise, barley may be grown and fed to pigs; no record of this internal transfer is shown in the account.

Analysis of the trading accounts cannot provide the information necessary for a critical examination of the individual enterprises. If this is required, a further breakdown of the resources used (variable costs) by the individual enterprises must be made. The gross margin for each enterprise can then be calculated by subtracting the variable costs from the output. When the information about individual enterprises has been assembled it can be compared with suitable standards.

The process of analysing a farm business can therefore be divided into two parts:
- general analysis based primarily upon financial accounts and other appropriate records;
- an analysis of the individual enterprises on the farm in the form of 'gross margins' for each enterprise.

If either or both of these stages are to be undertaken the work can be done much more easily and effectively if a system of accounting, recording and estimating designed to provide the necessary information is introduced from the start. The NFU/ADAS Farm Business Records Book provides, in its section 3—Gross Margins, specific pages for the individual recording and analysis of gross margin data for arable cash crops, dairy cows, other cattle, sheep, pigs, and poultry enterprises, as well as for recording variable costs of grassland and forage crops.

The annual trading account

The three main types of Annual Trading Account in diagrammatic form are illustrated in Table 6:
- Simple Account—No allocation of costs to enterprises;
- Gross Margin Account—Partial allocation of costs to enterprises;
- Full Cost Account—Complete allocation of costs to enterprises.

In each case the output is shown for the individual enterprises, the difference depending on the degree to which the costs are allocated. In the simple account the costs are divided into total 'variable' and total 'fixed' and a total gross margin is shown. In the Gross Margin account only the 'variable costs' are allocated to the individual enterprises and the 'fixed' costs remain as lump sums and the difference between the sum of the enterprise gross margins and the 'fixed' costs is the farm profit.

The generally used system of cash analysis described earlier shows the profit or loss of the farm business as a whole. Such accounts do not however indicate the source of the profit or loss. Even if the farm business as a whole shows a profit, that is no guarantee that all enterprises of the farm have contributed to the profit. The larger the number of departments on a farm, the greater the possibility is that one or more may be unprofitable.

In order to split up the farm accounts to show the profit or loss on separate departments, full cost accounts are necessary. In the full cost account all the costs are allocated so giving a profit for each enterprise, again gross margins for each enterprise are shown as well as the profit. The greater the degree of allocation required, the more detailed the records must be.

Cost accounting plays a vital part in the successful management of industrial operations, where the management wishes to estimate the costs of manufacture of the various products of the factory with considerable accuracy. In this way, management can concentrate production on the most profitable lines of output.

The same techniques may be applied to agriculture with the aim of producing a series of Enterprise Net Margins, each relating to one department or enterprise. To produce such an account it is necessary to keep records of all inputs used in the particular enterprise. If the enterprise is a crop, then it means recording all the labour and materials used; the cost of all fertilisers and seeds purchased are recorded and a careful record made of the quantities and value applied to each field. The allocation of men, machinery and tractor hours, of rent and general overhead expenses is necessary but difficult. Because of the complications of the weather, seasonality of operations, etc. the farm worker carries out a large variety of operations in several departments over a short period so a time sheet system is needed to charge labour to the proper account.

Each day a worker fills in the number of hours, the name of the field, the crop, and other necessary details such as tractor use. Each week the time sheet details of labour and machinery hours are charged to the appropriate department. The bulk of farm purchases and expenses can thus be charged to the various departments and to these will be added an appropriate split of the overhead expenses such as office expenses, car and telephone charges, according to acreage or size of department. The farmer can thus calculate the Net Margin of the enterprise.

The derivation of an enterprise Net Margin may be presented graphically as:

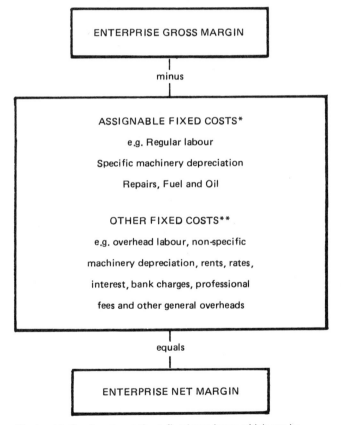

*Assignable fixed costs are those fixed cost items which can be identified as being fairly readily allocated to a consuming enterprise.
** Other fixed costs are those items of Fixed Costs incurred on the farm as a whole which are not readily associated with any particular enterprise.

To carry out cost accounting, a high degree of accounting skill is necessary and the large amount of clerical work required by cost accounting requires a first class farm office organisation.

At first sight, the full cost account would seem to be the most desirable. However, there are two main objections to this method:

- The allocation of general overhead costs can only be made on an arbitrary basis and in some cases it is virtually impossible to make a reasonable allocation.
- The allocation of labour and machinery costs to individual enterprises can be misleading. This is particularly true for arable crops, since it is not the annual cost of labour incurred by the enterprise that is important, so much as the use of labour at peak periods. A saving of labour and machinery costs may not be obtained simply by the elimination of an enterprise, because it may not be practical to make an actual reduction in men and machines.

The system of cost accounting, or complete enterprise costing can also be misleading in the following way. Labour used is costed at the going hourly rate and a share of costs such as buildings, regular labour, is charged, even if these resources and their costs were still there if the particular enterprise was discarded. So although an enterprise which uses spare resources may be shown by cost accounting to be making a loss, it is in fact contributing to total farm profit which would fall if the enterprise was discarded.

The conclusion drawn by most farm economists is that the value of cost accounting on the individual farm is very limited. It is a poor tool for replanning the farm. It is also troublesome, time-consuming and costly. Such costings however have a place for enquiries into the general trends in the profitability of an enterprise, eg. for research or price-fixing purposes.

The value of an analysis of a simple account for farm management purposes is limited to an examination of the outputs of individual enterprises and levels of costs in total and the profits obtained. While there is some correlation between output and profit, it is the relationship between output and costs that is important. Hence the allocation of costs in the gross margin account is the minimum necessary to make a useful assessment of the contribution made to the business by individual enterprises. The recording required is not excessive and of the three it is the most useful method of accounting for farm management purposes, since the gross margin information is very valuable for planning purposes.

Table 6 The three types of annual trading account layout*

	1 Simple Account used in Farm Management Survey and by ADAS								2 Gross Margin Account (Partial Cost Allocation) Used by the ADAS								3 Full Cost Account (Full Cost Allocation)						
	Total	Enterprises						Total	Enterprises							Total	Enterprises						
		A	B	C	D	E	F			A	B	C	D	E	F			A	B	C	D	E	F
OUTPUT																							
Variable Costs: (Allocated Costs)																							
Purchased food																							
Seeds																							
Fertilizers and sprays																							
Vet. and Medicine																							
Haulage																							
Contract work																							
Casual labour																							
Total																							
GROSS MARGIN																							
Fixed Costs: (Unallocated Costs)																							
Regular labour (incl. est of Farmer and Wife's)																							
Equipment depreciation																							
Equipment repairs																							
Fuel and power																							
Rent and rates																							
Building and other repairs																							
Miscellaneous																							
Total																							
PROFIT/LOSS																							

* *Farm Management Account.* Kerr, H. W. T., University of Nottingham, 1968.

3 Examination of the farm business

Analysis of Farm Accounts and Records

A study of past financial results helps to uncover any points of weakness in the organisation and management of a farm business. Ideally, the results for two or three years should be studied, since one year's results may have been influenced by exceptional circumstances. However, where only one year's results are available some guidance can obtained about a farm as a basis for tentative future plans.

Adjusting the accounts for inter farm comparison

Before calculating a number of simple performance measures, for comparison with standards published by the various universities, certain adjustments to the accounts must be made. These adjustments are as follows:

- a charge should be made for unpaid family labour, of sons and daughters, equivalent to the hired labour rate. Amounts included in the accounts for tax purposes of wife's wages, where no physical labour is done, should be deducted;
- in the case of an owner-occupied farm a rental charge should be entered, based on the level of rents paid in the locality;
- any cottage or other rents received should be excluded;
- exclude interest charges of all types;
- check the machinery depreciation item[1]—if the allowances claimed for income tax purposes have been given, these could be significantly different from the true decrease in value. This will be particularly noticeable when the maximum permitted first-year allowance has been taken and could materially upset the inter farm comparisons. It is essential, therefore, to eliminate this high figure from depreciation, or at least be aware that it is inflated. The best way of overcoming the problem is to record separately the new machinery purchased within the year;
- allowances should be made for the private use of farm vehicles and of the farmhouse, and for farm produce consumed in the farmhouse or supplied to workers for which no payment is made;
- exclude non-farming receipts and expenses;

---

[1] See also section 'Farm accounting in a period of inflation'. In the Farm Management Survey, machinery depreciation is now calculated using a 'replacement cost' basis. If the farmer's machinery depreciation is calculated on the usual 'historic cost' basis, it will require adjusting.

- specific subsidies received for livestock and crops should be added to the output of the appropriate enterprises;
- the revenue must be adjusted for any valuation changes in sale products and, in the case of livestock, for livestock purchases;
- expenditure on seeds, feedingstuffs and fertilisers must be adjusted for valuation changes, in order to express the true input of the resource in question during the year;
- expenses and receipts for produce bought for resale should be excluded;
- the value of purchased stores which are sold should be deducted from the cost of that item;
- exclude costs of paid management.

Small charges amounting to only a few pounds can be ignored but any that appear likely to affect the profit appreciably must be taken into account. Composite items, such as 'feeds, seeds and manures' and 'livestock purchases', must be broken down. A well prepared set of accounts makes a farm business analysis an easy undertaking. The valuation should be in sufficient detail to permit the output for each enterprise to be calculated separately.

The ADAS use a standard form MA1 'Information for Farm Business Analysis' (see Appendix 2) which is an expanded Trading and Profit and Loss Account. For management purposes additional information is necessary and facilities for this are provided.

Definition of input and output terms[1]

Once the above adjustments have been made, the outputs and inputs from the accounts can be recast into a management account. The definition of the relevant terms are given below.

Output Terms

These terms generally relate to the value of production of an enterprise[2], or of the whole farm, that is either consumed on the farm or sold, but can in certain cases include sundry revenue such as general subsidies and grants, contracting and wayleaves. Returns are Revenue adjusted for valuation changes. Enterprise Output is Returns for an enterprise plus transfers out and the value of produce used or consumed for which no cash is receivable (by the business) less expenditure on, and transfers in of livestock.

Enterprise Output of a Livestock Enterprise is its Returns, plus the market value of livestock and livestock products transferred to another enterprise (transfer out), plus the market value of any production from the enterprise consumed on the farm, less expenditure on livestock[3] and less the market value of livestock transferred in from another enterprise (transfers in). In general the Enterprise Output relates to the recording year, but where the production cycle differs, it relates to that period.

---

[1] Definitions of Terms used in Agricultural Business Management, MAFF; GFM 21, amended October 1978.

[2] Enterprise—an identifiable sector of the farm or horticultural business for which there are specific potential Returns.

[3] Before deducting any purchase grant.

Enterprise Output of a Sale Crop Enterprise is the total value of the crop produced; it equals Returns from the crop plus the market value of any part of the crop used. When (as is preferable) this is calculated for the 'harvest year', as distinct from the accounting year, there will be no opening valuation.

Enterprise Output from Forage consists primarily of the outputs of grazing livestock enterprises using the Forage Area. In addition it includes keep let and occasional fodder sales, as of hay, together with an adjustment for significant changes in the valuation of stocks of home grown fodder.

Total Farm Output is the sum of the enterprise outputs from sale crops, non-grazing livestock [1] and the Forage Area, plus sundry revenue.

Gross Output (where calculated) is total Returns, plus the value of produce consumed in the farmhouse or supplied to workers, less expenditure on livestock[2], livestock products and other produce bought for resale. Total Gross Output includes the Gross Output of Livestock, crops and items of sundry revenue.

Input Terms

Inputs are resources used in the production process, eg. feed, materials, labour and machinery, measured in physical or financial terms.

Costs. For gross margin and farm accounting purposes they are operating expenses with certain adjustments and be divided into two types:

- Variable costs, and
- Fixed costs.

Variable Costs are defined as those costs which can both be readily allocated to a specific enterprise and will vary in approximately direct proportion to changes in the scale of that enterprise. Examples of Variable Costs are fertilisers (excluding lime), sprays, seed and concentrate feedingstuffs (purchased or home-grown) and, where appropriate, casual labour and contract costs. Purchases of livestock are not treated as variable costs but deducted in the calculation of the appropiate enterprise outputs.

Fixed Costs are those costs which either cannot readily be allocated to a specific enterprise or will not vary with small changes in the scale of the individual enterprises. Examples of Fixed Costs are regular labour, machinery costs, rent and rates, and general expenses. Fuel and repairs are usually treated as Fixed Costs but glasshouse fuel is generally treated as a Variable Cost. Fixed Costs normally include:

- the value of payments-in-kind to workers (if not already included in the earnings figure used);
- the Depreciation on items of capital expenditure including machinery after adding losses and deducting gains arising from the sale of machinery;

and exclude:

- the value of purchased stores used in the farmhouse (eg. coal, electricity), or sold off the farm;

---

[1] eg. veal calves, barley beef, pigs and poultry.
[2] Before deducting any purchase grant.

- allowances for the private use of farm vehicles;
- the rental value of the private share of the farmhouse;
- any labour and materials used in capital projects.

In certain circumstances (eg. in the derivation of Management and Investment Income, see page 43), Fixed Costs are subject to various adjustments.

Adjusted Fixed Costs are Fixed Costs less Interest, Paid Management Costs and Ownership Charges, plus Notional Rent, Unpaid Family Labour and Farmer and Spouse Labour Costs.

Total Inputs equals total Variable Costs plus Adjusted Fixed Costs.

Calculating Outputs

The procedure for calculating enterprise output and costs is demonstrated by the following examples. Firstly the enterprise output of a dairy herd would be calculated as shown on pages 40 and 41.

Calculating Costs

Some of the items on the expenditure side of the account also have to be modified, to express the true cost of the resource during the year. For example, if there were stocks of purchased feed on hand at the beginning and end of the year, the cost of purchased feed would be arrived at as follows:

Expenditure on feed	22 727
Opening valuation	490
	23 217
less: Closing valuation	540
Cost of feed	22 677

Many items of expenditure, such as wages and rent, for which there are no valuations, can be transferred straight from expenditure to costs without any adjustment. Cost and gross output may then be summarised in Table 7 on page 42.

Calculation of performance measures

When the necessary adjustments have been made to the accounts a number of simple measures of performance can be calculated. Their purpose is to: (a) show what economic success is being achieved by comparing them with farms of a similar type and size in the locality; (b) uncover whatever weaknesses or strengths there may be in the present organisation and management of the farm; and (c) suggest ways in which greater financial success might be achieved. These measures of performance must not be studied in isolation from one another; the emphasis must be on the business as a whole. There are other measures which could be used in addition to those described below. The particular measures chosen will depend to some extent on the farm and type of farming, and the type of local efficiency standards available.

Enterprise Output of Livestock
Example Dairy Herd

| SALES | £40 400 (Milk) |
| | £ 7 740 (Stock) |

plus

| SUBSIDIES (including) headage payments) | £NIL |

plus

| TRANSFERS OUT TO OTHER LIVESTOCK ENTERPRISES | £1 050 Calves - Dairy Replacements |
| | £2 600 Calves - Beef Production |

plus

| PRODUCE CONSUMED IN HOUSE etc. | £280 |

plus

| CLOSING VALUATION | £20 800 |

minus

| LIVESTOCK PURCHASES | £1 740 |

minus

| TRANSFERS IN FROM OTHER LIVESTOCK ENTERPRISES | £8 000 Freshly Calved Heifers |

minus

| OPENING VALUATION | £20 000 |

equals

| ENTERPRISE OUTPUT | £43 130 |

Secondly, the enterprise output of a cereals enterprise would be calculated thus:

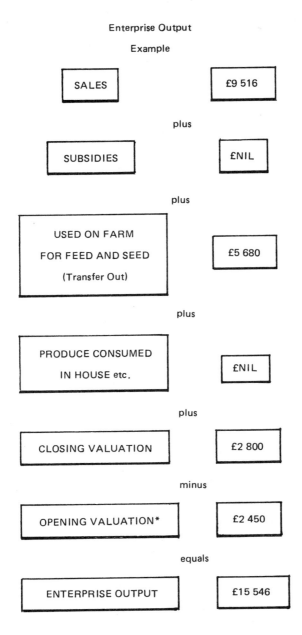

*Where a harvest year or production cycle basis is followed there will be no opening valuation.

SUMMARY OF FINANCIAL RESULTS TABLE 7

	Your Farm of ....100.0.....Adjusted hectares		STANDARD or YOUR FARM LAST YEAR £ per Adjusted ha
ENTERPRISE OUTPUTS	£ Total	£ per Adjusted ha	
Cereals	16,766	167.7	
Other Crops			
Dairy Herd and Milk	43,130	431.3	
Dairy Followers	6,922	69.2	
Beef Cattle	11,525	115.2	
Sheep and Wool			
Pigs			
Poultry and Eggs			
Other Livestock			
Change in Tillage Valuation			
Forage Sales and Valuation Change	250	2.5	
Other Receipts	1,410	14.1	
TOTAL FARM OUTPUT	80,003	800.0	
VARIABLE COSTS			
Feed: Homegrown	5,680	56.8	
Bought	22,677	226.8	
Seed: Homegrown			
Bought	1,504	15.0	
Fertilizer	5,268	52.7	
Casual Labour			
Contract Charges			
Other Variable Costs	5,040	50.4	
TOTAL VARIABLE COSTS	40,169	401.7	
TOTAL FARM GROSS MARGIN	39,834	398.3	
FIXED COSTS			
Regular Labour: Paid	7,200	72.0	
Unpaid	2,000	20.0	
Machinery: Depreciation	2,900	29.0	
Fuel, Oil, Electricity	1,120	11.2	
Repairs, Tax, Insurance, Leasing	2,050	20.5	
Rent and Rates – Paid			
Rent – Notional if Owner Occupier	4,000	40.0	
Other Fixed Costs	3,246	32.5	
TOTAL FIXED COSTS	22,516	225.2	
MANAGEMENT AND INVESTMENT INCOME	17,318	173.1	
Add Back: Notional Rent	4,000	40.0	
Unpaid Labour	2,000	20.0	
Subtract: Financing Charges	2,040	20.4	
Paid Management			
PROFIT	21,278	212.8	

(91434)

The first stage in the analysis is to compare the farm results with the average costs and output figures for comparable farms. These are usually available both as averages of all farms and as above average 'premium' results achieved by the more successful farms in the group. In order to offset differences in size of farm, these standards are usually expressed on a per adjusted hectare basis. When per unit area measures are used, the area basis should be clearly stated.

The **'Adjusted Agricultural Utilisable Area'** is calculated as follows:

To the area of crops and grass are added the 'grass equivalent' of any area of rough grazing land on the farm and the 'grass equivalent' of any area of Hill Rights. Where there is rough grazing land on the farm (or the benefit of Hill Rights) it is usually converted into a 'grass equivalent' on the basis of three hectares of rough grazing (or Hill Rights) equivalent to one hectare of grass. The accounts analysis can then be set out with average or above average results from similar farms shown alongside for comparison: see form MA2 Farm Management Report (Appendix 3). With this overall comparison in mind it is possible, using the MA2, to go through the various aspects of the farm business by considering in turn the following measures of performance.

Farm Profit or Loss is Total Farm Gross Margin less the sum of the Fixed Costs incurred. Alternatively it is Total Farm Output less the sum of Variable and Fixed Costs. It represents the surplus or deficit before imputing any Notional Charges such as Unpaid Family Labour Costs and Notional Rent.

Net Farm Income is Farm Profit or Loss after adding back Interest and Ownership Charges, minus Unpaid Family Labour Costs and Notional Rent. It represents the reward to the farmer and spouse for their own manual labour, management and interest on tenant-type capital invested in the farm, whether borrowed or not.

Management and Investment Income is Net Farm Income minus Farmer and Spouse Labour Cost plus Paid Management Costs. Alternatively it is Total Farm Gross Margin minus Adjusted Fixed Costs or Total Farm Output minus Total Inputs. It represents the reward to management, both paid and unpaid and the return on tenant-type capital invested in the farm, whether borrowed or not.

Return on Tenant's Capital. A measure of performance is the return on tenant's capital. This is calculated by expressing the management and investment income as a percentage of tenant-type (or operating) capital.

It is difficult to measure the total amount of tenant's capital invested in a farm business. A commonly accepted method is to take the average of the opening and closing valuations of machinery, stores (feed, seed, fertilisers, etc.), crops and livestock. This measure gives a relationship between the level of tenant's capital investment and the reward to capital and management. It should be remembered, of course, that the total capital invested is usually greater than the estimate as calculated here. But this calculation is simple and provides a useful guide to the return on the tenant's capital. More accurate estimates can be obtained by calculating the annual average from a number of valuations during the year or by constructing cash flows preferably on a monthly basis.

Total Farm Output per hectare. Since a high proportion of the farmer's costs are fixed, it is important that a sufficiently high level of production should be achieved, in order to obtain an adequate margin over all costs. A useful measure of the level of production, related to type of farm, is the Total Farm Output per hectare. The Total Farm Output of the farm is divided by the adjusted area to obtain the Total Farm Ouput per hectare.

Farm Output less all feed and seed per hectare. The purchase of feeds and seeds is virtually a process of buying additional land and not all farms depend to the same extent on purchased feed and seed. A truer reflection of production from the farm itself can be obtained by calculating the Farm Output less all feed and seed which is derived by subtracting the value of purchases of feed, livestock keep, seeds, bulbs, cuttings and plants for growing on, from Total Farm Output and dividing by the adjusted hectarage. This is a better measure of the use being made of the productive capacity of the land actually being farmed.

Factors Affecting the Level of Output. Output is influenced by three factors:
- Intensity of farming system;
- Yields obtained in individual enterprises;
- Prices realised for products sold.

The intensity of the farm system depends on the proportion of high output enterprises undertaken, for example dairy cows, potatoes and sugar beet have a higher level of output than cereals, beef and sheep. Output is increased as yields and unit prices rise.

An examination of each enterprise using these criteria will show which, if any of them, are responsible for depressing the total output of the farm. By comparing the way the level of output per hectare differs from that of the average farm, it will be possible to suggest ways and means of adjusting the farming system to raise the output and improve the net farm income. This might involve reducing the size of a 'low output enterprise', eg. beef, and increasing the size of a 'high output enterprise', eg. dairying.

Stocking Density. The stocking density is an important factor in determining the level of output on most farms. Because of the different classes of farm livestock, it is impossible to measure the relative density of stocking on farms, unless they are reduced to a common basis of measurement.

Grazing Livestock Units. It is usual to convert all the grazing livestock into 'livestock units' based on assessments of their dietary energy requirements. The average number of animals on the farm is calculated by recording the number on the farm each month and dividing by 12. The number of livestock units on a farm may be calculated from the values given in Table 8 as shown in Table 9.

Adjusted Forage Area. The forage area which includes kale, feed roots, etc. but not home grown cereals retained for feed, should be adjusted in the following ways: add the area equivalent of keep rented and deduct the area equivalent of keep let: deduct the area equivalent of occasional fodder and seed sales[1]: add or deduct the area equivalent of changes in the valuation of stocks

---

[1] Where sales of seed or fodder crops, such as hay, are a regular part of farm policy, they should be regarded as sale crops not forage crops.

of home grown fodder (but changes caused by yield variation attributable to weather conditions, the severity or length of the winter, etc. may be ignored): convert hectares of rough grazing to their grassland equivalent. The following adjustments may also be made where important:
- add the area equivalent of purchased fodder;
- add the area equivalent of catch crops and also of grazing from cash crops of hay or herbage seed.[1]

Table 8 Livestock units co-efficients

Type of Stock	Livestock Unit Co-efficient	Type of Stock	Recommended Livestock Units
CATTLE		SHEEP	
Dairy Cows*	1·00	Ewe and Ewe Replacements	
Dairy Bulls	0·65	(excluding suckling lambs)	
Beef Cows (excluding suckling calf)	0·75	light weight 40 kg	0·06
		medium weight 60 kg	0·08
Beef Bulls	0·65	heavy weight 80 kg	0·11
Other Cattle (excluding intensive beef systems)		Rams	0·08
		Lambs	
0–12 months	0·34	Birth to Store	0·04
12–24 months	0·65	Birth to Fat	0·04
over 24 months†	0·80	Birth to Hoggets	0·08
Barley Beef	0·47	Purchased Stores	0·04
HORSES	0·80	PIGS	
		Sows (including litters to weaning)	0·44
POULTRY			
Over 6 months	0·017	Boars	0·35
Under 6 months (excluding broilers)	0·0044	Pigs fattened per pig fattened during the year	0·09
Broilers	0·0017		
Turkeys (breeding stock only)	0·017	per pig on the farm at any one time	0·17

* The ratios are based on metabolisable energy requirements, with one Livestock Unit being defined as the feed energy allowance for the maintenance of a 625 kg Friesian cow and the production of a 40 kg calf, and 4 500 litres of milk at 3·6% butterfat and 8·6% solids—not fat.

† Reduced in proportion to time animal is on farm.

Adjusted Forage Area per Grazing Livestock Unit. The ratio obtained by relating the number of grazing livestock units (cattle, sheep, horses) to the adjusted forage area, indicates whether or not the farm is understocked in comparison with other farms in the neighbourhood. If the adjusted forage area on the farm illustrated in Table 9 was calculated to be 60 hectares the stocking density would be 1·7 Grazing Livestock Units per adjusted forage hectare.

Feed Conversion. Expenditure on concentrated feedingstuffs is the most important single item of cost on many farms and the extravagant use of concentrated feed is a common cause of low profits. In practice it has been shown that many farmers are either unaware that they are over-feeding, or unecessarily

---

[1] Where sales of seed or fodder crops, such as hay, are a regular part of farm policy, they should be regarded as sale crops not forage crops.

fearful of the effects of reducing the ration of concentrates. To relate livestock output to feed costs the quantity of feed consumed by the different classes of livestock must be known; this involves keeping records.

Example of the calculation of grazing livestock units on a dairy farm rearing replacements

Table 9

Type of Stock	Livestock Units per Animal	Annual Average Number of Animals	Total Livestock Units
Dairy Cows	1·0	80	80·0
Dairy Bull	0·7	Nil (AI)	—
Other Cattle (over 24 months)	0·80	4	3·2
Other Cattle (12 to 24 months)	0·65	19	12.4
Other Cattle (0–12 months)	0·34	19	6·5
Total Grazing Livestock Units			102·1

Note: this calculation does not take into account the farm's 52 Beef Cattle reared indoors all their life, which came under the heading of non-grazing livestock.

As well as the calculation of stocking density, the data on Grazing Livestock Units can be used in a calculation which indicates the actual production from the farm's grassland resources. This routine calculation applies to the dairy herd, dairy followers, beef cattle and sheep. From the Enterprise Output of each class of stock, the total allocated feed costs (including farm grain) are deducted, and the resulting figure is divided by the number of Grazing Livestock Units for that class of stock to give 'Enterprise Output less allocated feed costs per Grazing Livestock Unit'. It emphasises the importance of feed allocation and good physical records generally.

Physical records and feed allocation to the different classes of livestock provides the opportunity for the calculation of further important measures of efficiency, particularly of concentrated feed usage, for example:

Dairy Herd:
  Milk production   litres per cow
                                       value £   per cow
                                         pence per litre
  Concentrate inputs   kg per litre
                                         pence per litre
                                         £   per cow
  Milk Value less Concentrates   £   per cow
Other Livestock:
  Feed Costs per £100 output   Dairy followers
                                           Beef Cattle
                                           Sheep
                                           Pigs
                                           Poultry

Labour and Machinery. Efficiency factors relating to labour and machinery may be calculated and compared with standards, and these comparisons may indicate when labour and machinery use is excessive but give little indication of why it should be so.

The combined labour and machinery costs form the major part of the total costs on many farms, and it is essential that output should be high enough in relation to this expenditure.

Before dealing in more detail with efficiency factors relating to labour and machinery, it is necessary to consider a problem in definition of terms of output. The older approach to expressing farm output comprised two major definitions which in their briefest form may be expressed as:

Gross Output = Sales + increase in valuation, less purchase of livestock
  − decrease

Net Output = Gross Output less purchased feeding stuffs and seeds

Nowadays with the adoption of the terminology 'Enterprise Outputs', the term 'Total Farm Output' (ie. the sum of all the enterprise outputs from sale crops, non-grazing livestock, grazing livestock and any other output from the forage area, plus sundry revenue) is preferred in the official agricultural economists' 'Definitions of Terms used in Agricultural Business Management'. Although a definition of 'Gross Output' is provided, the term 'Net Output' no longer appears in the official list. The position is therefore that 'Total Farm Output (less Feed and Seed)' has taken the place of 'Net Output'. It is still, however, possible to find published standards using the 'Gross Output' and 'Net Output' terminology, and for the sake of brevity here these terms, which are also easier to remember, will be used. In any case there is not uniform consistency in the published standards from agricultural economics departments, and it is always necessary for the user of the standards to check exactly what standard is being quoted.

Labour

Commonly used measures of labour efficiency are the gross and net output per £100 labour cost including farmer and wife labour but excluding paid management. By comparing this figure with the average of farms of the same type and size, a useful indication of the productivity of labour on the farm is obtained. Within certain limits, a high output per £100 labour cost indicates efficiency in the use of labour, and a low figure indicates that labour is not productively employed. On small farms, an increase in the volume of output will usually be the only way of increasing the productivity of labour, since a reduction in the farm staff will seldom be possible.

But on larger farms, where the volume of ouptut is satisfactory, a low output per £100 labour cost may be due to an excessive amount of labour, and consideration should therefore be given to the possibility of reducing the size of the farm staff by better farm organisation.

Machinery

The gross and net output per £100 machinery costs are used to measure the efficiency with which machinery is being used. This figure can be compared with the average of similar farms to obtain an indication of the economy of machinery use. A high output per £100 machinery on the farm may indicate that the farm is under-mechanised. Conversely, a low figure suggests that poor use is being made of the machinery on the farm or that the farm is over-mechanised. If machinery costs are higher than the average, it is useful to examine the level of the individual items making up the total of power and machinery costs to see if any economies could be made.

Machinery costs are difficult to alter without a change in organisation. The biggest single item is often depreciation; this is a legacy of past expenditure which can only be reduced over a period of time and by resolving not to replace certain items of machinery. Farmers should obviously avoid over-capitalisation in machinery, since this results in a high depreciation charge. In considering a replacement programme for machinery, there are three points to be considered. Firstly, from a tax angle, investment incentives on new machines can be an important factor on large arable farms where regular trading-in is practised. Secondly, the physical life of some machinery is considerable and, on small farms in particular, machines may last for a large part of the farmer's working life. Thirdly, purchases in the secondhand machinery market have become a means of reducing costs on small farms.

An excessive repair bill is usually a sign of poor maintenance; the causes of high repair costs are technical, and expert investigation is required to discover ways and means of reducing them. High fuel costs are more amenable to reduction, but some system of recording may be necessary.

Labour and Machinery

These costs can be interacting and therefore it will often be necessary to consider them in combination. Gross or Net Output per £100 Labour and Machinery Costs is the measure normally used.

Additional Measures of Performance. The performance measures described above are sufficient for the preliminary analysis of most farm businesses. In some cases, however, it might be advantageous to calculate some additional measures of performance, for example the 'Farm Productivity' which is the 'Total Farm Output (£) per £100 Total Inputs (including Unpaid Labour Cost)'.

An example of a Farm Management Report (MA2) based on 'Information for Farm Analysis' (MA1—Appendix 2) is given at Appendix 3.

The comparison in this case is with a group of farms in the 'Milk and General Cropping' category. This comparison is a first step in the analysis of the farm business, using the performance measures listed above.

Sources and allocation of funds

In the past two decades considerable progress has been made in modifying the conventional profit and loss account so that it becomes a valuable tool for management. The ADAS has designed and used a standard form for this purpose (see MA1 and MA4 in Appendices 2 and 4). From these modified trading accounts a number of management indices can be calculated and provide some assessment of the management efficiency of the farmer concerned. These indices usually relate to return to land or return to labour and, with the exception of some crude assessment of the working capital, the various aspects of capital appraisal have been ignored. It is not easy to evaluate or to calculate returns so far as capital is concerned, since there is a continuous flow of capital and its use can only be obtained by examining the balance sheet as well as the trading account.

Until very recently records collected by the universities and other organisations have tended to look at the net farm income rather than the capital resources necessary to obtain a certain level of cash income. The balance sheet can and should perform a more fundamental role than that of merely indicating the capital position of a business at any particular point of time, since it can indicate, when used with the profit and loss account, the flow of funds in and out of the business. Pioneering work in this field[1] indicated the method that can be used for calculating sources of funds available and the allocation of these funds.

A farm business which is being expanded or intensified has a great appetite for capital. The cash position may be as important to such a business as the profitability which is indicated in the trading account. The farmer's own personal affairs are usually closely associated with those of the farm business and cash may be removed through payments of tax, personal expenditure, repayments of loans and the purchase of equipment not revealed in the trading account. Particularly in an expanding business this may lead to the situation known as 'over trading' where there is a shortage of cash to keep the business running even though the profitably as indicated by the trading account is satisfactory. Therefore it may often be prudent to determine exactly how the funds which become available during the year have been used. Cash which may have become available through the sale of equipment, by running down stocks, by recovery of loans (eg. from sundry debtors) or from outside sources and all in-goings and out-goings are taken into account when drawing up the balance sheet. This note shows how the information contained in the specimen balance sheet shown in Table 4 can be adjusted to show clearly what has taken place.

Because of the fact that the trading account includes opening and closing valuations of livestock, crops and stores; depreciation of machinery and buildings; and adjustments for debtors and creditors, the trading profit figure may bear little resemblance to the net cash flow for the same year. Furthermore, the trading account does not include outlays on capital assets, nor does it incorporate drawings for personal use, nor again either the repayment or raising of loans and the payment of tax. These are additional potentially powerful influences affecting the overall net cash flow position, which can cause further divergence between it and trading profit.

---

[1]Davies, G D D and Dunford, W J, The Farm Balance Sheet, Its Construction and Interpretation; Department of Agricultural Economics, University of Exeter, Report No. 167, September 1967.

The funds retained in or removed from the business during the year can be determined quite simply by deducting the opening from the closing capital (Owner Equity) as follows:

	£
Closing capital	282 530
Opening capital	272 684
Retained in the business	9 846

This sum can also be derived from the Farm Profit as follows:

Farm Profit		21 278
Income tax (on previous year's profit)	6 432	
Living expenses	5 000	
		11 432
Retained in the business		9 846

Both these calculations show the funds retained in the business but neither of them indicate how they have been used. This can be done by setting out the sources of funds available to the business and the allocation of these funds within it in a columnar form of presentation in Table 10 on pages 52 and 53.

Gross margin analysis

An alternative or complementary approach is to regard the farm business as a group of productive enterprises, each with its own output and direct or variable costs. These enterprises are centred upon the general farm, which provides common services for all the enterprises and the necessary co-ordination of processes. The main division of costs is, therefore, into variable costs, ie. those which can be allocated to an enterprise and which will alter in size as the scale of the enterprise is changed, and those overhead or fixed costs which cannot be easily allocated to an enterprise and therefore belong to the farm as a whole.

Costs

Farm costs may be divided into fixed and variable costs.

Variable Costs are defined as those costs which can both be readily allocated to a specific enterprise and will vary in approximately direct proportion to changes in the scale of that enterprise.

Examples of Variable Costs

Crops	Livestock
Seed	Stock purchases
Fertiliser	Concentrates
Casual labour	Casual labour
Contract services	Contract services
Sprays, etc.	Sundries (vet, etc.)
Sundries (twine, etc.)	

Fixed Costs are those costs which either cannot be readily allocated to a specific enterprise or will not vary with small changes in the scale of the individual enterprises.

> Examples of Fixed Costs
> Regular labour
> Unallocated casual labour
> Paid Management and secretarial
> Machinery repairs and leasing charges
> Fuel and electricity
> Machinery depreciation
> Rent and rates
> Depreciation of buildings and fixtures
> Sundries (water, insurance, office, phone, professional fees, etc.)
> Financing charges

Machinery repairs and fuel are included in fixed costs, because it is too complicated to allocate these to individual enterprises.

Output

In addition to the division of costs it is necessary to allocate output to the various enterprises. For cash crops this presents little difficulty, though grain and pulse crops fed on the farm are credited with the sale value of the product for analysis and planning purposes. For example, if barley is grown and some of the production is fed to fattening pigs, the sale value of the barley fed to the pigs would be included in the output of barley, and the value of that barley would be regarded as a variable cost to the fattening pig enterprise.

In calculating the output of crops, one is concerned with the 'crop year' rather than the accounting year. For instance, on a farm where accounts are kept on the basis of a Michaelmas valuation, most of the sales of barley recorded between October 1980 and September 1981 will relate to the harvest of August 1980 but the variable costs will relate to the crop being grown for August 1981. In such cases, it is important to relate costs and returns for the same crop, hence the concept of a crop year. The problem can best be met by keeping a separate record of sales for crops at each harvest. Where this is not possible, the best estimate should be made of the value of crops produced during the crop year under review and relate it to the appropriate variable costs. Forage crops have no direct cash value, and hence no direct output, but have only variable costs.

Output from livestock presents no problems, being a direct calculation from sales corrected for valuation changes or livestock purchases during the period concerned.

TABLE 10

Sources and Allocation of Funds

CALCULATION OF ANNUAL CASH FLOW

SECTION A : CALCULATION OF CASH FLOW FROM TRADING

	±	£
PROFIT/LOSS (from MA 1, MA 16 or Accounts)		21278
ADD: Depreciation of Machinery		2900
Depreciation of Buildings and Fixtures		246
Decrease in Valuation of: Livestock		
Harvested Crops		
Growing Crops and Tillages		
Stores and Materials		
Decrease in Debtors		
Increase in Creditors		210
Notional Payments		
Sub-Total (a)		24634
DEDUCT: Notional Receipts		1080
Increase in Valuation of: Livestock		857
Harvested Crops		600
Growing Crops and Tillages		
Stores and Materials		520
Increase in Debtors		154
Decrease in Creditors		
Sub-Total (b)		3211
CASH FLOW FROM TRADING (a)−(b)=(c)	±	21423

SECTION B : CAPITAL TRANSACTIONS

	±	£
ADD: Receipts - Disposals of Land and Buildings		
Receipts - Disposals of Machinery and Fixed Equipment		2000
Capital Grants		
Sub-Total (d)		2000
DEDUCT: Purchase of Land and Buildings		
Improvements to land and Buildings		2460
Purchase of Machinery & Fixed Equipment		8000
Sub-Total (e)		10460
CASH FLOW FROM CAPITAL (d)−(e)=(f)	±	(−) 8460
BALANCE AVAILABLE FOR TAXATION AND PRIVATE USE (c)+(f)=(g)	±	12963

Balance available (brought forward)(g) | +/− | 12963 |

SECTION C : PRIVATE

    ADD: Funds Introduced ....

        Private Receipts ....

            Sub-Total (h) | 0 |

    DEDUCT:
        Private Drawings ....5000..........

        Tax ....6432..........

        Off farm Investments ....

            Sub-Total (i) | 11432 |

            Total (h)−(i)=(j) | +/− | − 11432 |

TOTAL SURPLUS/DEFICIT OF FUNDS FOR YEAR (g)+(j)=(k) | +/− | 1531 |

SECTION D : FINANCIAL SUMMARY

    ADD: Increase in Loans ....

        Increase in Overdraft ....1637..........

        Reduction in Credit Balances and Cash ....

            Sub-Total (l) | 1637 |

    DEDUCT:
        Reduction of Loans ....106..........

        Reduction in overdraft ....

        Increase in Credit Balances and Cash ....

            Sub-Total (m) | 106 |

            Total (l)−(m)=(n) | +/− | 1531 |

Total (n) must equal Total (k).

Gross Margins

The Gross Margin of an enterprise is its Enterprise Output less its Variable Costs. They are calculated for each enterprise by subtracting the variable costs from the relevant output. The gross margin can be expressed on a per hectare or per head basis for analytical purposes, or left as the total for the enterprise as a whole. The fixed costs of the farm business are subtracted from the total of the gross margins for all the enterprises to give a profit or loss.

The Gross Margin of a Grazing Livestock Enterprise is the Enterprise Output of that enterprise less the Variable Costs, including the allocated variable costs of grass and other forage. [Note however that in some situations the Gross Margin of a Grazing Livestock Enterprise (before deduction of forage variable costs) is calculated where it is considered inappropriate to allocate forage variable costs.]

The treatment of forage-consuming livestock differs from that of the non-grazing livestock because, in addition to the usual variable costs of feed, vet. and medicines, they must also bear the variable costs of the forage crops they consume. As a result, the livestock gross margin can be regarded as the product of the forage area and the value of such gross margins can be expressed on a per hectare basis, to be directly comparable with data for the cash crop area.

The most precise definition of the Gross Margin from the Forage Area is the sum of the Grazing Livestock Enterprise Gross Margins (before deduction of forage variable costs) plus sundry revenue from keep let and occasional sales of fodder, eg. hay[1], together with an adjustment for significant changes in the valuation of stocks of home-grown fodder, less:

    the variable costs of grass and other forage crops;
    unallocated purchased roughages and rented keep.

The diagram below represents this allocation of costs and output, with the resultant gross margins and farm income.

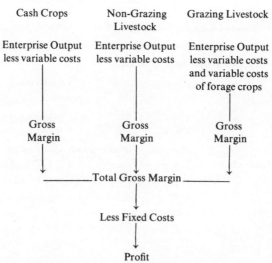

[1] Where sales of seed or fodder crops such as hay are a regular part of farm policy, they should be regarded as sale crop enterprises.

Adding together the Gross Margins for each enterprise gives the Gross Margin for the farm business as a whole. By deducting the total fixed costs from the total gross margins the Profit is obtained.

If fertiliser, seed, feed and casual labour costs can be allocated to each enterprise, the Annual Accounts can be drawn up in a Gross Margin form (MA4—see Appendix 4).

Records and preparation of gross margin data

The first step is to record the necessary data to allow for the calculation of the gross margin of each enterprise and the fixed costs of the farm.

For cereals and other cash crops the variable costs and outputs for each enterprise on the farm should be recorded on a harvest year basis. Often in a financial year's records, costs and outputs will relate to more than one harvest year. The costs of fertilisers, seeds, etc. incurred by crops still in the ground at the end of the financial year should be excluded from the gross margin records, but those costs incurred in the previous financial year which relate to the current harvest year must be included. The quantities fed to livestock and used for seed should also be valued at market price and added to the actual sales of the crop to arrive at the output for the crop year.

If the recording year ends before all the crops have been disposed of the remaining stocks should be valued at market price. However, almost invariably, when these stocks are sold or fed to livestock on the farm, their quantity or realisation value differs from the year end valuation. This difference between the valuation and realisation should be separately shown. Provision for this calculation is made in form MA4.

With grassland and forage crops it is not necessary to record the variable costs separately for each harvest year though, if the financial year should end, for example, at 31 March, it is more accurate to carry forward the fertiliser applied in spring for the following season's crops. In such cases this should be shown as part of the valuation of growing crops.

Where hay is grown for sale as part of the farm policy, the area of grassland so used should be treated as a cash crop in the same way as cereals and arable cash crops. When the aftermath is grazed by livestock, then a proportion of the variable costs should be charged to the forage.

For livestock enterprises it is essential to allocate purchased feedingstuffs and home grown cereals separately to each enterprise. Where dairy herd replacements are reared they should be treated as a separate enterprise from the milking herd. Downcalving heifers transferred from the rearing enterprise to the dairy herd should be valued at market price, credited to the output of the rearing enterprise and charged to the dairy herd. Likewise calves retained for rearing should be credited to the dairy herd and charged to the heifer rearing enterprise. Unless this is done, comparison of gross margins from the various enterprises will not be possible. Similarly with other types of livestock if more than one system is practised on a farm, each system should be treated as a separate enterprise. The variable costs of purchased feed and home grown cereals, livestock purchases or transfers (as referred to above), veterinary costs and sundries should be allocated to each enterprise separately.

The allocation of feedingstuffs to the various livestock enterprises can be a complicated process especially where home mixing is practised. The records required must be accurately kept to ensure that the total feedingstuffs allocated to the various livestock enterprises reconcile with purchased feed and home grown cereals consumed. The NFU Farm Record Book provides most of the forms necessary for recording gross margin data.

Where no special records have been kept in the past, it is possible to arrive at a rough estimate of enterprise performance from an ordinary set of accounts and the information a farmer could be expected to have available in his farm office. The normal variable costs of crops can be estimated by working out the usual use of fertilisers, seed, etc. and pricing the quantities. Similarly, the output from crops, based on normal yields and standard prices, can be estimated.

With livestock, the variation in the use of concentrates can be so great that such estimates are highly dangerous. Where one enterprise is very much larger than all the others, an arbitrary allocation can be made without much loss of accuracy. There are other cases where stock may be fed quite differently; for instance, pigs only on home-grown cereals and protein, and cows only on purchased balanced cake. In such cases an examination of the latest accounts, together with the relevant invoices, should enable a fair estimate to be made of the variable costs to be set against the relevant output.

Where such estimates cannot be made safely, it is better to admit that there is a serious gap in the management records kept by the farmer and to set about providing suitable records for the future, rather than to base the analysis and planning on suspect data.

A summary of gross margins and fixed costs has been prepared on Farm Management Analysis form MA4 and is shown in Appendix 4.

Interpreting the gross margin summary

Once the gross margins of each enterprise and the fixed costs of the farm have been calculated, an examination of the results for the existing organisation can be undertaken.

Cash Crops

If the gross margin per hectare of a cash crop is low, it is almost always due either to low yields or low price per unit rather than excessive variable costs, though the level of these should also be checked. Low physical yields are due to poor day-to-day management or a technical reason, such as poor cultivations, drainage, deficient soil, wrong seed or fertiliser. Low price will be due to poor marketing, grading or wrong variety.

Livestock

Here, low gross margins can be due not only to low output, the result of yield or price, as above, but also to excessive inputs, either high variable costs, especially feedingstuffs, or low stocking density. Low physical yields will again be due to breed or a technical fault, such as incorrect feeding. Low prices may be due to aiming at the wrong market, poor finishing or lack of selling ability. High variable costs are usually due to excessive concentrate feeding, though more rarely excess costs occur amongst the 'sundries', especially on the equipment and veterinary side. High feed costs may be due simply to a failure to keep a check upon the rate of feeding, or it may be due to some genetic cause or physical environment.

In grazing livestock enterprises stocking density (ie. grazing livestock units per adjusted forage hectare) can be the most important factor. Low stocking density may be the result of failure to obtain high production from forage crops, lack of capital to stock the land or a low standard of husbandry.

It must be borne in mind that concentrate feeding and stocking density are dependent on one another. A low gross margin per head, due to high feed costs, may be more than counter balanced by a high gross margin per hectare due to a high stocking rate. Whether this is the correct policy depends upon the price levels of feed and the relevant output, and also upon the possible alternative uses of the resources, especially the land and working capital which are involved.

Care should be taken in interpreting livestock gross margin per hectare because of the difficulty in allocating the area used to different classes of livestock. The usual method of allocating the area of forage is on a livestock unit basis, expressing the results as grazing livestock units per forage hectare. It may be more advisable to obtain the gross margin without allocating forage costs.

Once the cause of a low gross margin in any particular enterprise has been found, the farmer can consider whether it is possible and economic to remedy the fault, in order to bring the level of that enterprise up to the accepted standard. But this belongs more properly to the next section on reorganisation.

Fixed Costs

One further piece of analysis must be carried out to examine the general level of the fixed costs: these are: labour, consisting of paid regular labour and casual labour which cannot be allocated to an enterprise; paid management; power and machinery costs; property charges; sundries not appropriate to particular enterprises and financing charges. These are set out in more detail in MA4 (see Appendix 4).

The examination of labour and machinery should follow the same procedure as described under labour and machinery in the previous section. It is difficult to make a major alteration in these costs without changing the whole farm organisation. However, minor changes in the farm system do not usually result in changes in fixed costs. In the case of labour and machinery, which are described as 'lumpy', changes are usually in the size of the work force, or buying a new or larger machine. Such moves involve a considerable alteration to the farm systems if their extra costs are to be covered. Rent, rates and sundries do not usually vary without major changes in size of farm or system of farming. In many minor changes and substitutions of one enterprise for another these costs will not alter.

Fixed costs should nevertheless be scrutinised to ensure that the costs incurred are reasonable for the system of farming adopted.

4 Planning the farm business

The starting point for farm planning must be awareness on the part of the planner that an economic problem exists. At the outset the cause of the problem may not be clear, all that may be known is that profit is too low. The first task is to carry out an analysis of the current situation on the farm as described in Section 3, partly to throw light on the true nature and cause of the problem, and partly to provide the data required for re-planning. Planning, therefore, falls into two interrelated parts:
- Identification of the problem;
- Planning the solution.

Identification of the problem

Ideally, farm records for 2 or 3 years should be used to 'normalise' yield and price fluctuations. Simple averages of 2 or 3 years' yields and quantities of inputs may be used to provide the basis for normalising although planning should never be based solely on averages since these may still be biased by abnormal results. The objective of the normalising procedure is to provide the best possible estimate of what has happened to the farm in the past. Future projections will require the application of expected prices to these normalised yields and quantities of inputs. When the data have been normalised they can be presented in the form of a gross margin account.

When the normalised gross margin account has been established, it is possible to examine the levels of performance of the individual enterprises and the relationship between the enterprise gross margins and the fixed costs. The objective at this stage is to establish the true nature of the problem and to seek ways and means of improving the efficiency of the existing system. The examination should follow a pattern so that 'no stone is left unturned'. The criterion should be the comparison of these levels of gross margin and fixed costs with the potential levels which can be obtained by the particular farmer on his holding.

Other Information Required for Planning

Before proceeding from the analysis to the planning stage, further information is required in addition to enterprise gross margins and fixed costs:
- What physical resources are available and what are the limitations to the production of each enterprise, or combination of enterprises?

- Which enterprise, other than those undertaken at present, could be considered for the future?
- Are there alternative methods of production to be considered such as a switch from casual labour to machine lifting of potatoes?

If new enterprises or new methods of production are contemplated information on these can be obtained from local data collected from similar type and size of farms, but it is important to remember that these standard figures should be adjusted to meet the particular circumstances of the individual farm being planned.

The resources and limitations should be set down under broad headings as follows:

Land—are there areas which differ in the potential use to which they may be put for reasons of soil texture, topography, location etc.?

Rotation—what limits on cropping exist for disease or risk minimisation?

Institutional limits—what quotas or contracts exist or can be made available?

Equipment, machinery and labour—the capacity of major items of machinery and fixed equipment must be noted and the availability of regular and casual labour. In addition it is necessary to know:

- The rate at which the major farm operations are carried out in terms of hectares per day.
- The size and composition of the gang which is normally used. The time of year during which the operation must be completed.

Capital and credit restrictions—the opportunities for extending existing credit limits and the attitude to further borrowing should be explored.

Limitations should not be accepted blindly, otherwise perfectly feasible plans can be excluded. Personal likes and dislikes may be irrational in terms of a particular goal and need careful examination to ensure that profitable opportunities are not being overlooked.

Planning the solution

All planning methods depend upon the reliability of the data on which they are based. This principle applies as much to linear programming as it does to partial budgeting.

Each planning method has its place, and it is up to the user to decide which method to use. The simpler methods of planning are easier to understand and interpret but are not so useful for complex situations. Linear programming can handle planning situations where large numbers of alternative enterprises are possible—cases with up to 200 alternatives are being processed.

The procedure of assessing the present situation, obtaining planning data for forward budgets and selecting the combination of enterprises to form the whole farm plan is the same for all planning methods. The advent of mechanical aids such as calculating machines and the computer, only widens the scope of planning methods by taking the burden of mathematical calculation off the shoulders of the farmer or his advisers.

The planning methods used are:
- Partial Budgeting;
- Whole Farm Budgeting;
- Programme Planning;
- Linear Programming.

Partial Budgets

Although not strictly 'planning', many small reorganisation problems involve partial budgeting. These are undertaken when the basic farm plan is not changed and the farmer is concerned with the marginal costs and returns resulting from peripheral changes to the farm plan. In cases where the farm is already well planned, partial budgeting for minor adjustment may be all that is required. Care must be exercised that changes which at first glance appear peripheral do not, in fact, seriously alter some of the fixed costs eg. financial charges. Examples of situations where partial budgeting is appropriate are the following:

What would be the effect of increasing the dairy herd from 60 to 80 cows and reducing the cereal acreage?

Would it pay to eliminate 10 ha of potatoes and replace by barley if one man could be saved?

A clear answer may be derived from a partial budget by considering 4 questions via.:

Debit	Credit
1. The extra Costs	3. Costs Saved
2. Revenue Foregone	4. Extra Revenue
Total	Total

Partial budgets may be approached by setting out all the costs and returns involved or more simply by using the Gross Margins, see Table 11. In the more

Table II Examples of partial budgets

FULL METHOD						
DEBIT			CREDIT			
Extra Costs			Costs Saved			
Barley 10 hectares			Potatoes 10 hectares			
	£	£			£	£
			Seed		2 350	
			Fertilisers		1 650	
			Sprays		750	
Seed	230		Casual Labour		1 700	
Fertilisers	570		Sundries		1 450	
Sprays	200		Wages/man		3 500	
	—	1 000			—	11 400
Revenue Foregone			Extra Revenue			
Potato Sales			Barley Sales			
			40 Tonnes		3 600	
			Reduction in Net			
			Farm Income		4 000	
300 Tonnes		18 000			—	7 600
		19 000				19 000

GROSS MARGIN METHOD

		£
Loss in Gross Margins 10 hectares of Potatoes at £1 010 per hectare		10 100
Gain in Gross Margins 10 hectares of Barley at £260 per hectare		2 600
		7 500
Less savings in Fixed Costs—one man at £3 500 pa		3 500
Reduction in Net Farm Income		4 000

involved substitutions great care must be exercised when using the Gross Margin method because changes in fixed costs may easily be overlooked.

Quite complicated proposals may be tackled using the partial budgeting technique, but it is usually advisable to spell out the final answer in a whole farm budget. The full implications of the changes on fixed costs can then be better assessed and the total costs and total returns can be seen.

Whole Farm Budgets

Budgets for the whole farm should be used when a major re-organisation of the farming business is contemplated, or when a new farm is being planned. The main principle of whole farm budgeting is that all items of income and expenditure are calculated so that a complete picture is obtained of the financial implications of the new plan.

Traditional Method

The traditional method of preparing a whole farm budget involves the detailed calculation and presentation of the output and costs for a particular plan.

Gross Margin Method

In the past decade it has been appreciated that the chores involved in whole farm planning and budgeting can be simplified by using Gross Margin data. The procedure is as before to establish the resources available, the opportunities open to the farmer, the limitations that apply and then to seek an increase in farm profit in one of the following ways:

- Increase gross margins at the same level of fixed costs.
- Increase gross margins at an increased level of fixed costs.
- Maintain existing gross margins but decrease fixed costs.
- Decrease gross margins, at still further decreased fixed costs.

Proceeding in this order, the farmer can first establish whether the financial output of the business can be increased with no change in the present resources of land, labour and capital. This entails a search for some combination of enterprises which gives a higher total gross margin than is being obtained at present.

The next step is to examine the possibility for improving the farm profit by increasing the total gross margin even if this entails an increase in the fixed costs.

Example

The farm is a 162 hectare dairy and arable farm. There are 80 cows and 60 followers in the dairy herd. 40 hectares of wheat and 40 hectares of barley are grown together with 8 hectares of potatoes. The labour force consists of the farmer and 3 men.

Table 12 Existing organisation

Gross Output	£	Costs	£
Milk Sales	46 000	Purchased Foods	20 390
Cattle Sales	8 400	Seed	5 500
Wheat	20 400	Ferts	8 840
Barley	15 600	Sprays	4 000
Potatoes	13 200	Vet and Med	960
Miscellaneous	1 400	Labour Regular	11 350
		Labour Casual	1 200
	105 000	Machinery Depreciation and Repairs	13 770
		Fuel and Power	3 369
		Rent and Rates	9 687
		Other Repairs	2 075
		Miscellaneous	4 000
			85 141
		Net Farm Income	£19 859

A plan to increase dairy cows to 120 buying in replacements was considered. Cropping would be 80 hectares cereals, 8 hectares potatoes and the remaining 74 hectares in grass.

Table 13 Budget for proposed plan

Gross Output	£	Costs	£
Milk and cattle	69 000	Purchased foods	24 960
Wheat	20 400	Seed	5 500
Barley	15 600	Fertilisers	9 600
Potatoes	13 200	Sprays	4 000
Miscellaneous	1 400	Vet and Med	1 280
		Labour Regular	12 500
	119 600	Casual Labour	1 200
		Machinery Depreciation and Repairs	13 970
		Fuel and Power	3 969
		Rent and Rates	9 687
		Other Repairs	2 075
		Miscellaneous	4 375
			93 116
		Net Farm Income	£26 484

The third stage is to investigate whether the existing system could be maintained at a lower level of fixed cost, ie. by employing less labour or less machinery or by altering the balance between these two resources.

The final stage is to examine whether a less intensive farm system, resulting in a fall in the total gross margin accompanied by a greater reduction in fixed costs, is more profitable. It usually means reducing the number of men employed.

Arable Crop Enterprises

On arable farms the objective of planning is to select a combination of crops which give the optimum total gross margin within the resources available on the farm.

Grazing Livestock Enterprises

Planning to include Grazing Livestock introduces a new element into the procedure in that, if grazing livestock occur at all, they may do so only at certain levels eg. 40 cows.

Where the dairy herd appears on an arable farm then the land needed for the cows may be subtracted from the total acreage and the residue treated as an arable unit, using where possible the grass as a break crop in the arable rotation.

If the inclusion of cows is in question, then a whole farm budget comparing a dairy/arable system with an 'arable only' system is essential.

Where stock are kept purely to use rotational grass—eg. beef, sheep or dairy heifers, then their performance may be reduced to a Gross Margin per ha basis, and they may be entered into the calculation as ha of beef, sheep or dairy heifers, and be selected or discarded according to their merits as compared with other possible break crops.

On the all grass farm, where an enterprise such as a herd of dairy cows must be included, it should be entered first in the plan and the remaining land utilised by other crops or stock selected on the basis of their Gross Margins.

In the same way, marginal substitution of one enterprise for another can be tested. Thus—if one cow requires 0·5 hectares and has a Gross Margin excluding forage variable costs of £373, and sheep stocked at 10 ewes per hectare have Gross Margin excluding forage variable costs of £340 per hectare, it is easy to demonstrate the effect on total Gross Margins of substituting sheep for cows. To reduce the herd by one cow reduces the Gross Margin by £373 and allows on the 0·5 hectares released 5 ewes to come into the plan with a Gross Margin of £170. Thus the substitution would reduce the overall gross margin by £203.

Non Grazing Livestock Enterprises

On many farms there will be intensive livestock enterprises (ie. non land using). On farms where such enterprises occur their performance will be measured in the initial analysis and, if satisfactory, they should be included at that level in the first plan when the land using enterprises are selected.

This plan may then be modified as marginal checks are made to show the effect of using some of the farm's resources—labour and capital—to permit the expansion of another enterprise, and also of releasing these resources by its elimination.

Marginal Capital

It is important to establish whether capital is a limiting factor. If this is so, it should be remembered that neither partial nor whole farm budgeting methods may always take this fully into account, in such circumstances different planning techniques are required.

Marginal capital is that extra capital required to fund a project. Funds may be required to purchase additional fixed assets such as land, fixed equipment, buildings, dairy cows and other breeding stock.

However it should be remembered that extra funds are usually required to finance the operation of the project ie. working capital. This may be either extra variable or fixed costs and in the simplest case may be the accumulated total cost of these items. In more complex instances, a cash flow (see page 76) may be required to calculate working capital requirements.

Marginal capital planning consists of two main considerations:

- calculation of the feasibility of the project. By this, is meant the ability of the project to generate sufficient extra funds to repay the marginal capital required (whether borrowed or not) within the constraints of time or interest associated with the source of the capital;
- calculation of the worthwhileness of the project. For a project to be worthwhile the extra funds generated must be capable of:
repaying the capital required;
paying any interest charges;
leaving a sufficient sum to cover any risk associated with the project;
rewarding the farmer for his managerial skills.

There are a number of techniques which may be used to undertake a capital appraisal of a project. The simpler methods such as calculation of simple rate of return or an annual capital charge are easily applied. A more sophisticated but more comprehensive approach involves the use of Discounted Cash Flows.

Labour Planning

The allocation of labour and machinery to individual enterprises requires detailed time sheets and is not necessarily helpful—indeed it may be positively misleading. This is because the demand which enterprises make on labour at peak periods during the year is more important than their total labour demand. Total Annual Labour Requirements for a farm can be calculated using the Annual Labour Requirements, shown in Table 14 for each enterprise undertaken.

Seasonal Labour Requirements

On predominantly arable farms, man days may give a misleading answer due to the seasonal nature of the labour demand.

Table 14 Average annual labour requirements (man hours)

Crops	Per Hectare	Livestock	Per Head
Arable:		Dairy Cows:	
Cereals—straw burnt	15	Cowshed	65
Cereals—straw carted	20	Parlour (under 70 cows)	46
Potatoes (Mechanised)		Parlour (over 70 cows)	34
Main Crop	170	Calves:	
Second Early	150	0–6 months	16
Early	120	6–12 months	12
Potatoes (Non Mechanised)		Veal/Place	16
Main Crop	270	Other Cattle:	
Second Early	250	1–2 years	12
Early	220	2 years +	12
Sugar Beet	90	Barley Beef (6 months—slaughter)	9
Vining Peas	32	Suckler Herds (per cow):	
Oilseed Rape	20	Single Sucklers	20
Field Beans, Dried Peas	20	Multiple Sucklers (3 calves)	
Herbage Seed	26	(+ 8 hrs per extra calf)	36
Forage:		Sheep:	
Grass —Grazed	8	Breeding Ewes	4
—Conserved, 1 cut	24	Other Sheep (6 months +)	2
2 cuts	40	Pigs:	
Kale—Grazed	20	Sows	28
Turnips, Swedes, Mangolds		Boars	28
—Harvested	120	Others (over 2 months)	4
—Grazed	48	Poultry:	
		Laying/bird housed	0·32
		Broilers/bird housed	0·24
		Turkeys/bird housed	0·40

In these circumstances, it is useful to calculate the seasonal labour requirements of the farm programme, to enable a comparison to be made month by month between the estimated labour requirements and the labour actually available on the farm to do the work. This will show when the peak labour requirements occur, and also whether or not the farm's labour force can cope with the estimated work load. In Table 15 seasonal labour requirements have been expressed as percentages of total requirements; thus, after the annual labour requirements have been calculated for a farm, the seasonal distribution can be estimated.

Adjustments to these figures should be made if the seasonal distribution of labour requirements for the enterprises on a particular farm is known to be different from that shown in Table 15; for instance, where early lambs are produced or store cattle purchased for fattening.

On the whole, unless there is a sheer physical excess of labour in the current organisation, reduction of labour costs usually come about only as a result of reorganisation of the pattern of enterprises or methods of production.

To focus attention on the seasonal labour and machine requirements of different enterprises; to avoid overlooking the resources in the planning process as may happen with intuitive enterprise selection followed by partial budgeting, a labour profile should be prepared. (See Figure 1). It is also valuable in forcing the planner to consider gang size requirements for field operations. A further advantage is that the pattern of labour distribution shown on the profile for an existing system will often indicate the opportunities for changing the balance of enterprises.

Labour peaks can also be examined by preparing a profile diagram in which the requirements of each enterprise in every month (or even half month) of the year is determined. An example is shown in Figure 1.

Figure 1.

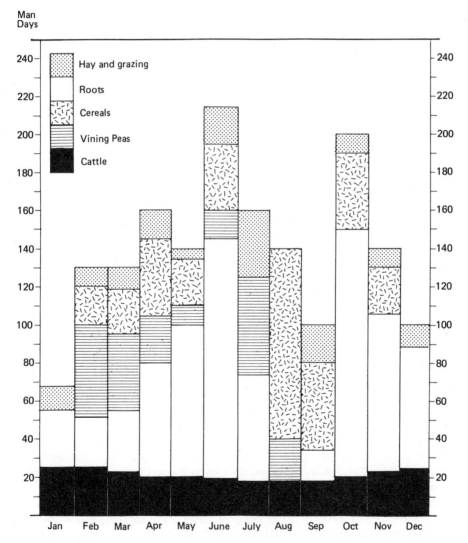

Fig 1: Example of Labour Profile for a 165 hectare farm (Standard man days)
Source: Camamile, G. H. and Theophilus, T. W. D., *Records for Profitable Farming*.
Hutchinson & Co. (Publishers) Ltd., London, 1964.

Although the peaks and the distribution of labour between enterprises is clearly shown there are several disadvantages to this method. If it is accepted that time sheets are unlikely to be kept generally, standard figures, known to be a mean of widely fluctuating performance, have to be used, instead of actual performance. It is not possible to see how the men are deployed or the composition of a gang working on a job. It is difficult to incorporate the machine times.

Table 15 Percentage labour requirements*

	Jan	Feb	Mar	Apr	May	June	July	Aug	Sept	Oct	Nov	Dec
	%	%	%	%	%	%	%	%	%	%	%	%
Cereals:												
Combined Winter	—	8	8	8	—	—	—	17	17	17	17	8
Combined Spring	—	—	20	20	—	—	—	20	20	10	10	—
Potatoes:												
Main Crop	3	4	6	12	3	3	1	—	12	41	12	3
Sugar Beet	—	—	1	7	27	13	5	—	7	27	13	—
Kale: Folded	—	—	12	44	32	12	—	—	—	—	—	—
Hay	—	—	10	10	—	40	30	10	—	—	—	—
Silage	—	—	10	10	40	30	10	—	—	—	—	—
Grazing	—	—	17	17	—	16	17	17	16	—	—	—
Dairy Cows:												
Cow Shed	10	10	10	10	6	6	6	6	6	9	10	11
Parlour	10	10	9	8	7	7	7	7	8	8	9	10
Other Cattle:												
Over 1 year	15	15	14	5	3	3	3	3	5	10	10	14
Under 1 year	11	9	9	6	6	3	3	3	11	14	14	11
Sheep	5	5	18	11	11	16	6	5	6	5	6	6

Note: Pigs and Poultry have a fairly even distribution depending on the system.
*Source: *Farm Business Data*—University of Reading, May 1967.

Gang Work Days

Profile diagrams drawn up on the basis of 'Gang Work Days' can go some way to reduce these disadvantages. A 'Gang Work Day' (GWD) is similar to a standard man day, consisting of eight hours, but it is a measure of the time taken by the gang needed to carry out a particular job. Thus, if it takes eight hours (ie. 1 GWD) for a gang of three men to complete a job the work content would be 1 GWD on the Gang Work Day basis, but three man days on the standard man day basis.

The average number of days available in each month of the year is given in Table 16, after allowance has been made for weather, sickness, holidays, etc, balanced by reasonable overtime.

Table 16 Work days available

Operation	Soil Type	Jan	Feb	Mar	Apr	May	Jun	Jul	Aug	Sep	Oct	Nov	Dec
Ploughing	Heavy	11	8	20	24	25	27	29	27	27	27	22	16
	Light	21	17	26	26	27	28	30	27	27	28	24	24
Heavy	Heavy	3	3	13	22	22	25	27	26	25	24	15	6
Cultivations	Light	11	10	23	25	25	26	28	25	25	25	20	23
Light	Heavy	2	2	11	20	20	23	26	24	22	21	10	2
Cultivations	Light	7	7	20	24	23	26	27	24	23	23	16	11
Drilling	Heavy			10	20	19	21	23	21	20	20	10	
	Light			18	23	23	26	26	22	21	21	13	
Potato	Heavy			9	20	7							
Planting	Light			16	24	9							
Top Work	Heavy		2	12	17								
	Light		4	19	22								
Crop Spraying					14	16	16	18	15				
Silage Making						18	24	26	24	12			
Cereal Harvesting								15	19	15	7		
Potato	Heavy						3	5	16	8	7	2	
Harvest	Light						14	23	25	24	24	9	
Root	Heavy	8									8	3	3
Harvest	Light										24	21	18

A form, Table 17, based upon this scale can be prepared, from which profile diagrams can be drawn. The information needed to make up the profile is as follows:

Column

1. Activity — Ploughing, combine harvesting, etc.
2. Period — Dates between which the job should be completed.
3. Hectares of each separate crop.
4. Hectares per GWD — Speed of working by gang.
5. GWD Required — Number of GWD to complete job, obtained by dividing (3) by (4).
6. Days available to complete job.
7. Number of Gangs required to complete work in time available. If the days available are less than the days required to do the job by one gang, another gang will be required.
8. Composition of Gang — The number of regular and casual workers and the number of tractors required to make up the gang.

An example is shown (Figure 2) of a GWD labour profile and the data used to construct the profile for a 150 hectare dairy farm growing the following crops is given in Table 17.

Winter wheat	32 ha
Spring cereals	8 ha
Potatoes	8 ha

Figure 2.

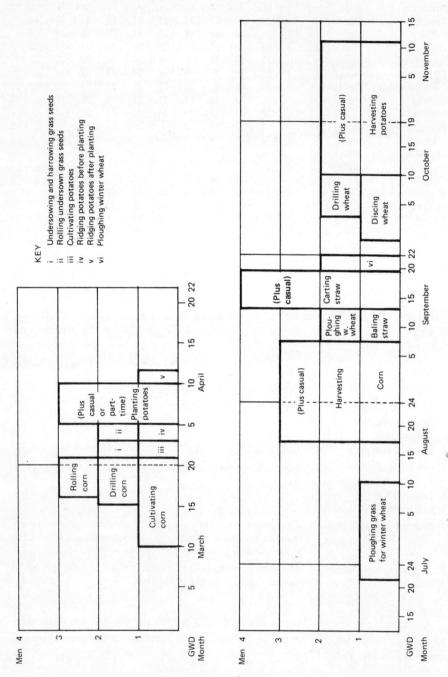

Fig 2: Labour Profile using Gang Work Days
Source: Kerr, H. W. T. and Thomas, H. A., *An Exercise in Planning*, University of Nottingham, March 1967. FR. No. 166

Table 17 Field labour required in peak periods

Activity (1)	Period (2)	ha (3)	ha per GWD (4)	GWD required (5)	Days available (6)	No. of gangs (7)	Composition of gang (8) Regular labour	Casual labour	Tractors	Notes
SPRING										
(1) Corn Cultivation Cross harrowed x2	15 March— 30 April	32	4	8	30	1	1		1	
Drilled and harrowed	15 March— 30 April	32	5	6½	30	1	1		1	
Rolling	Within 1 week	32	11	3	30	1	1		1	
(2) Undersowing grass seeding and harrowing	Before corn (1) comes through or after it is established (2)	16	11	1½	30	1	1		1	
Rolling	Within 1 week	16	11	1½	30	1	1		1	
(3) Potatoes Cultivate x3	Early April	8	4	2	7	1	1		1	
Planting	1 April— 15 May	8	2	4	30	1	1	2	1	
AUTUMN										
(1) Harvesting corn	20 August— 30 September	72	9	8	28	1	3		1	
Straw baling (1 T/cow = 1 acre/cow)		40	8	5		1	1		1	
Carting		40	6	7		1	2	2	2	
(2) Winter Wheat Ploughing	½ from 15 July	16	2	8		1	1		1	
	½ after harvest	24	2	12		1	1		1	
Discing	After grass x 3	16	3	6						
	After corn x 2	24	4	6		2	1		1	
Drilling and harrowing	October	40	4	10	20	1	1		1	
(3) Potatoes Burning off haulm	End of September	8	9	1	7	1	1		1	
Harvesting and storing	1 October— 15 November	8	½	16	27	1	3	5	3	
Riddling	Winter	8	½	16		1	3	4		33 t Crop

Advantages of the GWD Method are:

● The information relevant to a particular case can usually be obtained with reasonable accuracy without much difficulty.

Simple recording may be needed but there is no need to keep detailed time sheets. Where actual information is available it should be used in preference to 'standards'. The latter can be used as guides to performance levels achieved or where new enterprises are to be introduced into the farm plan.

● Attention is focused on the composition of the gang and speed of working. The efficiency of the gang can be examined and if necessary a work study assessment of specific jobs can be made.

● It is easy to see how the labour force is deployed at any particular time.

It will not be necessary always to go into the detail shown here. Where the total labour force can be determined quite simply by merely considering peak requirements of labour and machinery no profile diagram is necessary, but where the problem is complicated, it can be of great help.

It will frequently be found that the tentative plan has one or two labour peaks of relatively short duration. There are three ways of dealing with the problem: (a) eliminate the peaks, (b) fill in the troughs or (c) replan with returns to labour at those critical periods as the dominating influence.

(*a*) To eliminate peak labour requirements without modifying the farm plan either employ a contractor, employ casual or part-time workers, or use a different technique or mechanise the operation. On many farms peaks occur in the autumn and late spring. Employing a contractor to plough, a casual gang to shear sheep, or mechanising the beet crop are common methods or reducing labour peaks.

(*b*) Filling the troughs in the labour chart may be self defeating as an end in itself. The actions taken should be shown to add to farm income and not simply be an exercise in keeping men employed. It will either be done by adding an enterprise, such as fattening pork pigs in empty cattle yards in the summer, or by substituting enterprises which have higher labour demands in the periods when labour is available. Thus, second early potatoes or grass seed, both harvested in July, are favoured on arable farms because they do not demand autumn labour.

(*c*) When labour is the most restricting resource the best farm income is achieved by selecting enterprises with the highest returns to labour at the critical periods. Better methods are now being sought of assessing the time available for field work at various periods of the year. Until improved data are available, the planner must use his judgement in completing existing labour charts to take account of differences between farms with different soil types. It is often the practice to include an element of 'float' in planning labour use, ie. artificially to limit the period available for critical jobs so as to allow for adverse soil or weather conditions. It is difficult to give guidance on the amount of float that should be allowed on a labour chart for particular situations, the precise amount will vary with the time of year, with soil type and farming system. For example, more float should be allowed on heavy land in the autumn period than on light land. A subjective decision is required on each farm and, with experience, the appropriate figures can be fairly closely judged for each situation.

An example is given of three possible alternative systems for a 150 hectare farm:

Present system   53 ha forage, 89 ha cereals, 8 ha potatoes, 80 cows and followers

Plan I   Increase cows to 120 with followers
Increase forage area to 78 ha, grow 8 ha potatoes, 64 ha cereals

Plan II   Eliminate dairy herd, grow 20 ha potatoes, 130 ha cereals

Plan III   Replace dairy herd with beef, annual output 120 head 60 ha forage, 8 ha potatoes, 82 ha cereals

Table 18   Assumptions made in compiling budgets.

	Present System	Plan I	Plan II	Plan III
Crops	ha	ha	ha	ha
Wheat	45	34	70	42
Barley	44	30	60	40
Potatoes	8	8	20	8
Total crops	97	72	150	90
Forage	53	78	—	60
	150	150	150	150
Livestock numbers				
Dairy cows	80	120	—	—
Replacement units	20	30		
Beef (annual output)	—	—	—	120
Stocking density				
(Livestock units per hectare)	2·03	2·07	—	2·0
Nitrogen, kg/ha	250	250	—	180
Labour				
Cowman @ £5 750	1	1	—	—
Stockman @ £4 850	1	1	—	1
Tractor driver @ £4 600	1	1	2	1

Gross margins used in the budgets are as follows:

    Wheat   £350/ha
    Barley   £270/ha
    Potatoes   £860/ha
    Dairy cows   £350/head
    Dairy heifers   £250/calved heifer
    Beef   £160/finished beast
    Forage costs   Dairy £100/ha
                             Beef £80/ha

Table 19 Present position and budgets for plans I to III

	Present Position	Plan I	Plan II	Plan III
GROSS MARGINS from:				
Wheat	15 750	11 900	24 500	14 700
Barley	11 880	8 100	16 200	10 800
Potatoes	6 880	6 880	17 200	6 880
Total crops (a)	34 510	26 880	57 900	32 380
Cows	28 000	42 000	—	—
Replacement units	5 000	7 500	—	—
Beef	—	—	—	19 200
	33 000	49 500	—	19 200
Less Forage Costs	5 300	7 800	—	4 800
Total livestock (b)	27 700	41 700	—	14 400
Miscellaneous Income (c)	750	750	750	750
TOTAL GROSS MARGIN (a + b + c = d)	62 960	69 330	58 650	47 530
FIXED COSTS				
Regular labour	15 200	15 200	9 200	9 450
Machinery repairs and depreciation	7 600	7 800	8 000	7 600
Fuel and Power	3 000	3 300	2 500	2 800
Rent and Rates	9 000	9 000	9 000	9 000
Other repairs	2 750	2 900	2 600	2 750
Miscellaneous	4 200	4 500	4 000	4 200
Total (e)	41 750	42 700	35 300	35 800
NET FARM INCOME (d—e)	21 210	26 630	23 350	11 730

Labour

Labour profiles, constructed for the peak periods by the Gang Work Day method indicated that the farmer would need the assistance of three men for Plan I and two men for Plans II and III (Table 19).

Capital

Increasing the size of the dairy herd in Plan I would require capital expenditure on improving the milking parlour, extra cattle accommodation and more cows. In Plans II and III capital would be released by selling the dairy herd, but larger machinery would be required for potato planting and harvesting and corn harvesting in Plan II and more working capital would be required in Plan III before the first group of beef cattle were ready for sale (Table 21).

Table 21 Capital budget

	Plan I		Plan II		Plan III	
	Extra Capital Required	Capital Released	Extra Capital Required	Capital Released	Extra Capital Required	Capital Released
Buildings	16 000				4 000	
Parlour	7 000					
Cattle	18 000			31 500		31 500
Machinery			25 000			
Working Capital					28 000	
Total	41 000		25 000	31 500	32 000	31 500

Decisions

Plans I and II both show an increase in budgeted net farm income over the present system. Plan III shows a lower income. Plan I would require a substantial injection of capital. The farmer would want to satisfy himself that the increase in income would be sufficient to justify the capital expenditure. Before taking any decision it would be prudent to test the sensitivity of the plans to changes in costs and prices.

Sensitivity

When Gross Margin budgets are drawn up, the best estimates of future prices, costs and yields, should be used. The actual outcome may be different from that expected, however. It is, therefore, a wise precaution to see what effect changes in these factors would have on the Gross Margins and the Whole Farm Budget.

In the previous example, the dairy cows were budgeted to produce 5 200 litres of milk per head head at an average price of 11·1p per litre. A variation of 10 per cent in milk yield or milk price would cause a change in the Gross Margin per cow of ± £57·70. This carried through to the Whole Farm Budget would cause a variation in the Net Farm Income in Plan I of ± £6 924. If milk yield or price were 10 per cent below expectations, Plan I would show a Net Farm Income below that of Plan II but still higher than Plan III.

Variations in cereal prices would affect all three plans. An increase in price would increase cereal Gross Margins. It would also cause an increase in concentrate cost for dairy cows and beef cattle and thereby lower the Gross Margins on these enterprises. If cereal prices were £10 per tonne more than expected and concentrate prices were also £10 per tonne higher, the increase in Wheat Gross Margin at a yield of 5 tonnes per hectare would be £50 and barley Gross Margin at a yield of 4 tonnes per hectare would be increased by £40. The reductions in Gross Margins of dairy cows and beef cattle at concentrate usages of 1·8 tonnes and 1 tonne per head respectively, would be £18 and £10.

The effect of these changes on the Whole Farm Budgets would be:

		£
Plan I		
	34 ha wheat x 50	1 700
	30 ha barley x 40	1 200
		2 900
	less 80 dairy cows x 18	1 440
	increase in Net Farm Income	1 460
Plan II		
	70 ha wheat x 50	3 500
	60 ha barley x 40	2 400
	increases in Net Farm Income	5 900
Plan III		
	42 ha wheat x 50	2 100
	40 ha barley x 40	1 600
		3 700
	less 120 beef x 10	1 200
	increase in Net Farm Income	2 500

A reduction in cereal prices, of a similar amount, would have the opposite effect on the Net Farm Incomes in each plan.

These calculations show that Plan I would be less sensitive than Plans II and III to changes in cereal prices. Plan I would be highly sensitive to changes in milk yields or prices, which would have no effect on the other two plans.

Cash Flow Budgets

The Cash Flow sets out the actual cash receipts and expenditure over a period of time indicating when particular transactions are expected to occur together with the effect they will have on the cumulative balance.

The period covered by the cash flow can be as short as a few weeks or it may be for several years ahead. The longer the period covered the more speculative the cash flow becomes. The information can be set out on a weekly, monthly, quarterly, annual or any other convenient time basis. The balance can be made for each time period and the cumulative balance carried forward. The bank balance or overdraft existing at the commencement of the calculations is used as the starting point.

It is possible to calculate the amount of cash available after appropriate deductions have been made for meeting taxation, the amount required for living expenses, cash required for reinvestment in the business, the cash required to meet current bank charges and any repayments of capital, the repayment of annual debtors and to obtain surplus or deficit of cash that will be available or required for either reinvestment in the farm or for investment off the farm.

A cash flow can be used for the following purposes:
To calculate the expected future level of a bank balance or overdraft.

To forecast the future financial pattern of a farm plan. The detailed forecast would include farm receipts and expenditure, personal spending, tax payment, planned investment etc.

To assess the feasibility of alternative investment projects.

The alternatives would include the possibility of continuing the present farm plan.

To form the basis for discounting future returns from investment projects.

Cash flows have also been called 'capital profiles' and 'capital diaries' but in recent years the term cash flow has been more generally used, particularly in relation to discounting methods for appraising long term capital investments.

As more capital is likely to be borrowed from outside the family resources, more attention will have to be given to the production of cash flow projections and a realistic attempt made to calculate or allocate the cash available after meeting living expenses, taxation, repayment of capital and the interest charges associated with borrowed capital. At current prices, particularly where land investment is concerned, there is likely to be a deficit of cash available in many farm businesses.

It may be necessary to refinance the business on a longer-term basis and in other cases to dispose of some of the capital assets or, alternatively, to reduce some of the capital expenditure in new projects. Any farmer interested in undertaking this particular exercise can contact his agricultural advisory officer, who can advise on this appraisal.

Undue emphasis has been given to the calculation of the net worth of the business; farmers, bankers and other suppliers of credit have tended to lend money on this factor of net worth, as well as the integrity of individuals. There is an abundance of capital resources available within the agricultural industry measured in net worth terms, but this capital can only be released if and when a person disposes of his business. On many farms there is a shortage of cash available. It is necessary to calculate the cash available and the allocation of this cash to meet the present commitments, before investing further capital. An example of a Cash Flow is shown on form MA7 (Appendix 5). Further details of using a Cash Flow as a method of Budgetary control are discussed in Section 5—Financial Control. (See page 90 and Table 28).

Programme Planning

A more refined process of Gross Margin planning is referred to as programme planning.

The first stage in programme planning is to list the resources available on the farm, the resource requirements of possible enterprises, and the gross margins of those enterprises. A two-way table is then constructed with each enterprise having a column and each resource having a row. The body of the table therefore shows the resource requirement of each enterprise. (See Table 22).

The next stage is to calculate the gross margin per unit of each scarce resource and note the order in brackets behind this figure. This is done simply by dividing the gross margin by the resource requirement of that unit. (See Table 23).

Having done this the table is redrawn with resources available listed along the top. (See Table 24).

Potatoes, having the highest gross margin per hectare, are shown first at the maximum level and the resource requirements of eight hectares are deducted from those available to give a new list of resources now available. Dairy cows have the next highest gross margin per hectare, and come in at the maximum number. Winter wheat, the next enterprise tried, can only feature at 28 hectares because October-November labour is now exhausted. Barley occupies the residual land.

Table 22 Available resources and enterprise requirements

Resource	Available (ie maximum)	W Wheat 1 hectare	Barley 1 hectare	Potatoes 1 hectare	Dairy (1 cow + ¼ replacement unit)
Land	150	1	1	1	0·8
Cereal land	150	1	1		
Potatoes	8			1	
Wheat	32	1			
Dairy cows	100				1
Mar-Apr labour	540	2·5	5	12·5	0
Aug-Sept labour	680	10	5	2·5	0
Oct-Nov labour	460	5	0	40	0
Gross margin		38	32	70	100

(eg. 1 hectare of W Wheat requires 2·5 hours of Mar-Apr labour, 10 hours of Aug-Sept labour, 5 hours of Oct-Nov labour, and produces a gross margin of £350)

Table 23 Gross margins to land and labour

	W Wheat	Barley	Potatoes	Dairy
Per hectare of land	350(3)	270(4)	860(1)	515(2)
Per hour of Mar-Apr labour	140(1)	54(3)	69(2)	—
Per hour of Aug-Sept labour	35(3)	54(2)	344(1)	—
Per hour of Oct-Nov labour	70(1)	—	21(2)	—

Table 24

	Land	Mar-Apr labour	Aug-Sept labour	Oct-Nov labour	Gross Margin £
Available	150	540	680	460	—
Potatoes (8 ha)	8	100	20	320	6 880
Dairy (100)	142	440	660	140	6 880
	80	—	—	—	41 200
W Wheat (ha)	62	440	660	140	48 080
	28	70	280	140	9 800
Barley (ha)	34	370	380	—	57 880
	34	170	170	—	9 180
Unused resources	—	200	210	—	—
Total Gross Margin					67 060

From the total gross margins of £67 060 must be deducted the fixed costs:

	£
Regular labour	14 350
Power and machinery	11 200
Rent and rates	9 300
Sundries	4 450
	39 300

to give a net farm income of £27 760

In this example the resources have been taken in descending order of returns to land. An alternative approach would have been, after deducting the first enterprise giving the best returns to this and so on. A third approach is to treat each resource in turn as limiting and introduce enterprises in order of returns to this resource until each runs out. Here we are left with a series of plans, the most promising of which can be developed.

Linear Programming*

This is a technique which normally requires the use of a computer. Information about the farm concerned is arranged in a special way to form a mathematical model. The computer looks at all possible combinations of enterprises which are feasible. Not only does it select the farming system which makes the very maximum profit but it provides most useful supplementary information. For enterprises not selected it indicates to what extent their profitability must rise before they are worth while considering. For those already in the selected system it indicates how much their profitability can fall before a system change is likely to be needed. It also identifies the bottlenecks to expansion, and indicates how much an extra unit of a limiting factor (such as land, labour) is worth.

As in other planning methods, the enterprises possible and resources available need to be set out. In linear programming a table (normally called a 'matrix') is constructed. (See Table 25).

Preparation of Matrix

Four types of information are required.

ACTIVITIES (or enterprises) which are possible for the farm business.

NET REVENUES (normally gross margins) for each of the activities.

AVAILABLE RESOURCES AND CONSTRAINTS within which the farm plan must operate.

RESOURCE REQUIREMENTS of each activity.

EXAMPLE

Background

A farm comprises 162 ha of good sandy loam, suitable for a wide range of crops including cash roots and cereals. Adequate grain drying and storage facilities exist for a farm of this type.

---

[1]For detailed information on Linear Programming see MAFF/ADAS Book—Farm Planning by Computer.

There is a 700 ton potato store. Potato and sugar beet harvesting machinery are available along with a full range of tractors and cultivating equipment.

A covered yard is suitable for 100 mature beef cattle but there are no facilities for dairy cows. Sheep handling facilities exist for 600 ewes.

Three regular men are employed.

ACTIVITIES

Activities which are to be considered are as follows:

ACTIVITY	NET REVENUE per hectare or per unit £
Sugar Beet	550
Potatoes	900
Winter Wheat	440
Barley	340
Grass Seed	300
Oil Seed Rape	320
Winter Beans	250
Barley Beef	20
18 Month Beef	80
Ewes	27
Ley	−138
Hay	−150

This list includes:

- All enterprises which the farmer is willing to consider, including those being operated on the farm at present and also other enterprises which may be introduced in the future.
- Although not included in this simple example similar lists often incorporate other activities such as hiring casual labour, buying straw, selling straw, selling hay. There is also usually a wider range of both livestock and crop enterprises. Quite often a crop like barley would be split into 1st year, 2nd year, 3rd, 4th and perhaps 5th year, with separate enterprises representing each.

NET REVENUE

Net revenue is the contribution which each activity makes to the business as a whole, and is expressed as output less variable costs. For many activities the net revenue is therefore the gross margin.

For the activities which supply resources, eg. ley for grazing, hay for stockfeed, the net revenue will be negative, consisting of the variable costs of the home grown crop.

The method of obtaining the net revenue (or gross margin) for linear programming is the same as in any other planning method.

Gross margin information needs to be collected for each enterprise. These figures then need to be 'normalised' (to offset the effects of abnormal weather and prices in particular years). As the linear programme will be used for future planning, the gross margins also need to be adjusted for any technical improvements or known price changes which are likely to occur. Fixed costs also need to be calculated for the whole farm. These are deducted from the total net revenue (gross margin) calculated by the computer, in order to obtain net farm income.

In Table 25 the activity names (abbreviated to eight characters or less) form column headings. The first row, called 'margin', contains the net revenues (margins) for each enterprise.

AVAILABLE RESOURCES AND CONSTRAINTS

The activities proposed can only be carried out within the limits of the resources of the farm. These 'resources' include not only land, labour, capital, buildings and management, which are normally thought of as farm resources, but also rotational limits, institutional and personal limitations, such as the potato and sugar beet quotas, and personal preferences about enterprise size.

Land

In our example the farm consists of 162 ha of good sandy loam. In Table 25 this is represented by the second row 'LAND' followed by 162. The symbol '+' in the next column indicates to the computer that the total area of crops must be equal to or less than 162 ha.

On many farms there is more than one type of land and possibly some permanent pasture. If the land types offer different cropping possibilities they can be entered on different rows in the matrix, eg. there could be light land suitable for potatoes, sugar beet, cereals, etc. and perhaps heavy land suited only to leys, beans, grass seed and cereals. These different land types would have their own set of crop activities.

Labour

Three regular men are employed. Casual labour is not available but the men are prepared to work overtime as required, ie. evenings and weekends.

Work on the land is restricted by weather and soil conditions, particularly during the winter. After making allowances for these factors, available hours including overtime, grouped into periods of the year are as follows:

	Hours
October/November (root harvesting)	1912
December/January (winter ploughing)	1502
February/March/April (spring sowing)	2867
May/June (crop spraying/haymaking)	2185
July/August/September (cereal harvest)	3112

Code words of eight characters or less are allocated to these labour periods and each is given a row in Table 25, eg. Labour in October/November is shown as LABON 1912. Again the symbol + is used to indicate to the computer that labour required by activities in October/November must not exceed 1912 hours.

For obvious reasons our example case must be simple. It would be preferable to have the overtime labour 'hired' as needed by having separate activities for overtime October/November, overtime December/January, and so on. If casual labour had been available this could also have been accommodated in a similar manner. Sometimes it is preferable to have labour for livestock distinct from labour used for arable enterprises. For alternative ways of handling labour see MAFF/ADAS Book-Farm Planning by Computer.

Capital

In the example some working capital is required to pay for wages, rent and other fixed costs. In addition there is £90 000 available to purchase seeds, fertilisers, sprays, feeds, livestock and miscellaneous enterprise costs. The bottom row of Table 25 represents this aspect, ie. CAPITAL 90 000.

A single capital constraint such as this tends to overestimate capital needs because it does not take account of cash flow. More sophisticated ways of handling capital are covered in Reference Book 419. A single capital row is usually preferable to ignoring capital altogether.

Buildings and Fixed Equipment

Our farm has existing covered yard accommodation for 100 adult cattle. Represented in Table 25 by the row YARD 100. Similarly sheep handling facilities for 600 ewes are covered by the row SHEEPLIM 600.

To accommodate 18 month beef in *new* buildings, we could have an additional 18 month beef enterprise, with an amortisation charge for the new building deducted from the gross margin. An extra amount representing the unit cost of the building would also be added to the capital requirement for this new beef enterprise.

Institutional Constraints

In our example case the sugar beet quota limits the area of this crop to 36 ha; represented by the row BEETLIM in Table 25. Similarly the potato quota restricts this crop to 18 ha and this is done by the row SPUDSLIM. Other institutional limits which can sometimes be encountered include vegetable contracts.

Rotational Limits

In the example case the following rotational limits apply:

Sugar beet is limited to 1 year in 4 (40 ha). Potatoes are limited to 1 year in 6 (27 ha). As the quotas for these crops are less than the rotational limits the crop areas would be controlled by the institutional constraints BEETLIM and SPUDSLIM. No additional rotational contraints have, therefore, been built into this matrix. If it was felt that there was a possibility of quotas being increased beyond the rotational limits at some time in the future, it would be advisable to include rotational constraints in the matrix in addition even though they would be redundant initially.

Other rotational constraints are:
Beans limited to 1 year in 5 by BEANLIM (32 ha).
Winter wheat must follow a break crop or a ley by means of the row WHEATCAP.

The row GRAZING makes sure that grazing is sufficient for grazing livestock. It is usually preferable to have 2 grazing constraints, one for the first half of the year and one for the second when larger areas of grass are needed per livestock unit.

The row HAYREQ makes sure that sufficient hay is grown for livestock.

Care must be taken not to have too many rotational constraints, or make them too rigid, otherwise the only solution to the matrix may be the present farm plan.

Personal Constraints

Being new to the oil seed rape crop the farmer wishes to restrict its area to 18 ha at this point in time. This is done by row RAPELIM (18).

Grass seed limited to 16 ha due to combine capacity.

Other personal assessments of risk and uncertainty for various enterprises can, to a limited extent, be catered for in this way.

RESOURCE REQUIREMENTS

Each activity requires a certain amount of resource, eg. wheat requires land, labour, capital and needs to follow a break crop or ley. These are winter wheat's resource requirements. If we look at Table 25 and read down the winter wheat column we can see that for each unit of winter wheat (in this case 1 hectare):

- There is a net revenue (gross margin) of £440.
- One hectare of land is required.
- One hectare of wheat 'capacity' is needed, ie. we have said that wheat must follow a ley or a break crop, thus one or both of these must have been grown before capacity for winter wheat is created.

A glance along the WHEATCROP row indicates that SPUDS have a negative—1 entry here. This means that one hectare of potatoes *create the capacity* for one hectare winter wheat. Similarly capacity is created by grass seed, oil seed rape and winter beans.

As ley is only ploughed out every three years the capacity for wheat created by a ley is only 33 per cent of the total area of ley, hence -0.33 under ley in row WHEATCAP.

- Wheat needs 7·41 hours of October/November labour.
- It needs 1·48 hours of February/March/April labour.
- 3·46 hours of July/August/September labour are required.
- Capital needed for seed, fertiliser, sprays and miscellaneous costs totals £160.
- The SUM row at the bottom is a check total of all the entries, The computer adds the entries it receives and checks that its total and that from the SUM row correspond. This helps to identify any errors made in getting data into the computer.

Obtaining facts and figures for this part of linear programming can be quite time consuming. This particularly applies to the labour requirements for arable crops at different times of the year, and perhaps to grass and concentrate requirements for livestock at different rates of stocking. In other planning methods the resource requirements are often conveniently overlooked or estimated vaguely, whereas in linear programming they do need to be stated with more precision. Because these coefficients are often not readily available in written form the technique is not as widely used as it might be. It is by no means an impossible task to assemble the data required. Once it has been prepared for one farm situation it is that much easier to modify it for the next. Whilst farmers do not have their enterprise labour requirements written down, they certainly have them in their heads. Sit down around the table with any good farmer and he can tell you, for example, what operations he carries out on sugar beet, when he ploughs, how long it takes, when he starts to work down and how long that operation will be. Similarly he will deal with fertiliser application, herbicide, drilling, subsequent actions and harvesting. The resultant figures can be cross checked for feasibility against the John Nix handbook or other published information.

Processing

Table 25 represents the matrix as it would be submitted for the current ADAS Linear Programming procedure. In this format the user can see the logical relationship and interactions between activities and resources. It is advisable to check these carefully before processing.

Computation can, in fact, be done on a calculating machine and this process does not require more than a modest knowledge of arithmetic. To solve a problem the size of that in the example matrix could, however, take several days of solid work, involving a large number of simple calculations. The Ministry's computer takes 2 minutes to do the same work even though it is also doing several other jobs at the same time.

The formuli used to solve an LP problem are often presented in a rather fearsome mathmematical form. In actual fact the procedure is simple and logical and very practical.

In simple outline it consists of:
- Selecting the enterprise with the highest net revenue.
- Introducing that enterprise until one of the resources is used up.
- This is then solution one.
- It then looks to see if any other enterprise could make better use of that limiting resource.
- If so that enterprise would be substituted for some or all of the first enterprise until some other resource becomes limiting.
- The substitution procedure continues until no further improvements in total net revenue are obtained.
- During the process the opportunity cost of giving up one or more enterprises as another is introduced is adequately considered.

Interpretation of results

An appropriate amount of time must be spent on the interpretation of results if the linear programming technique is to be used to its full advantage. Solutions should never be accepted blindly since the reasons for the exclusion or inclusion of a particular activity may be as important as the fact itself.

Feasibility. The first task in the interpretation of results must be to check that the constraints have held and that the solution is feasible, ie. that no weaknesses or mistakes occurred in the design or content of the matrix. To do this it is necessary to check each constraint to find out whether each has operated as intended. It may be that an essential constraint was omitted from the matrix and this will have to be included before proceeding further.

Reasons for the Solution obtained. After checking the feasibility and consistency of the solution, the reasons for obtaining that particular solution must be worked out. The starting point is to look first of all at the list of marginal value products, to find out which resources have been limiting. Only those in short supply will have influenced the plan obtained. For example in the solution to our problem labour in October/November is very limiting with an extra hour at that time of year shown to be worth £12·25. Although potatoes have the highest net revenue per hectare (£900) they require 57·33 hours of labour in that period. Net revenue is thus £15·70 per hour. Sugar beet on the other hand with a net revenue of £550/ha needs only 22·24 hours of labour in October/November and so gives a return of £24·73/hour. Thus in the solution sugar beet reaches the maximum allowed by its constraint, ie. 36 hectares. Potatoes in the solution are 10 hectares and well below the quota of 18 hectares.

The shortage of labour in October/November also influences the choice of other break crops. Although oil seed rape and grass seed have much lower net revenues per hectare they are also selected in preference to potatoes as their return to labour in the autumn is much higher. The list of marginal value products indicates that if it is possible to relax the constraints on both oil seed rape and grass seed, then increases in total net revenue are possible. An extra hectare of oil seed rape being shown to be worth £116 and an extra hectare of grass seed £102.

In contrast winter beans are not selected as a break crop, not only because their net revenue/hectare is low but also because of their relatively high demand for October/November labour.

Land is also a limiting factor and an extra hectare is shown to be worth £273. This also has an influence on the choice of enterprises. Ewes give a return per hour of October/November labour of £27 ÷ 0·45 hours = £60. This is of course higher than that from sugar beet (£24·73). On the other hand ewes require 0·09 ha grazing and 0·08 tonnes hay which is equivalent to 0·011 ha, making a total of 0·101 ha/ewe. With a net revenue of £27/head this represents a return of £267/ha. Sugar beet has a return of £550 and oil seed rape a return of £320. It is of course when there are several limiting factors operating at the same time that Linear Programming's ability to give the correct weighting is all important.

The solution consists of cash crops only because in this case they give the best returns to land and October/November labour.

A rise of £44 is needed in the net revenue from 18 month beef and £11 from barley beef before these livestock enterprises are worth considering. Because livestock are not selected there is a surplus of capital.

Necessity for Re-runs. If the relationships of enterprises as shown in the matrix are checked carefully before computation the chances are that the initial solution will be satisfactory. Re-runs may be necessary, however, to eliminate keying errors. It may also be necessary to make alterations to the constraints if it is felt that the answer does not in fact reflect the true position on the farm concerned, eg. after studying the results it might be thought worthwhile to make the cereal rotation less rigid or to remodel the livestock feeding constraints and coefficients.

If the example case was 'live' it would be desirable to have a re-run with the oil seed rape constraint relaxed to the rotational limit (32 ha). To ease the October/November bottleneck it might be possible to improve beet or potato harvesting techniques. In this case re-runs with appropriately lower labour requirements for these crops would be sensible. The resulting difference in total gross margin could then be used in an investment appraisal exercise to determine whether the cost of a new machine is justified.

Use of Linear Programming in ADAS

When Linear Programming was first used it was confined to individual farm cases. The tendency these days is to use it to build a model of a typical farm in a given area. The results are then used as background for giving advice to similar farms.

As every farm is different results from such a model cannot be applied to other farms without adjustment. It is also important to test that the recommendations are indeed applicable to other cases. This can be done by partial budgeting.

A model indentifies the enterprises which are likely to be most important and those which are non starters for a particular locality. It also identifies possible bottlenecks and gives an idea as to how stable a particular situation is likely to be in these dynamic times.

Integer Programming[1]

In ADAS, Linear Programming is now used in conjunction with Integer Programming. The optimum LP solution is found and then a branch and bound technique (Integer Programming) employed to overcome the weaknesses of LP.

This additional process allows many more practical features to be built into models and includes the following advantages:
- Choice of solution—there is usually one or more near optimum solutions as well as the optimum one.
- Labour can be hired in one man units.
- It is possible to make activities mutually exclusive, eg. cows on hay, *or* silage, but not some of each.
- It is also possible to specify that *if* an activity is selected then it will be of at least a minimum size.
- Set up costs can be taken into account, eg. if oil seed rape has not been grown on the farm before one can insist on the purchase of a windrower before rape is selected for the solution.

Other Developments—Micro Computers

The arrival of the micro computer is likely to make other computer programmes more easily available for farm management. Programs are already on the market for financial control of the farm business, dairy herd and other livestock monitoring. We are likely to see enterprise forecast programmes, already available on some main frame computers, becoming widely used on micro computers. In addition, programmes for whole farm budgeting, cash flow forecasts and livestock rationing are very much on the cards.[2] Some of these programmes are in fact already available and abbreviated versions are possible on programmable calculators.

Further Reading

[1] Butterworth, K.—Practical Application of Integer Programming to Farm Planning—Farm Management Vol. 2, No. 3, Autumn 1972.

[2] ADAS Publication—'Introduction to Micro Computing'. Micro computers in the Farm Office series.

Table 25—Example Matrix

ACTIVITIES

RESOURCES		Symb GM	1 Beet	2 Spuds	3 Wheat	4 Barley	5 Grass-Seed	6 Rape	7 W-Beans	8 Barl-Beef	9 Beef-18	10 Ewes	11 Ley	12 Hay
Margin		Z	550	900	440	340	300	320	250	20	80	27	−138	−150
Land	162	+	1	1	1	1	1	1	1	0	0	0	1	1
Wheatcap	0	+	0	−1	1	0	−1	−1	−1	0	0	0	−0·33	−0·33
Beetlim	36	+	1	0	0	0	0	0	0	0	0	0	0	0
Spudslim	18	+	0	1	0	0	0	0	0	0	0	0	0	0
Gseedlim	16	+	0	0	0	0	1	0	0	0	0	0	0	0
Rapelim	18	+	0	0	0	0	0	1	0	0	0	0	0	0
Beanlim	32	+	0	0	0	0	0	0	1	0	0	0	0	0
Yard	100	+	0	0	0	0	0	0	0	0·64	1	0	0	0
Sheeplim	600	+	0	0	0	0	0	0	0	0	0	1	0	0
Labon	1 912	+	22·24	57·33	7·41	5·44	0	0·49	8·16	2·55	7·75	0·45	0	0
Labdj	1 502	+	4·20	37·07	0	0	0	0	0	2·55	7·5	0·5	0	0
Labfma	2 867	+	10·63	20·75	1·48	4·69	1·98	0·99	0	3·83	9	1·7	1·98	1·97
Labmj	2 185	+	18·53	0·99	0	0·49	1·48	0	0·49	2·55	0·5	0·7	1·48	17·3
Labjas	3 112	+	0·49	39·78	3·46	3·46	7·41	14·08	4·20	3·83	0·75	0·65	1·73	17·3
Grazing	0	+	0	0	0	0	0	0	0	0	0	0·09	−1	0
Hayreq	0	+	0	0	0	0	0	0	0	0	1·5	0·08	0	−7·5
Capital	90 000	+	280	829	160	150	175	220	150	300	270	77	138	150
Sum			888·59	1885·92	614·35	505·08	486·87	556·56	413·85	335·95	378·12	109·17	4·86	29·74

Table 26 Solution to matrix example

TOTAL NET REVENUE £71 597

Activity	Size	Net rev/unit	Decrease in total net revenue if 1 unit introduced into solution
Beet	36·00	550	0
Spuds	9·93	900	0
Wheat	43·93	440	0
Barley	38·13	340	0
Grasseed	16·00	300	0
Rape	18·00	320	0
Wbeans	0	250	−47·47
Barlbeef	0	20	−11·25
Beef 18	0	80	−43·62
Ewes	0	27	0
Ley	0	−138	−147·57
Hay	0	−150	−398·30

ROW INFORMATION

Constraint	Amount unused	Original limit	Marginal value product of 1 unit of resource
Land	0	162	273
Wheatcap	0	0	75
Beetlim	0	36	4
Spudslim	8·07	18	0
Gseedlim	0	16	102
Rapelim	0	18	116
Beanlim	32	32	0
Yard	100	100	0
Sheeplim	600	600	0
Labon	0	1 912	12·25
Labdj	982·5	1 502	0
Labfma	1 984·8	2 867	0
Labmj	1 465·7	2 185	0
Labjas	2 043·2	3 112	0
Grazing	0	0	238
Hayreq	0	0	0
Capital	52 175	90 000	0

STABILITY OF SOLUTION

Activity	Net rev/unit	Lower limit of net rev	Upper limit of net rev
Beet	550	546	00
Spuds	900	662	913
Wheat	440	361	453
Barley	340	314	350
Grasseed	300	197	00
Rape	320	203	00
Ewes	27	5·5	40

5 Financial control

The function of a manager of any business is to achieve the desired profit by making the best possible combination of resources and skills available to him. Since the 1950s, considerable progress has been made in establishing methods of recording physical and financial data on farms, and great strides made in the process of analysing, interpreting and planning farm businesses to improve profits and resource use.

Any plan which evolves from the critical examination of a business remains a paper plan unless it is put into operation and the objectives achieved. The manager will want to know how the plan is progressing at various stages and to satisfy himself that the potential outcome is in fact being realised. It is the implementation of the plan and its day-to-day control which creates the major problem for management. The objectives of control measures should therefore be:

- The achievement of the control of performance of individual enterprises.
- The achievement of financial control of plans for the whole farm business.

Methods of Control

There are two types of control measures appropriate to the agricultural industry:

short period enterprise control techniques;
budgets or forecasts for the business as a whole.

In both types, the objective is to compare actual with planned levels of performance; any deviations from the plan can be investigated and put right as soon as they become apparent.

Farming has been slower than some industries to adopt business control measures due to inherent difficulties; the problems of weather, certain natural phenomena beyond the control of the farmer and the extended production cycle in many enterprises. Once a plan has been accepted, on most of our farms certain major inputs or fixed costs like labour and power and machinery are predetermined, are used in a continuous flow and are not storable. It is also true that when a decision has been taken to grow a given acreage of cereals in the plan, expenditure on seeds, fertilisers and sprays (ie. variable costs) are predetermined, and little needs to be done to control such expenditure. The return obtained from an enterprise or business as a whole is influenced more by technical and marketing considerations than by trying to control the production cycle once it is under way. In practice, therefore, there is little point in trying short period control measures of arable crops, although monthly or quarterly overall financial control may still be worthwhile on a large arable farm.

Short Period Control Measures of Individual Enterprises

Nevertheless, there are certain factors of production which do justify a serious attempt at control measures. Factory type enterprises like pigs, poultry and veal calves readily lend themselves to periodic checks. So too does milk production. The control of feed inputs in a pig fattening enterprise, or concentrate feed in milk production, justify careful control since they amount to a major part of costs. Measures of kg meal per kg liveweight gain, or feed costs per £100 pig output, are the really important tests of efficient pig production. In milk production, concentrates fed per litre, calving index, percentage of dry cows and milk yields are critical factors in a profitable enterprise.

There are numerous enterprise check schemes designed by the ADAS, Meat and Livestock Commission, Milk Marketing Board and commercial organisations, to provide farmers with the means of controlling these enterprises. The ADAS offer a poultry laying flock scheme which enables a monthly check on egg sales and feed to be made. The MMB Dairy Management Scheme not only enables a monthly check on performance to be made, but can also incorporate a forecast against which the farmer can measure actual achievement. The main function of these types of scheme is to provide a comparison between actual and planned performance, so that management can take appropriate steps to stem deviations from the plan before these erode the profit margin. There is little point in the exercise unless this comparison of actual and planned performance can be made.

Budgetary Appraisal for the Whole Farm Business

There are two distinctly different systems of whole farm control measures.

System 1

This, the simpler one, which might be called Budgetary Appraisal, sets out a detailed budget for next year's output and costs, and at the end of the year compares them with the actual performance. This is a logical sequence of farm planning. It involves the comparison of budgeted input and output in physical and financial terms with actual performance. Differences will be due to deviations of one or more of the following factors:

- Number of livestock or hectares.
- Yields.
- Prices realised.
- Quantities of inputs.
- Costs of inputs.

The control procedure should measure the extent to which each of the above factors has been responsible for the differences between plan and actual performance. Table 27 gives an example of this type of comparison, setting out an analysis of variance under the above headings to indicate where the differences lie. It is imperative that the budget is prepared in sufficient detail to allow comparison of these factors.

A careful examination of this example clearly shows the major reasons for not reaching the planned profit level. These lie in the failure of the farmer to keep the 90 cows he planned to do, to grow the area of barley planned, or to achieve the milk yield planned. The price of input items was higher than expected and power and machinery costs were particularly high. Quite obviously no forecast can be absolutely accurate; it is nevertheless valuable in clarifying these deviations which can readily be explained from those which need further investigation, as for example in this case of failure to keep the number of cows planned. The reason may be that the original plan was over optimistic, or, as is often the case, the farmer failed to appreciate the effect the extra cows would have on the final profit. The technique can be a useful measure of control on an annual basis. The explanation of the differences can be illustrated by taking the example of the cows.

Output Variance. Ninety cows, averaging 5 200 litres at an average price of 11·0p per litre were budgeted. In practice, the farmer only kept 87 cows which averaged 5 100 litres at an average price of 10·9p per litre. Output was £3 117 below the budget. The variation from the planned output is explained as follows. The farmer kept 3 cows less than planned, so the plan fell short on programme by 3 cows × 5 100 litres × 10·9p = £ 667·70.

The explanation due to programme is minus £1 667·70.
The variance due to yield is 90 cows × 100 litres × 10·9p = £981·00.
The variance due to price is 90 cows × 5 200 litres × 0·1p = £468·00.
The sum of these variances add up to £3 116·70 (rounded to £3 117): so how the difference arose has been explained.

In working out the variance of Output for the Programme and Yield, the rule is to multiply by the actual price received and not the forecast price.

Input Variance. If feed for the cows is used to demonstrate the analysis on the costs side, the planned budget was for 162 tonnes of feed at £130 per tonne costing £21 060. In practice the farmer kept 87 cows instead of 90, used 147·9 tonnes of feed and had to pay £135 per tonne giving an actual cost of £19 966·50. Cost of feed was £1 093·50 less than budgeted.

The explanation of this variance is partly due to the failure to reach the programme, partly due to the increased price and partly due to the reduction in the amount used. Thus the variance due to programme is 3 cows × 1·7 tonnes at £135 = £688.
Variance due to reduced input of feed = 9 tonnes at £135 = £1 215.
Variance due to price increase = 162 tonnes × £5 = £810.
The variation from the plan of feed costs is:
 −£688 −£1 215 + £810 = −£1 093

System 2

A more complete technique can be described as Budgetary Control. The purpose is to help management maintain a continuous scrutiny over the flow of cash in and out of the business. It involves greater effort on the part of the farmer, together with a more sophisticated budgeting method. The objective is to provide a cash flow control of the business over relatively short periods, usually monthly. In practical terms, there are considerable difficulties in making forecasts in such detail, but the attempt may often be very worthwhile to provide management with the facility to explore any deviations from plan, even where some can logically be explained.

The need for financial control is most acute in two main types of business situation, large businesses and those where the availability of capital is very critical.

The chief essentials required for financial control are the preparation of a list of capital expenditure and the recording of payments and receipts as they occur. Included in these statements must be any capital income as well as expenditure, private drawings, tax payments, interest charges and any repayments of capital. An example of the two combined in one, a cash flow budget on a monthly basis, is included in Table 28. In some farm situations a cash flow budget on a quarterly basis will be adequate. Much depends upon when the major cash flows in and out of the business take place. On a mainly cereal-growing farm, the cash out-flows occur primarily in the spring, and the cash in-flows in the autumn or winter, depending upon when the produce is sold. In these circumstances, a quarterly cash flow would suffice unless there was an overriding need to control all items of cash flow over shorter periods to keep within an overdraft ceiling.

Table 28 is a cash flow budget prepared for a year on a monthly basis. As can be seen from the headings, this layout can be used to provide a comparison between budget and actual by leaving alternative columns to be used for actual results.

A Commitment Budget. The system of cash flow control outlined above is perfectly satisfactory where bills are paid promptly. If payments are delayed or made on account, the system only adequately covers the receipts and expenditure, but fails to give any control of the quantities of resources used in the period. Under these circumstances, it is desirable to undertake in addition a Commitment Budget which as its name implies, indicates what commitments have been entered into. It can take the same form as a cash flow budget, except that entries would be made from delivery notes or invoices so that a control can be operated on the resources used rather than paid for.

Integration with the Recording System. Having drawn up the cash flow budget, some system of obtaining the actual performance for the selected control periods, eg. monthly or quarterly, is necessary. Providing there is a cash analysis recording system similar to the NFU Record Book, it is relatively easy to use one page per month, totalling the analysed columns to provide the actual cash in-flows and out-flows for the month. Ideally, the forecast payments and receipts should be transferred from the cash flow budget to the top of the recording sheets for comparison with the actual. It is also desirable to have cumulative totals as well as period totals. An example on the payments side of the cash analysis is shown in Table 29.

To obtain an overall picture, a single summary sheet could be used to show the budgeted and actual net in-flow and out-flow of cash at the end of each period. A suggested cash flow budget is in Table 30 on a quarterly basis. It would be possible to adopt the same format for a monthly period.

Table 27 Annual Budgetary Appraisal

Analysis of variance between actual and budgeted results on a 100 ha farm

Farm Plan	Cropping and Stocking (Budgeted)	Cropping and Stocking (Actual)
	90 Cows	87 Cows
	20 Sows	20 Sows
	55 ha Barley	53 ha Barley
	45 ha Grass	47 ha Grass

Budget for output		Actual output results		Measure of variance		Explanation of variance		
				By enterprise	By item	Programme	Performance Yield	Price
	£	£	£	£	£	£	£	£	£
MILK								
90 Cows×5 200 litres @ 11·0p		51 480		87 Cows×5 100 litres @ 10·9p		48 363	−3 117	−3 117	−1 668	−981	−468
CATTLE								
20 Cull cows @ £300	6 000		22 Cull cows @ £285	6 270			+270	+570		−300
85 Calves @ £60	5 100		82 Calves @ £62	5 084			−16	−186		+170
	11 100			11 354					
Less 10 Heifers @ £500	5 000		Less 9 Heifers @ £490	4 410					
		6 100			6 944	+844	−590 (+)	−490 (+)		−100 (+)
SOWS								
20 Sows×18 Weaners @ £22	7 920		20 Sows×19 Weaners @ £23	8 740			+820	+460		+360
6 Cull Sows @ £65	390		5 Cull Sows @ £59	295			−95	−59		−36
	8 310			9 035					
Less 6 Gilts @ £80	480		Less 6 Gilts @ £80	480					
		7 830			8 555	+725				
BARLEY								
55 ha×4·5 tonnes @ £90		22 275	53 ha×4·2 tonnes @ £92		20 479	−1 796	−1 796	−773	−1 518	+495
TOTAL GROSS OUTPUT		87 685			84 341	−3 344	−3 344	−1 626	−2 039	+321

Table 27 (cont)

Budget for input			Actual input results		Measure of variance		Explanation of variance		
					By input	By item	Programme	Input	Cost
	£	£		£	£	£	£	£	£
PURCHASED FEED									
Cows 90×1·8 tonnes @ £130	21 060		Cows 87×1·7 tonnes @ £135	19 967		−1 093	− 688	−1 215	+ 810
Pigs 20×1·8 tonnes @ £140	5 040		Pigs 20×1·9 tonnes @ £146	5 548		+ 508		+ 292	+ 216
		26 100		25 515	− 585				
SEED									
Barley 55 ha×155 kg @ 22p	1 875		Barley 53 ha×155 kg @ 22p	1 807		− 68	− 68		
Grass 10 ha @ £50	500		Grass 10 ha @ £53	530		+ 30			+ 30
		2 375		2 337	− 38				
FERTILISERS									
Barley 55 ha×0·5 tonnes @£110	3 025		Barley 53 ha×0·5 tonnes @£114	3 021		− 4	− 114		+ 110
Grass 25 tonnes @ £110	2 750		Grass 26 tonnes @ £114	2 964		+ 214	+ 126	− 12	+ 100
20 tonnes @ £100	2 000		20 tonnes @ £102	2 040		+ 40	+ 87	+ 87	− 40
		7 775		8 025	+ 250				
PAID LABOUR									
1 Herdsman @ £105/week	5 460		1 Herdsman @ £103/week	5 356		− 104			− 104
1 Other Worker @ £80/week	4 160		1 Other Worker @ £81/week	4 212		+ 52			+ 52
Casual labour	1 000		Casual labour	745		− 255			− 255
		10 620		10 313	− 307				
RENT									
100 ha @ £70/ha		7 000	100 ha @ £70/ha	7 000					
POWER AND MACHINERY									
100 ha @ £120/ha		12 000	100 ha @ £127/ha	12 700	+ 700	+ 700			+ 700
SUNDRIES									
100 ha @ £40/ha		4 000	100 ha @ £37/ha	3 700	− 300	− 300			− 300
TOTAL INPUTS		69 870		69 590	− 280	− 280	− 657	−1 022	+1 399
NET FARM INCOME		17 815		14 751	−3 064	−3 064	− 969	−1 017	−1 078

Table 28 Example of a monthly cash flow budget

		Period actual/budget APRIL			Period actual/budget MAY			Period actual/budget JUNE			Period actual/budget JULY		
	Date	No.	Val.	Total	No.	Val.	Total	No.	Val.	Total	No.	Val.	Total
RECEIPTS													
Trading													
Milk				4 150			4 100			3 620			2 680
Cattle:													
Calves													
Cull cows										2 100			1 200
Sheep							790						
Pigs				400						530			400
Poultry													
Crops:													
Wheat													
Barley				8 000			7 275						
Potatoes													
Roots													
Sundries													
Grants (CDP)													
Other receipts													
Capital													
Grants													
Mach. sales				65									
Personal receipts													
Sub Total (a)				12 615			12 165			6 250			4 280

Table 28 (contd.)

PAYMENTS				
Trading				
Feed:				
D. cows	350	50	50	220
Pigs	500	310	270	480
Vet. & Med.	100	100	100	100
Livestock purchases:			GILTS 240	
Seed				
Fertilisers			4 000	
Sprays				
Wages—Permanent	750	755	900	750
Casual	50	50	250	50
Power & Mach.—Fuel		400		
Resp Ins.	500	400	400	400
Con. Hire			800	
Rent				
Sundries	300	250	280	260
Interest, HP & Mort. payment			1 840	
Capital				
Buildings etc.				
Machinery				
Other incl. capital repayments				
Personal				
Drawings	280	280	280	280
Tax				
Any other payments			400	
Sub Total (*b*)	2 830	2 595	9 810	2 540
Net Cash Flow for period (*a–b*)+	9 785	9 570		1 740
Net Cash Flow for period (*a–b*)−			3 560	
Cumulative Balance	−3 715	−4 145	−7 705	−5 965
Annual Net Cash Flow B/F—£23 500				

Table 28 (contd.)

Date	Period actual/budget AUGUST			Period actual/budget SEPTEMBER			Period actual/budget OCTOBER			Period actual/budget NOVEMBER		
	No.	Val.	Total	No.	Val.	Total	No.	Val.	Total	No.	Val.	Total
RECEIPTS												
Trading												
Milk			2 350			4 230			5 000			5 340
Cattle:												
Calves			500			1 020			1 760			1 010
Cull cows			1 500			300						
Sheep												
Pigs			790			530			910			400
Poultry												
Crops:												
Wheat												
Barley												
Potatoes												
Roots												
Sundries:												
Grants (CDP)			120			150			1 800			
Other receipts			800									
Capital												
Grants												
Mach. sales									65			
Personal receipts												
Sub Total (a)			6 060			6 230			9 535			6 750

98

Table 28 (contd.)
PAYMENTS
Trading

Feed:					
D. cows		650	2 150	2 890	3 650
Pigs		340	510	490	460
Vet. & Med		100	100	100	100
Livestock purchases:	HEIFERS	5000			
Seed			1 425		
Fertilisers				750	750
Sprays		900	900	800	
Wages—Permanent		250	50	50	50
Casual		400			
Power & Mach.—Fuel		550	450	400	400
Resp Ins.		750	50	150	50
Con. Hire			3 500		
Rent		290	340	360	360
Sundries					
Interest, HP & Mort. payment					
Capital					
Buildings etc.		2 300	4 500		
Machinery					
Other incl. capital repayments					
Personal					
Drawings		280	280	280	280
Tax					800
Any other payments		250			
Sub Total (*b*)		12 060	14 255	6 270	6 900
Net Cash Flow for period (*a*–*b*)+			8 025	3 265	
Net Cash Flow for period (*a*–*b*)–		6 000			150
Cumulative Balance		–11 965	–19 990	–16 725	–16 875
Annual Net Cash Flow					

Table 28. (contd.)

	Date	Period actual/budget DECEMBER			Period actual/budget JANUARY			Period actual/budget FEBRUARY			Period actual/budget MARCH		
		No.	Val.	Total	No.	Val.	Total	No.	Val.	Total	No.	Val.	Total
RECEIPTS													
Trading													
Milk				5 680			5 270			4 760			4 300
Cattle:													
Calves				520			60			110			120
Cull cows				300						600			
Sheep													
Pigs				620			500			530			910
Poultry													
Crops:													
Wheat													
Barley													7 000
Potatoes													
Roots													
Sundries:													
Grants													
Other receipts													
Capital													
Grants													
Mach. sales											3 500			500
Personal receipts													
Sub Total (a)				7 120			5 830			9 500			12 830

Table 28 (contd.)
PAYMENTS
Trading

Feed:					
D. Cows	3 400		3 120		2 510	2 020
Pigs	430		460		510	280
Vet. & Med	100		100		100	100
Livestock purchases:			240	GILTS		
Seed		3 775			850	
Fertilisers						100
Sprays					250	400
Wages—Permanent	750		750		750	865
Casual	50		50		50	50
Power & Mach.—Fuel	400		400		400	450
Resp Ins.	500					
Con. Hire	40					
Rent						3 500
Sundries	350		390		320	350
Interest, HP & Mort. payment	1280					
Capital						
Buildings etc.						
Machinery					8 500	
Other incl. capital repayments						
Personal						
Drawings	280		280		280	280
Tax						
Any other payments	400					
Sub Total (*b*)	11 755		5 790		14 520	8 395
Net Cash Flow for period (*a*–*b*)+			40			4 435
Net Cash Flow for period (*a*–*b*)–	4 635				5 020	
Cumulative Balance	–21 510		–21 470		–26 490	–22 055
Annual Net Cash Flow						

Table 29 Payments analysis record for month of April 1980

Date	Detail	Amount £	Cheque No.	Contra £	Petty Cash £	Livestock Purchase £	Feed £	Seed £	Ferts £	Sprays £	Wages (Perm) £	Wages (Casual) £	Fuel & Oil £	Mach Repairs £	Contract & Hire £	Insurance £	Rent & Rates £	Sundries £	Drawings £	Etc.
	Brought Forward	—		—	—	—	—	—	—	—	—	—	—	—	—	—	—	—	—	
	Budget—March	2 830					850				750	50		500				300	280	
2/3	Cam Feeds Ltd	320					320													
5/3	Post Office (Tel. a/c)	57																57		
5/3	J Howarth	103									103									
5/3	M Munslow	77									77									
12/3	Watsons Ltd	625												625						
13/3	A & B King	287																287		
16/3	Cam Feeds Ltd	595					595													
17/3	Farm Supplies Ltd	179								130								49		
19/3	J Howarth	214									214									
19/3	M Munslow	164									164									
19/3	CASH (Drawings)	203																	230	
26/3	J Howarth	106									106									
26/3	M Munslow	79									79									
	TOTAL	3 036					915			130	743			625				393	230	
	CUM TOTAL carried fwd	3 036					915			130	743			625				393	230	

Table 30 Cash flow summary

Month	BUDGET						ACTUAL					
	Total Receipts £	Total Payments £	Net Inflow £	Net Outflow £	Bank Balance		Total Receipts £	Total Payments £	Net Inflow £	Net Outflow £	Bank Balance	
					At Bank £	Overdrawn £					At Bank £	Overdrawn £
Balance at 1/4/80						23 500						23 500
April	12 615	2 830	9 785		13 715							
May	12 165	2 595	9 570		4 145							
June	6 250	9 810		3 560	7 705							
July	4 280	2 540	1 740		5 965							
August	6 060	12 060		6 000	11 965							
September	6 230	14 255		8 025	19 990							
October	9 535	6 270	3 265		16 725							
November	6 750	6 900		150	16 875							
December	7 120	11 755		4 635	21 510							
January	5 830	5 790	40		21 470							
February	9 500	14 520		5 020	26 490							
March	12 830	8 395	4 435		22 055							
TOTALS	99 165	97 720										

103

APPENDIX I
IN CONFIDENCE

Farm Management No. | 9 9 8 8 7 7 7

Year Ended | Month MARCH | 19 79

Total Farm Acreage (including Roads, Buildings, Woodlands) | 1 0 0

MINISTRY OF AGRICULTURE, FISHERIES AND FOOD

AGRICULTURAL DEVELOPMENT AND ADVISORY SERVICE

FARM FINANCIAL ANALYSIS

Calculation of Annual Cash Flow
Balance Sheet

Note: For Farm Business Recording Scheme purposes it is only obligatory to complete page 2

MA 9 (Revised May 1972)

2610 D.926842 8800cps. 8/72 Cr.P.C. Gp.800

CALCULATION OF ANNUAL CASH FLOW

SECTION A : CALCULATION OF CASH FLOW FROM TRADING

	£
PROFIT/LOSS (from MA 1, MA 16 or Accounts)	+/− 21278
ADD: Depreciation of Machinery	2900
Depreciation of Buildings and Fixtures	246
Decrease in Valuation of:	
Livestock	
Harvested Crops	
Growing Crops and Tillages	
Stores and Materials	
Decrease in Debtors	
Increase in Creditors	210
Notional Payments	
Sub-Total (a)	24634
DEDUCT:	
Notional Receipts	1080
Increase in Valuation of:	
Livestock	857
Harvested Crops	600
Growing Crops and Tillages	
Stores and Materials	520
Increase in Debtors	154
Decrease in Creditors	
Sub-Total (b)	3211
CASH FLOW FROM TRADING (a)−(b)=(c)	+/− 21423

SECTION B : CAPITAL TRANSACTIONS

	£
ADD: Receipts - Disposals of Land and Buildings	
Receipts - Disposals of Machinery and Fixed Equipment	2000
Capital Grants	
Sub-Total (d)	2000
DEDUCT:	
Purchase of Land and Buildings	
Improvements to land and Buildings	2460
Purchase of Machinery & Fixed Equipment	8000
Sub-Total (e)	10460
CASH FLOW FROM CAPITAL (d)−(e)=(f)	+/− −8460
BALANCE AVAILABLE FOR TAXATION AND PRIVATE USE (c)+(f)=(g)	+/− 12963

Balance available (brought forward)(g) | ± 12963 |

SECTION C : PRIVATE

    ADD: Funds Introduced ....

         Private Receipts ....

                           Sub-Total (h) | 0 |

    DEDUCT:
         Private Drawings ....5000........

         Tax ....6432........

         Off farm Investments ....

                         Sub-Total (i) | 11432 |

                         Total (h)−(i)=(j) | ± −11432 |

TOTAL SURPLUS/DEFICIT OF FUNDS FOR YEAR (g)+(j)=(k) | ± 1531 |

SECTION D : FINANCIAL SUMMARY

    ADD: Increase in Loans ....

         Increase in Overdraft ....1637........

         Reduction in Credit Balances and Cash ....

                         Sub-Total (l) | 1637 |

    DEDUCT:
         Reduction of Loans ....106........

         Reduction in overdraft ....

         Increase in Credit Balances and Cash ....

                         Sub-Total (m) | 106 |

                         Total (l)−(m)=(n) | ± 1531 |

Total (n) must equal Total (k).

SECTION E   BALANCE SHEET

	As at 19......	19.79....	19..........
FIXED ASSETS			
Land (1)		250000	
Buildings and Fixtures		2214	
Machinery		9300	
Orchards and Other Long Term Crops			
Breeding Livestock		20800	
TOTAL FIXED ASSETS (a)		282314	
CURRENT ASSETS			
PHYSICAL WORKING ASSETS			
Other Livestock		15189	
Harvested Crops		3150	
Growing Crops and Tillages		900	
Stores and Materials		5060	
Sub-Total (b)		24299	
LIQUID ASSETS			
Debtors and Prepayments		3100	
Bank Balance and Deposits		2880	
Cash in Hand		120	
Sub-Total (c)		6100	
TOTAL CURRENT ASSETS (b)+(c)=(d)		30399	
TOTAL ASSETS (a)+(d)=(e)		312713	
CURRENT LIABILITIES			
Creditors (including HP outstanding)		2870	
Tax owed		8419	
Bank overdraft		6500	
TOTAL CURRENT LIABILITIES (f)		17789	
NET ASSETS (e)−(f)=(g)		294924	
LONG TERM LIABILITIES			
Bank Loans			
Private Loans			
Other Long Term Loans Loan AMC £12500 Fixed Rate of Interest 12½% over 20 years		12394	
SUB TOTAL LOAN CAPITAL (h)		12394	
OWNER EQUITY (2) (g)−(h)=(i)		282530	
CAPITAL EMPLOYED (h)+(i)=(j)		294924	

1. Use present values (ie within 4 years)
2. Owner Equity — including appreciation of fixed assets

SECTION F — BALANCE SHEET RATIOS

		19.......	1979.....	19.......
OWNER EQUITY	$\dfrac{\text{Owner Equity}}{\text{Total Assets}} \times 100$			
	$\dfrac{\text{(i) } 282530}{\text{(e) } 312713} \times 100$	%	90.4 %	%
FIXED ASSETS	$\dfrac{\text{Fixed Assets}}{\text{Total Assets}} \times 100$			
	$\dfrac{\text{(a) } 282314}{\text{(e) } 312713} \times 100$	%	90.3 %	%
GEARING	$\dfrac{\text{Loan Capital}}{\text{Owner Equity}} \times 100$			
	$\dfrac{\text{(h) } 12394}{\text{(i) } 282530} \times 100$	%	4.4 %	%
CURRENT	$\dfrac{\text{Current Assets}}{\text{Current Liabilities}} \times 100$			
	$\dfrac{\text{(d) } 30399}{\text{(f) } 17789} \times 100$	%	170.9 %	%
LIQUIDITY	$\dfrac{\text{Liquid Assets}}{\text{Current Liabilities}} \times 100$			
	$\dfrac{\text{(c) } 6100}{\text{(f) } 17789} \times 100$	%	34.3 %	%

SECTION G — RETURNS ON CAPITAL

RETURN ON OWNER EQUITY

$\dfrac{^{*}\text{Profit} \times 100}{\text{Owner Equity}}$

$\dfrac{21278}{282530} \times 100$

19.......	19.......	19.......
%	7.5 %	%

RETURN ON NET ASSETS

$\dfrac{^{*}\text{Profit plus Long Term Interest} \times 100}{\text{Net Assets}}$

$\dfrac{21278 + 1548}{294924} \times 100$

%	7.7 %	%

RETURN ON TOTAL ASSETS

$\dfrac{^{*}\text{Profit plus all Interest} \times 100}{\text{Total Assets}}$

$\dfrac{21278 + 2040}{312713} \times 100$

%	7.5 %	%

*Profit is the difference between Gross Output and Costs, Costs include all Interest but NOT Tax.

APPENDIX 2
Ministry of Agriculture, Fisheries and Food

IN CONFIDENCE

AGRICULTURAL DEVELOPMENT AND ADVISORY SERVICE

INFORMATION FOR FARM BUSINESS ANALYSIS

	FOR OFFICE USE	
1.		1
2. I.C. (2=normal; 3=replacement)		2
3. Region and Group		1 3 0
	County / District/Adviser / Farm	
4. F.M. No.	9 9 8 8 7 7 7	

For a management analysis of the farm business, please complete this document as fully as possible. Notes for guidance on this document appear overleaf. If you are in difficulty, require clarification of some items or advice generally, please contact your Local Agricultural Advisory Officer.

PLEASE HELP US TO GIVE YOU A BETTER SERVICE BY MAKING YOUR ENTRIES CLEARLY IN INK.

5. Accounting Year End (Month 01=Jan. Month 12=Dec.) .... Month `0 3` Year `7 9`

6. Area Used for Agriculture (to nearest whole hectare—unit values on extreme right) .... `1 0 0`

7. Data Mark. `1`

8. Altitude of Main Part of Farm (1=0 to 100m; 2=101 to 200m; 3=201 to 300m; 4=301 to 400m; 5=over 400m) .... `1`

9. Average Annual Rainfall (1=under 750mm; 2=751 to 999mm; 3=1000 to 1249mm; 4=1250 to 1499mm; 5=over 1500mm) .... `2`

10. Predominant Soil Texture (1=heavy; 2=medium; 3=light; 4=peaty; 5=organic) .... `2`

11. Form of Business (1=Sole Trader; 2=Family Partnership; 3=Non-Family Partnership; 4=Farming Company; 5=Other) .... `1`

12. Percentage of Farm Rented (1=None; 2=1 to 24%; 3=25 to 49%; 4=50 to 75%; 5=75 to 99%; 6=all) .... `1`

13. *Breed of Dairy Cows (1=Friesian; 2=Ayrshire; 3=Channel Islands; 4=Mixed; 5=Other) .... `1`

14. *Average Weight of Ewes (1=Under 50 kg; 2=50 to 70 kg; 3=Over 70 kg) .... `0`

MA 1 (Rev. June 1976)
 (Reprinted 1977)

*If not applicable, please enter O in the box.

NOTES FOR GUIDANCE

This MA1 document is standardised to assist in the collection of management information about the farm business during one financial year. It can be completed from normal Trading and Profit and Loss account data plus a limited amount of additional information.

For a full evaluation of the business please complete the form as fully as possible by providing financial and physical information in the spaces provided.

The form is designed for computer processing of data and as a basis for discussion between a farmer and his adviser. To enable computer processing to proceed smoothly please observe the following rules:

(1) DO TRY AND FIT THE FARM INFORMATION TO THE LINE DESCRIPTION.
DO NOT BRACKET ENTRIES.

(2) DO NOT CROSS OUT ANY PRINTED LINES,
DO NOT INSERT ALTERNATIVES.

(3) WHOLE NUMBERS ARE REQUIRED FOR MOST BOXES BUT
WHERE DECIMAL ENTRIES ARE INDICATED PLEASE FILL THE DECIMAL SPACE WITH A VALUE OR A ZERO.

(4) INCLUDE ITEMS OF EXPENDITURE AND REVENUE ONCE ONLY.
(i.e. avoid double counting.)

AMOUNT OF INFORMATION REQUIRED. The form can be processed through to a management report if only a small amount of information is available. Naturally, however, the provision of a small amount of information will only produce a limited summary of the business. It follows, therefore, that the more detail you supply in each section, the better the management report and the greater its value to you.

In cases where an accurate division between groups is impossible (e.g. Dairy Cows and Beef Cows) make an arbitrary split between them. It is important, however, to have consistency of treatment on each page i.e. if you have elected to class animals as Dairy Cows on Page 3 please ensure that they are also classed as Dairy Cows on Pages 4, 5, 7 and 11.

SEE "NOTES FOR SPECIFIC ENTRIES" ON PAGE 114.

VALUATIONS
(See Note 7)

	Item	OPENING VALUATION at (date) 03/78		CLOSING VALUATION at (date) 03/79	
LIVESTOCK (AT REALISTIC VALUES)		No.	£	No.	£
Dairy Cows	1	80	20000	80	20800
Bulls for Dairy Herd	2				
Dairy Heifers	3	39	8112	40	8084
Beef Cows	6				
Bulls for Beef Herd	7				
Other Beef Cattle	8	52	7020	49	7105
Ewes (incl. shearlings) and Rams	11				
Other Sheep	13				
Sows and Boars	14				
Other Pigs (incl. in-pig gilts)	16				
Laying Hens (incl. breeders)	17				
Other Poultry (specify)	18				
Other Livestock (specify)	19				
Livestock Products: Wool	20				
Other Livestock Products (specify)	21				
(a) Total Livestock	22		35132		35989

	Item	Tonnes (if possible)	£	Tonnes (if possible)	£
HARVESTED CROPS					
Wheat	23	35	2450	35	2800
Barley	24				
Oats	25				
Potatoes	26				
Sugar Beet	27				
Other Cash Crops (specify and code: see Page 8) Crop Code	28				
	29				
	30				
	31				
	32				
	405				
	406				
Feed Roots	33				
Other Forage Crops (specify)	34				
Hay (Home Grown)	35				
Silage (Home Grown)	36				250
Straw (Home Grown)	37		100		100
(b) Total Harvested Crops	39		2550		3150
GROWING CROPS (see Note 8)					
Annual	407		900		900
Perennial	408				
(c) Total Growing Crops	40		900		900
GOODS IN STORE					
Fertilisers	41		3600		4000
Seeds and Plants (purchased)	42		400		440
Sprays	43				
Concentrates (purchased)	44		490		540
Hay (purchased)	45				
Straw (purchased)	46				
Other Purchased Feed	47				
Fuel, Oil, Gas etc.	48		50		80
Others (specify)	49				
(d) Total Goods in Store	51		4540		5060
(e) Total Valuation (a)+(b)+(c)+(d)	52		43122		45099

TRADING ACCOUNT — EXPENDITURE

Category	Item	No. bought (if possible)	A	£		Sub-Total	
LIVESTOCK PURCHASES	Dairy Cows and Calved Heifers		53	1740			
	Bulls for Dairy Herd		54				
	Dairy Heifers (including calves)		55				
	Beef Cows		56				
	Bulls for Beef Herd		57				
	Other Beef Cattle (including calves)		58				
	Ewes for Breeding and Rams		59				
	Store Lambs and other Sheep		62				
	Sows and Boars		63				
	Other Pigs (incl. in-pig gilts)		65				
	Poultry		66			Sub-Total	
	Other Livestock (specify)		67		68	1740	
FEED	Bought Concentrates		69	22727			
	Bought Hay and Feeding Straw		70				
	Other Feed		71			Sub-Total	
	Keep Rented		72		73	22727	
OTHER VARIABLE COSTS	Fertilisers (£) Lime (£)		74	5668			
	Purchased Seeds and Plants		75	1544			
	Crop Sprays		76	600			
	Crop Sundries (incl. P.M.B. Levy)		77	200			
	Veterinary and Medicines		78	3020			
	Contract Rearing: Cattle		79				
	Contract Rearing: Sheep		80				
	Livestock Sundries (incl. purchased bedding)		81	1220		Sub-Total	
	Other		82		83	12252	
WAGES AND SALARIES	Regular Labour (incl. National Insurance)		84	7200			
	Casual Labour		85				
	Wife (if paid)		86				
	Paid Management		87			Sub-Total	
	Secretarial		88		89	7200	
POWER AND MACHINERY	Machinery and Vehicle Repairs		90	2050			
	Contract and Transport		91				
	Leasing Charges		92				
	Fuel and Oil		93	1150			
	Electricity, Coal, Gas etc		94				
	Vehicle Tax and Insurance		96			Sub-Total	
	Depreciation of Cars, Vans, Lorries, Tractors and other machinery		97	2900	99	6100	
PROPERTY CHARGES	Rent		100				
	Rates		101				
	Depreciation of Buildings, Improvements and Fixed Equipment	Tenant's Items Only See Note 10	102	246		Sub-Total	
	Repairs to Buildings, Fences etc.		104	1090	105	1336	
SUNDRIES	Water		106	430			
	General Insurance		107	520			
	Office and Telephone		108	350			
	Professional Fees (Accountant, etc.)		109	380		Sub-Total	
	Others (incl. Subscriptions etc.)		110	230	112	1910	
FINANCING CHARGES	Bank Charges and Interest		113	492			
	Mortgage and other Loan Interest		114	1548		Sub-Total	
	Hire Purchase Interest		115		116	2040	
	Total Expenditure (box 68+73+83+89+99+105+112+116)				117	55305	
	Opening Valuation (box 52)				118	43122	
	Total (box 117+118)				119	98427	
	PROFIT (box 185 less box 119)				120	21278	
SUPPLEMENTARY INFORMATION	Value of Unpaid Manual Labour (incl. farmer)		121	2000			
	Estimated Rental Value of Owner-occupied Land		122	4000			
	Area of Owner-occupied Land (Hectares)		123	100			

TRADING ACCOUNT — REVENUE

		No. Sold (if possible)	£		
LIVESTOCK SALES (include associated premia and subsidies)	Dairy Cows	124	7740		
	Bulls from Dairy Herd	125			
	Calves from Dairy Herd (0 to 3 months)	126			
	Dairy Heifers (over 3 months)	127			
	Beef Cows	128			
	Bulls from Beef Herd	129			
	Other Beef Cattle Fat £ Store £	130	14040		
	Ewes and Rams Ewes £ Rams £	131			
	Lambs: Fat £ Store £	133			
	Sows and Boars: Sows £ Boars £	134			
	Other Pigs (incl. in-pig gilts)	136			
	Poultry	137			Sub-Total
	Other Livestock (specify)	138		139	21780
LIVESTOCK PRODUCE (include associated premia and subsidies)	Milk	140	40400		
	Cream, Cheese, Butter	141			
	Wool	142			
	Eggs	143			Sub-Total
	Other Produce (specify)	144		145	40400
LIVESTOCK SUBSIDIES	Cattle e.g. Hill Cow See Note 11.	146			
	Sheep e.g. Hill Sheep	147			Sub-Total
	Other Livestock Grants e.g. Brucellosis incentives	148		151	
CROP SALES AND SUBSIDIES	Wheat	152	7452		
	Barley	153	2064		
	Oats	154			
	Potatoes	155			
	Sugar Beet	156			
	Other Cash Crops (specify and code: See Page 8) Crop Code 157, 158, 159, 160, 161, 409, 410				
	Straw	162	1220		
	Hay and Silage	163			Sub-Total
	Feed Roots	164		165	10736
SUNDRIES	Lime Subsidy	166			
	Grant for Keeping Accounts	167			
	Other Grants (exclude Capital Grants) Specify:	168			
	e.g. FHDS Guidance Premium	169			
		170			
	Keep Let	172			
	Contract Work	173	610		
	Wayleaves, Rents received etc.	174			Sub-Total
	Others (specify)	175		176	610
NOTIONAL RECEIPTS	Rental Value of House	177	400		
	Private Use of Car, Electricity, etc.	178	400		
	Produce Consumed: Milk	179	280		
	Poultry and Eggs	180			Sub-Total
	Other	181		182	1080
	Total Revenue (box 139+145+151+165+176+182)			183	74606
	Closing Valuation (box 52)			184	45099
	Total (box 183+184)			185	119705
	LOSS (box 119 less box 185)			186	

NOTES FOR SPECIFIC ENTRIES

(6) **Crop Codes** — Blank lines are provided on Page 3 (Valuations), Page 5 (Revenue) and Page 9 (Farm Area) for crops additional to those listed. These crops will be identified by the computer by a code number and will subsequently be shown on the management report by name.

A list of crops codes is given on page 8 and the user is asked to enter these as appropriate in the code boxes provided. PLEASE ENSURE THAT CODING IS CONSISTENT BETWEEN PAGES 3, 5 and 9 i.e. use the same code number for the same crop appearing on different pages.

(7) **Valuations** — Realistic values are required for all entries. Numbers **must** be given for livestock valuations but tonnage figures for harvested crops are optional. Note that Hay, Silage, and Straw appear in two places, depending on whether they are homegrown (enter as Harvested Crops) or purchased (enter as Goods in Store).

(8) **Growing Crops** — Separate lines are provided for Annual crops and Perennial Crops (such as Trees, Bush crops, biennials etc). Both lines can, if necessary, include an element of valuation for tillages.

Valuations of Growing crops can be on various bases but are usually based on (a) expected sale value less expected costs to date of sale, or (b) Variable costs (seed, fertiliser etc) incurred to date. These and other methods of valuation are dealt with in more detail in "Definition of Terms used in Business Management" published by MAFF.

TRADING ACCOUNT ITEMS

(9) **Handling of VAT** — all entries in Expenditure and Revenue sections should normally not include VAT (i.e. entries should be net). If it is not possible to show items net, particularly in the expenditure section, contra entries of VAT should be made. VAT refunds on items of expenditure should be entered in boxes 169 or 170 and separately identified as VAT refunds. VAT in respect of Revenue items should not be entered at all.

(10) **Property Charges** — These should be for the **farm business** and therefore should not include items relating to the ownership of land. Entries in boxes 102 and 104 in particular should reflect this.

(11) **Livestock Sales and Subsidies** — There are two possible areas for the entry of livestock subsidies and grants. Subsidies, deficiency payments, quality premia and the like should be included with livestock sales; the Livestock subsidies section should be used for items not linked directly to sales.

F.H.D.S. guidance premia and similar grants related to farm plans should be entered in the Sundries section.

TENANT'S CAPITAL

		B Value at Start of Year (£)	Purchases (less Grants) (£)	Sales (£)	Depreciation* (£)	Value at End of Year (£)
Buildings and Improvements	Tenant's Items only	187	188 2460	189	190 246	191 2214
Fixed Equipment		192	193	194	195	196
Cars, Vans, and Lorries		197	198	199	200	201
Tractors and other machinery		202 6200	203 8000	204 2000	205 2900	206 9300
Sub-Total		207 6200	208 10460	209 2000	210 3146	211 11514
Closing Valuation of Livestock (from box 22)						212 35989
Crops and Goods in Store (from boxes 39, 40 and 51)						213 9110
Total Tenant's Capital (box 211+212+213)						214 56613

(*Depreciation should also be entered in boxes 97 and 102)

PHYSICAL INFORMATION

GRAZING LIVESTOCK (to nearest whole number)

		Annual Average
Dairy Cows	215	80
Bulls for Dairy Herd	216	
Dairy Replacements: 0—12 months	217	19
12—24 months	218	19
Over 24 months	219	4
Beef Cows	220	
Bulls for Beef Herd	221	
Other Beef Cattle: 0—12 months	222	
12—24 months	223	
Over 24 months	224	
Horses and other equines	225	
Ewes	226	
Ewe Replacements	227	
Rams	228	Throughput
Lambs: Birth to Store	229	
Birth to Fat	230	
Birth to Hoggetts	231	
Purchased Store Lambs	232	

NON-GRAZING LIVESTOCK

Calves Reared for Sale at under 3 months	234	
Beef Cattle Reared Indoors all their life	235	52
Sows	236	
Boars	237	
In-Pig Gilts	238	
Other Pigs for Sale Fat	239	
for Sale as Stores	240	
Laying Hens (incl. Breeders)	241	
Other Poultry (specify)	242	
Other Livestock (specify)	243	

TRANSFERS BETWEEN CATTLE ENTERPRISES

		Number	Value £
Calves from Dairy Herd retained for Dairy Replacements	244	21	1050
Calves from Dairy Herd retained for Beef Production	245	52	2600
Freshly Calved Heifers transferred to Dairy Herd	246	20	8000

LAMBING RESULT

Ewes put to Ram	247	
Lambs Reared	248	

CROP CODES

Crop code entries are required on Pages 3, 5 and 9 for crops other than WHEAT, BARLEY, OATS, POTATOES and SUGAR BEET. Please ensure that the coding is consistent on all three pages for the same crop.

CROP	CODE
CEREALS:	
MIXED CORN	1
RYE	11
OIL SEED RAPE	12
MUSTARD	21
MAIZE (for grain)	22
LEGUMES:	
PEAS (dried)	15
PEAS (green)	72
FIELD BEANS	18
BEANS (green)	76
GRASS PRODUCTS:	
HERBAGE SEED	23
CLOVER SEED	35
OTHER SEED CROPS	29
HAY FOR SALE	30
ROOT CROPS:	
CARROTS	53
PARSNIPS	57
SWEDES (for sale)	58
CHICORY	80
BRASSICAE:	
BRUSSELS SPROUTS	61
BROCCOLI	64
CAULIFLOWER	65
CABBAGE	68

CROP	CODE
OTHER VEGETABLES:	
ONIONS	56
BEETROOT	71
LETTUCE	81
LEEKS	82
CELERY	83
OTHER VEGETABLES	84
RHUBARB	60
OTHER CROPS:	
APPLES	209
PEARS	211
CHERRIES	213
PLUMS	215
OTHER TOP FRUIT	217
BLACKCURRANTS	218
STRAWBERRIES	220
RASPBERRIES	222
GOOSEBERRIES	224
OTHER SOFT FRUIT	226
HOPS	227
TULIPS	235
DAFFODILLS & NARCISSI	236
ANY OTHER CROP	90

PHYSICAL INFORMATION (continued)

FARM AREA (enter a decimal value or a zero behind decimal point)

		Hectares (To 1 decimal place)	Yield-Tonnes per Hectare
Wheat	249	18 . 0	5 . 4
Barley	250	22 . 0	4 . 4
Oats	251	.	.
Potatoes	252	.	.
Sugar Beet	253	.	.
Other Cash Crops (specify and code)	Crop Code		
	254 (	.	.
	255 (	.	.
	256 (	.	.
	257 (	.	.
	258 (	.	.
	259 (	.	.
	260 (	.	.
	261 (	.	.
	262 (	.	.
	263 (	.	.
	264 (	.	.
Kale	265	.	
Roots for Stockfeed	266	.	
Other Forage Crops for Livestock	267	.	
Temporary Grass	268	60 . 0	
Permanent Grass	269	.	
Fallow and Bare land	271	.	
Crops and Grass Total (boxes 249 to 271)	411	100 . 0	
Rough Grazing	270	.	
Area used for Agriculture (box 411+270)	272	100 . 0	
Buildings, Roads, Woods etc.	273	.	
Total Farm Area (box 272+273)	274	100 . 0	

FORAGE AREA OF FARM

		Equivalent Hectares (to 1 decimal place)
Grass (incl. hay and silage)	275	60 . 0
Kale, Roots, Forage, Maize etc.	276	.
Rough Grazing (grass equiv.)	277	.
Hill Rights (grass equiv.)	278	.
Grass Keep — Hired	279	.
LESS Grass Keep — Let	280	.
LESS Sales of Hay, Roots etc.	281	.
Opening stock minus closing stock of forage (if significant)	282	.
Adjusted Forage Area	283	60 . 0

ALLOCATION OF FEED

C	DAIRY COWS		DAIRY HEIFERS		BEEF CATTLE		SHEEP	
	tonnes	£	tonnes	£	tonnes	£	tonnes	£
Home Grown Wheat	284		285		286		287	
Home Grown Barley	288		289		290	71	5680	291
Home Grown Oats	292		293		294		295	
Other Home Grown Cereals	296		297		298		299	
Purchased Cereals	300		301		302		303	
Purchased Concentrates	304	152	16720	305	20	2400	306	22	3557	307
Wet Grains	312		313		314		315	
Wet Beet Pulp	316		317		318		319	
Total	308	152	16720	309	20	2400	310	93	9237	311

	PIGS		POULTRY	
	tonnes	£	tonnes	£
Home Grown Wheat	344		345	
Home Grown Barley	346		347	
Home Grown Oats	348		349	
Other Home Grown Cereals	350		351	
Purchased Cereals	352		353	
Purchased Concentrates	354		355	
Skim Milk (litres)	362		363	
Potatoes/Wet Grains	364		365	
Total	356		357	

	tonnes	Value per tonne £
Home Grown Potatoes used for Feed	412	

FARM CROPS RETAINED FOR SEED

	tonnes	Value per tonne £
Wheat	374	
Barley	375	
Oats	376	
Other Cereals	377	
Potatoes	378	
Other (specify)	379	

LABOUR

	Number	Hours worked during year*
Regular Male	380	2	4400
Regular Female	381		
Casual Male	382		
Casual Female	383		
Salaried Management	384		
Farmers and Partners	385		
Farmers' and Partners' Wives	386		
Total	387		

*Enter approximate time spent on manual work only
DO NOT include managerial or clerical work.

MONTHLY MILK PRODUCTION (LITRES)

Sales:			
April	388	34350
May	389	32004
June	390	32040
July	391	29760
August	392	26940
September	393	25706
October	394	28600
November	395	36430
December	396	38610
January	397	34810
February	398	34700
March	399	32690
Total Sales	400	386640
Farmhouse	401	1400
Employees	402	1400
Livestock	403	4100
Total Production	404	393540

APPENDIX 3

Farm Management No | 9 | 9 | 8 | 8 | 7 | 7 | 7 |

IN CONFIDENCE

MINISTRY OF AGRICULTURE, FISHERIES AND FOOD

AGRICULTURAL DEVELOPMENT AND ADVISORY SERVICE

FARM MANAGEMENT REPORT

for year ending ... 31 March 19.79

LAND AREAS

Total Area	100.0	ha
Agricultural Utilisable Area	100.0	ha
Adjusted Agricultural Utilisable Area	100.0	ha
Adjusted Forage Area	60.0	ha

MA2 (NR) (Revised 1978)

SUMMARY OF FINANCIAL RESULTS

	Your Farm of ....Adjusted hectares		STANDARD or YOUR FARM LAST YEAR
ENTERPRISE OUTPUTS	£ Total	£ per Adjusted ha	£ per Adjusted ha
Cereals	16766	167.7	
Other Crops			
Dairy Herd and Milk	43130	431.3	
Dairy Followers	6922	69.2	
Beef Cattle	11525	115.2	
Sheep and Wool			
Pigs			
Poultry and Eggs			
Other Livestock			
Change in Tillage Valuation			
Forage Sales and Valuation Change	250	2.5	
Other Receipts	1410	14.1	
TOTAL FARM OUTPUT	80003	800.0	
VARIABLE COSTS			
Feed: Homegrown	5680	56.8	
Bought	22677	226.8	
Seed: Homegrown			
Bought	1504	15.0	
Fertilizer	5268	52.7	
Casual Labour			
Contract Charges			
Other Variable Costs	5040	50.4	
TOTAL VARIABLE COSTS	40169	401.7	
TOTAL FARM GROSS MARGIN	39834	398.3	
FIXED COSTS			
Regular Labour: Paid	7200	72.0	
Unpaid	2000	20.0	
Machinery: Depreciation	2900	29.0	
Fuel, Oil, Electricity	1120	11.2	
Repairs, Tax, Insurance, Leasing	2050	20.5	
Rent and Rates – Paid			
Rent – Notional if Owner Occupier	4000	40.0	
Other Fixed Costs	3246	32.5	
TOTAL FIXED COSTS	22516	225.2	
MANAGEMENT AND INVESTMENT INCOME	17318	173.1	
Add Back: Notional Rent	4000	40.0	
Unpaid Labour	2000	20.0	
Subtract: Financing Charges	2040	20.4	
Paid Management			
PROFIT	21278	212.8	

(91434)

EXAMINING YOUR FARMING RESULTS

ARABLE CROPS

	YOUR FARM			STANDARD or YOUR FARM LAST YEAR	
	HECTARES	YIELD TONNES/HA	OUTPUT £/HA	YIELD TONNES/HA	OUTPUT £/HA
Wheat	18	5.4	433.4		
Barley	22	4.4	352.0		
Oats					
Potatoes					
Sugar Beet					
....					
....					
....					
....					
....					

LIVESTOCK ENTERPRISES

GRAZING LIVESTOCK	ENTERPRISE OUTPUT £	TOTAL ALLOCATED FEED COSTS (including Farm Grain) £	ENTERPRISE OUTPUT LESS ALLOCATED FEED COSTS £	GRAZING LIVESTOCK UNITS	ENTERPRISE OUTPUT LESS ALLOCATED FEED COSTS PER GRAZING LIVESTOCK UNIT	
					YOUR FARM £	* £
Dairy Herd	43130	16720	26410	80.0	330.1	
Dairy Followers	6922	2400	4522	22.2	203.7	
Beef Cattle	11525	9237	2288	–	–	
Sheep						
....						
Total Grazing Livestock Units				102.2		

OTHER PERFORMANCE DATA

			Your Farm	*
Stocking Density (Grazing livestock units per adjusted forage hectare)			1.7	
Dairy Herd:	Milk Production	Litres per Cow	4919	
		Value £ per Cow	505.0	
		Pence per litre	10.5	
	Concentrate Inputs	kg per litre	0.39	
		Pence per litre	4.3	
		£ per cow	209.0	
	Milk value less concs – £ per cow		296.0	
Other Livestock	Feed Costs per £100 output	Dairy Followers	34.7	
		Beef Cattle	80.2	
		Sheep		
		Pigs		
		Poultry		
Lambing Percentage (lambs reared/100 ewes put to ram)			%	%
Farm Productivity – Total Farm Output per £100 total Inputs (including Unpaid Labour Cost)			128	

*Insert STANDARD or LAST YEAR

(91434)

USE OF LABOUR AND MACHINERY

	FARM OUTPUT LESS ALL FEED AND SEED		GROSS MARGIN	
	Your Farm	*	Your Farm	*
Per £100 Labour Costs (Paid and Unpaid)	545.0		433.0	
Per £100 Machinery Costs	884.3		702.5	
Per £100 Labour and Machinery Costs	337.2		267.9	

RETURN ON TENANT'S CAPITAL

		Your Farm	
	£ Total	£ per Hectare	* £ per Hectare
Management and Investment Income	17318	173.2	
Tenant's Capital (Investment in Machinery, Tillages Live and Deadstock)	56613	566.1	
Total Investment in Machinery	9300	93.0	
Management and Investment Income as a Percentage of Tenant's Capital		30.6	

TRADING CASH FLOW

	Your Farm £
Trading Receipts	73526
Trading Expenses	52159
TRADING NET CASH FLOW	21367
Plus Increase in Valuation	1977
Less Depreciation	3146
Plus Notional Receipts	1080
PROFIT (See page 2)	21278

ADVISORY OFFICER'S COMMENTS:

*Insert STANDARD or LAST YEAR

Ncle 91434/1/171 300 8/79 TLB

APPENDIX 4

IN CONFIDENCE

Farm Management No. | 99 : 88 : 777 |

Month
Year Ended | March 19 79 |

Ministry of Agriculture, Fisheries and Food

AGRICULTURAL DEVELOPMENT AND ADVISORY SERVICE

Information For

FARM MANAGEMENT ANALYSIS

ENTERPRISE FINANCIAL INFORMATION

WHOLE FARM FINANCIAL SUMMARY

PHYSICAL INFORMATION

MA4 (Rev. 1978)

ENTERPRISE FINANCIAL INFORMATION

LIVESTOCK

Class of Stock			DAIRY COWS		DAIRY REPLACEMENTS		BARLEY BEEF	
OUTPUT			No.	£	No.	£	No.	£
Closing Valuation:	Breeding stock		80	20800				
	Other livestock				40	8084	49	7105
Sales:	Cull Breeding stock		22	7740				
	Other livestock (1)						48	14040
	Produce (2)		393540	40400				
Transfers out:	Breeding stock		21	1050	20	8000		
	Other livestock		52	2600				
	Produce (2)							
Home Consumption (incl. Employees)				280				
Hill Cow and Hill Sheep subsidies								
Other subsidies and miscellaneous output								
		Sub-Total (a)		72870		16084		21145
Opening Valuation;	Breeding stock		80	20000				
	Other livestock				39	8112	52	7020
Purchases:	Breeding stock			1740				
	Other livestock							
Transfers in:	Breeding stock		20	8000				
	Other livestock				21	1050	52	2600
		Sub-Total (b)		29740		9162		9620
LIVESTOCK OUTPUT (a) − (b)				43130		6922		11525
VARIABLE COSTS			tonnes		tonnes		tonnes	
Bought Concentrates			152·0	16720	20·0	2400	22·0	3557
Home Grown grain			·		·		71·0	5680
Other Feed (excl. forage and straw) (3)								
Contract Rearing								
Veterinary and Medicines				2114		332		574
A.I. Fees and bull hire								
Bedding and litter				484		356		480
Casual Labour and contract work								
Marketing Expenses								
Transport and Sundry costs								
TOTAL VARIABLE COSTS				19318		3088		10291
GROSS MARGIN (excl. forage costs)				23812		3834		1234
Average Number or Throughput				80		42		48
GROSS MARGIN PER HEAD (to 2 decimal places)				297·65		91·29		25·71
Yield (4)								
Enterprise Code				117		121		146

NOTES: (1) Include fatstock subsidies here.
(2) Milk in litres, wool in kilograms, eggs in dozens.
(3) e.g. Wet Grains, wet beet pulp, potatoes (purchased and home grown), milk fed to calves etc.
(4) Milk yield per cow, or lambing percentage, or pigs weaned per sow, or eggs per bird, etc.

ENTERPRISE FINANCIAL INFORMATION

LIVESTOCK

No.	£	No.	£	No.	£	No.	£	No.	£
tonnes		tonnes		tonnes		tonnes		tonnes	

ENTERPRISE FINANCIAL INFORMATION

CASH CROPS

CROPS HARVESTED DURING RECORDING YEAR (Note: Sales and Transfers of Opening Stocks should be entered on Page 11)

Crop		WHEAT	BARLEY					
PHYSICAL OUTPUT		tonnes	tonnes	tonnes	tonnes	tonnes	tonnes	tonnes
Main Product:	Closing stocks	35.0	.	.	.	.	.	.
	Sales	62.2	25.8	.	.	.	.	.
	Transfers out - feed	.	71.0	.	.	.	.	.
	Transfers out - seed	.	.	.	.	.	.	.
	Home Consumption	.	.	.	.	.	.	.
Total Yield		97.2	96.8	.	.	.	.	.
FINANCIAL OUTPUT		£	£	£	£	£	£	£
Main Product:	Closing Valuation	2800						
	Sales	5002	2064					
	Transfers out - feed		5680					
	Transfers out - seed							
	Home consumption							
	Sub-Total (a)	7802	7744					
Secondary Product:	Closing Valuation		100					
	Sales Straw	450	770					
	Transfers out							
	Sub-Total (b)	450	870					
Subsidies (c)								
CROP OUTPUT (a) + (b) + (c)		8252	8614					
VARIABLE COSTS								
Fertilizer		918	726					
Seed:	Purchased	648	594					
	Home Grown							
Sprays		270	330					
Casual Labour								
Contract Work								
Levies								
Sundry costs Baler Twine		90	110					
TOTAL VARIABLE COSTS		1926	1760					
GROSS MARGIN		6326	6854					
Hectares		18.0	22.0	.	.	.	.	.
GROSS MARGIN per Hectare (£) (to 2 dec. places)		351.44	311.55	.	.	.	.	.
Yield per Hectare (Tonnes) (to 2 dec. places)		5.40	4.40	.	.	.	.	.
Enterprise Code		3	7					

126

ENTERPRISE FINANCIAL INFORMATION

CASH CROPS

tonnes	tonnes	tonnes	tonnes	tonnes	tonnes	tonnes	tonnes	tonnes
£	£	£	£	£	£	£	£	£

FORAGE

FORAGE VALUATION CHANGE (VALUE AT VARIABLE COST OF PRODUCTION)

Home-Grown Crops only	Opening Valuation		Closing Valuation	
	Tonnes	£	Tonnes	£
Hay	.		.	
Silage	.		.	250
Harvested Roots	.		.	
Growing Roots				
TOTAL				250
VALUATION CHANGE	INCREASE (a)	250	DECREASE (b)	

MISCELLANEOUS FORAGE OUTPUT

Sales:	Tonnes	£
Hay	.	
Silage	.	
Harvested Roots	.	
Grass and Growing Roots		
Revenue for Agisted Stock		
Valuation Increase	(a)	250
TOTAL MISCELLANEOUS FORAGE OUTPUT	(c)	250

MISCELLANEOUS FORAGE COSTS

Transfer from Crops:	Tonnes	£
Feeding Straw	.	
Other	.	
Purchases:		
Hay	.	
Silage	.	
Harvested Roots	.	
Feeding Straw	.	
Grass and Growing Roots		
Expenditure on Agistment		
Valuation Decrease		(b)
TOTAL MISCELLANEOUS FORAGE COSTS		(d)

FORAGE VARIABLE COSTS (CROPS GROWN DURING RECORDING YEAR)

Crop	TEMP GRASS	£	£	£	£	£	£	£
Fertilizer	3624							
Seed	262							
Sprays								
Casual Labour								
Contract Work								
Sundry Costs								
TOTAL VARIABLE COSTS (e)	3886							
Actual Hectares	60.0	.	.	.	.	.	.	.
Adjusted Hectares	60.0	.	.	.	.	.	.	.
Enterprise Code	99							

SUMMARY

	£
Miscellaneous Forage Output (c)	250
Less Miscellaneous Forage Costs (d)	
Sub-total (c – d) to page 7	250
Forage Variable Costs (Total of e) to page 7	3886

WHOLE FARM FINANCIAL SUMMARY

FIXED COSTS

	Code	998 £
Wages		
Regular and National Insurance		7200
Produce consumed by Employees (1)		
Unallocated Casual		
Wife (if paid)		
Sub-Total		7200
Management		
Paid Management		
Secretarial		
Sub-Total		
Power and Machinery		
Machinery and Vehicle repairs		2050
Fuel and Oil		1120
Contract Hire, Transport (2)		
Leasing Charges		
Electricity and Coal		
Machinery depreciation		2900
Vehicle Tax and Insurance		
Sub-Total		6070
Property Charges		
Rent		
Rates		
Depreciation on Buildings and Improvements		246
Depreciation on Tenants Fixtures		
Property Repairs		1090
Sub-Total		1336
Sundries		
Water		430
General Insurance		520
Office and Telephone		350
Subscription and Professional Fees		380
Other		230
Sub-Total		1910
Financing Charges		
Bank Charges and Interest		492
Mortgage and Other Loan Interest		1548
H.P. Interest		
Sub-Total		2040
TOTAL FIXED COSTS		18556

GROSS MARGINS / OTHER RECEIPTS

ENTERPRISE	Code	999 £
Cash Crops		
Wheat	3	6326
Barley	7	6854
Sub-Total (a)		13180
Intensive Livestock		
Barley Beef	146	1234
Sub-Total (b)		1234
Grazing Livestock		
Dairy Cows	117	23812
Dairy Replacements	121	3834
Sub-Total		27646
Miscellaneous Forage Output/Costs ±		+ 250
Less Forage Variable Costs −		3886
GROSS MARGIN FROM FORAGE AREA (c)		24010
Other Receipts		
Contracting Work		610
Wayleaves, Rents received etc.		
Sundries		
Sub-Total (d)		610
TOTAL GROSS MARGINS/OTHER RECEIPTS (a) + (b) + (c) + (d)		39034
LESS FIXED COSTS		18556
SURPLUS/DEFICIT		20478

NOTES: (1) Where this has been credited to an enterprise (pages 2 to 5).
(2) The share of this item that has not been allocated to individual enterprises.

PHYSICAL INFORMATION

CROPPING

	Hectares
Wheat	18.0
Barley	22.0
Oats	.
Potatoes, early	.
Potatoes, main crop	.
Sugar Beet	.
Other Crops (Specify)	.
	.
	.
	.
	.
	.
	.
	.
	.
	.
	.
Kale	.
Roots for Stockfeed	.
Other Forage Crops	.
Temporary Grass	60.0
Permanent Grass	.
Fallow	.
Area of crops and grass	100.0
Rough Grazings	.
Area used for agriculture	100.0
Buildings, roads, woods, lakes etc.	.
Total Farm Area	100.0

Area rented	.
Area owner / occupied	100.0
Adjusted Farm Area *	100.0

* Area of Crops and Grass plus equivalent area of Rough Grazings and Hill Rights.

GRAZING LIVESTOCK ON THE FARM

		Annual Average or Throughput	Grazing Livestock Units Factor	Grazing Livestock Units
Dairy Cows		80	1.0	80.0
Bulls for Dairy Herd			0.7	.
Dairy Replacements	0–12 months	19	0.4	7.6
	12–24 months	19	0.6	11.4
	over 24 months	4	0.8	3.2
Beef Cows			0.6	.
Bulls for Beef Herd			0.7	.
Other Beef Cattle	0-12 months		0.4	.
	12–24 months		0.6	.
	over 24 months		0.8	.
Ewes			♂	.
Ewe Replacements			♂	.
Rams			0.1	.
Lambs:	Birth to Store	*	0.04	.
	Birth to Fat	*	0.06	.
	Birth to Hoggets	*	0.08	.
Store Lambs (Purchased and Tack)		*	0.04	.
Horses			+	.
			TOTAL	102.2

* Throughput

♂ Ewes and ewe replacements: under 50 kg — 0.08 G.L.U.
  50–70 kg — 0.10 G.L.U.
  over 70 kg — 0.12 G.L.U.

+ Horses: under 13.2 h — 0.6 G.L.U.
  13.2 - 15 h — 0.8 G.L.U.
  over 15 h — 1.0 G.L.U.

LESS FAVOURED AREAS

Tick if applicable

(a) Farm is wholly within the LFAs
(b) Farm is partly within the LFAs

PHYSICAL INFORMATION (CONTD.)

CALCULATION OF ADJUSTED FORAGE AREA

	Equivalent Hectares
Crops grown during the year: Grass (incl. hay and silage)	60.0
Kale, Roots, Forage Maize etc.	.
Rough Grazing (grass equiv.)	.
Hill Rights (grass equiv.)	.
Rented Keep, Purchased Grass and Growing Roots	.
Opening Valuation: Hay and Silage	.
Roots etc.	.
Forage value of Cash Crops (e.g. Herbage Seed)	.
Sub-Total (a)	60.0
Sales: Let Keep, Grass and Growing Roots	.
Hay and Silage	.
Harvested Roots	.
Closing Valuation: Hay and Silage	.
Roots etc.	.
Sub-Total (b)	0.0
TOTAL ADJUSTED FORAGE AREA (a) – (b)	60.0

MONTHLY MILK PRODUCTION (LITRES)

Sales			
April	34350	October	28600
May	32004	November	36430
June	32040	December	38610
July	29760	January	34810
August	26940	February	34700
September	25706	March	32690
Summer Total	180800	Winter Total	205840
		Total Sales	386640
		Farmhouse	1400
		Employees	1400
		Livestock	4100
		TOTAL PRODUCTION	393540

TO BE COMPLETED BY FARMER OR ADVISER

1. FORM OF BUSINESS — Tick as applicable
   (a) Sole trader or farmer and wife partnership — ✓
   (b) Family Partnership
   (c) Non family Partnership
   (d) Farming Company
   (e) Other

2. NOTIONAL VALUES

	£
(a) Rental value of owner-occupied land	4000
(b) Rental value of farmhouse(s)	400
(c) Private use of car, electricity, phone, etc.	400

3. Altitude range of major part of the farm (metres) — 50–75

4. Average annual Rainfall (mm) — 800

5. PREDOMINANT SOIL TEXTURE — Tick as applicable
   (a) Heavy – Clay loam, silty loam, silty clay loam
   (b) Medium – Very fine sandy loam, silty loam, loam, sandy clay loam — ✓
   (c) Light – Loamy sands and sandy loam except VFSL
   (d) Peaty – Peaty loam, loamy peat, light peat, sandy peat
   (e) Organic – Organic-sandy loams, loams, clay loams and silty loams

6. LABOUR

	Number	Hours worked during year*	Value of unpaid labour
(a) Regular male	2	4400	
(b) Regular female			
(c) Casual male			
(d) Casual female			
(e) Salaried managers			
(f) Farmer and partners	1	1250	2000
(g) Farmer's and partners' wives			

*Enter approximate time spent on manual work only, do NOT include managerial or clerical work.

7. CAPITAL INVESTED (At end of Accounting Year)

	£
(a) Livestock	35989
(b) Harvested Crops	3150
(c) Growing crops and tillages	900
(d) Long term crops (e.g. orchards, hops, etc.)	
(e) Goods in store	5060
(f) Cars, vans and lorries	
(g) Tractors and other machinery	9300
(h) Fixed equipment	
(i) Tenant's buildings and fixtures	2214
TOTAL	56613
(j) Owner-Occupiers value of holding and improvements	250000

Farm Management No. | 99: 88 : 777 |

REALISATION OF HARVESTED CROPS CARRIED FORWARD FROM PREVIOUS YEAR

Crop	WHEAT					
Harvest Year	1977					
PHYSICAL DATA (Tonnes) Main Product						
Opening Stocks (a)	35 0	.	.	.	.	.
Disposals: Sales	35 0	.	.	.	.	.
Transfers - feed	.	.	.	.	.	.
Transfers - seed	.	.	.	.	.	.
Home Consumption	.	.	.	.	.	.
Total Disposals (b)	35 0	.	.	.	.	.
Difference (b) − (a) +/−	.	.	.	.	.	.
FINANCIAL DATA (£)						
Opening Valuation: Main Product	2450					
Secondary Product	100					
Total (c)	2550					
Disposals						
Main Product: Sales	2450					
Transfers - feed						
Transfers - seed						
Home Consumption						
Secondary Product: Sales						
Transfers	100					
Subsidy Adjustments +/−						
Total Disposals (d)	2550					
Difference (d) − (c) +/−	0					
Additional Variable Costs						
Casual Labour						
Contract Work						
Sundries						
Levy Adjustments +/−						
Total						
Enterprise Code	3					

NOTE: This information will be used to amend the data which was submitted before these crops had been completely disposed of. The "Opening Stocks" and "Opening Valuation" figures should be the same as those entered as "Closing Stocks" and "Closing Valuation" on the previous year's return. These figures are unlikely to reconcile with the disposals and no attempt should be made to do so.

MA1 AND MA4 EXAMPLES

RECONCILIATION

	£
MA1 PROFIT	21 278
Less:	
Rental Value of House	400
	20 878
Less:	
Private Use of Car	400
MA4 SURPLUS	20 478

Appendix 5

Farm Management No 09:50:001

IN CONFIDENCE

AGRICULTURAL DEVELOPMENT AND ADVISORY SERVICE

MULTI STAGE
CASH FLOW PROJECTION

MINISTRY OF AGRICULTURE, FISHERIES AND FOOD

MA7 (Rev. 1972)

TRADING

Starting date		Date	PERIOD ACTUAL/BUDGET APR – JUN			PERIOD ACTUAL/BUDGET			PERIOD ACTUAL/BUDGET JUL – SEP			PERIOD ACTUAL/BUDGET			PERIOD ACTUAL/BUDGET OCT – DE		
			No.	Val.	Total	No.	Val.	Total	No.	Val.	Total	No.	Val.	Total	No.	Val.	Tot
R E C E I P T S	Milk				2100						2300						33
	Cattle: Calves				150						200						3
	Cull Cows				160						190						2
	Sheep																
	Pigs / Poultry																
	Crops: Wheat																7
	Barley																12
	Potatoes																22
	Sundries: Grants																
	Other Receipts																1
	Sub Total (a)				2410						2690						82
P A Y M E N T S	Feed D. Cows				410						160						8
	Y. Stock				20												1
	Vet. & Med.										90						
	Livestock Purchases: Cows																
	Seed				1180												
	Fertilisers				1370						200						3
	Sprays																
	Wages – Permanent				800						850						8
	– Casual																6
	Power & Mach. – Fuel				200						300						2
	– Reps., Ins.				150						200						1
	– Contract Hire																
	Rent & Rates				900												9
	Sundries				270						270						2
	Building Repairs										500						
	Sub Total (b)				5300						2570						43
	Net Cash Flow for period (a–b)				–2890						120						38
	Annual Net Cash Flow (for discounting purposes)																
	Interest (incl. H.P. & Mort.) on prev neg cum bal				– 300												– 3
	Trading Net Cash Flow for period (i)				–3190						120						35

PERSONAL

Receipts Sub Total (c)															
Drawings			350						350						3
Tax									320						
Any other payments															
Sub Total (d)			350						670						3
Personal Net Cash Flow for period (c–d) (ii)			– 350						– 670						– 3

CAPITAL

Grants															
Mach. Sales			300						800						
Sub Total (e)			300						800						
Buildings etc.															
Machinery			1000						3600						
Other incl. Capital Repayments															
Sub Total (f)			1000						3600						
Capital Net Cash Flow for period (e–f) (iii)			– 700						–2800						
Net Cash Flow for period (i + ii + iii)			–4240						–3350						3170
Cumulative Balance (Op Bal –6000) –			10240						– 13590						– 10420
Annual Net Cash Flow															

136

	PERIOD			PERIOD			PERIOD			PERIOD			PERIOD			PERIOD			PERIOD			PERIOD		
UAL/BUDGET	ACTUAL/BUDGET			ACTUAL/BUDGET			ACTUAL/BUDGET			ACTUAL/BUDGET			ACTUAL/BUDGET			ACTUAL/BUDGET			ACTUAL/BUDGET			ACTUAL/BUDGET		
	JAN – MAR						YEAR																	
Val.	Total	No.	Val.	Total	No.	Val.	Total	No.	Val.	Total	No.	Val.	Total	No.	Val.	Total	No.	Val.	Total	No.	Val.	Total	No.	Val.	Total
				3100						10800															
				250						910															
				220						850															
				3120						3840															
				1950						3200															
				1200						3400															
				150						300															
				9990						23300															
				920						2320															
				200						340															
				90						180															
										1260															
										1920															
				800						3300															
										600															
				200						900															
				150						650															
										1800															
				270						1080															
										500															
				2630						14850															
				7360						8450															
										– 640															
				7360						7810															
				350						1400															
				320						640															
				670						2040															
				– 670						–2040															
										1100															
										1100															
										4600															
										4600															
										–3500															
				6690						2270															
				–3730						–3730															

137

CALCULATION OF TRADING PROFIT

ITEM	Year 1	Year 2	Year 3	Year 4	Year 5
Trading Net Cash Flow for period (i) (Overleaf) Annual Total....					
National Receipts (Produce consumed, House rent, Private Use of car etc.)					
Closing Valuation – Livestock					
– Harvested Crops....					
– Tillages and Growing Crops					
– Stores					
Closing Debtors					
Opening Creditors					
SUB TOTAL (a)					
Opening Valuation – Livestock					
– Harvested Crops					
– Tillages and Growing Crops					
– Stores....					
Closing Creditors....					
Opening Debtors					
SUB TOTAL (b)					
PROFIT (a–b) before Depreciation					
Depreciation on Buildings and Fixtures					
Machinery Depreciation					
TRADING PROFIT after Depreciation					
Tax @					

Net Liquid Assets @		
	(starting date)	
Bank Balance	(+) £	
or overdraft	(–) £	
Debtors	(+) £	
Creditors	(–) £	
TOTAL	£	

Terminal Value @		
	(ending date)	
Buildings		£
Equipment		
TOTAL		

BIRDS AND BERRIES

*About the whitethorn's berried bush
The Fieldfare and the Redwing Thrush
Flit in unnumbered throngs, or speed
to rushy fen or plashy mead,
Impatient for their insect fare,
And darken with their flight the air.
What! do your northern banquets fail?
And, bound upon the autumnal gale,
Seek ye beneath our milder sky
And warmer sun a fresh supply?
Feed on, while yet the hedge-girt field
Rich store of scarlet haws shall yield!*

                    Bishop Mant

BIRDS AND BERRIES
A study of an ecological interaction

by BARBARA and DAVID SNOW

Illustrated by
JOHN BUSBY

T & A D POYSER

Calton

© *Barbara and David Snow 1988*

ISBN 0 85661 049 6

First published in 1988 by T & A D Poyser Limited
Town Head House, Calton, Waterhouses, Staffordshire, England

All rights reserved. No part of this book may be reproduced,
stored in a retrieval system, or transmitted in any form or
by any means, electrical, mechanical, photocopying or otherwise,
without the permission of the publisher

British Library Cataloguing in Publication Data

Snow, Barbara,
   Birds and berries.
   1. Seeds. Dispersal by fruit-eating birds
   I. Title   II. Snow, David, 1924–
   582′.016
   ISBN 0-85661-049-6

Printed and bound in Great Britain by
Butler & Tanner Ltd, Frome and London

Contents

Introduction 11
Acknowledgements 16

1 THE FRUITS 17
   The seasonal pattern of fruiting   18
   Holly *Ilex aquifolium*   22
   Yew *Taxus baccata*   28
   Ivy *Hedera helix*   31
   Mistletoe *Viscum album*   35
   Dog rose *Rosa canina*   41
   Other roses   44
   Haws *Crataegus* spp.   45
   Rowan *Sorbus aucuparia*   49
   Whitebeam *Sorbus aria*   51
   Wild service tree *Sorbus torminalis*   54
   Wild cherry *Prunus avium*   55
   Bird cherry *Prunus padus*   56
   Blackthorn *Prunus spinosa*   58
   Blackberry *Rubus fruticosus*   60
   Wild raspberry *Rubus idaeus*   63
   Crab-apple *Malus sylvestris*   64
   Elder *Sambucus nigra*   65
   Wayfaring tree *Viburnum lantana*   68
   Guelder rose *Viburnum opulus*   70
   Honeysuckles *Lonicera* spp.   72
   Dogwood *Cornus sanguinea*   75
   Privet *Ligustrum vulgare*   77
   Buckthorn *Rhamnus catharticus*   79
   Spindle *Euonymus europaeus*   81
   Wild currant *Ribes rubrum*   84
   Wild gooseberry *Ribes uva-crispi*   85
   Woody nightshade *Solanum dulcamara*   86
   Black nightshade *Solanum nigrum*   88
   Deadly nightshade *Atropa belladonna*   89
   White bryony *Bryonia dioica*   90
   Black bryony *Tamus communis*   92
   Spurge laurel *Daphne laureola*   94
   Mezereon *Daphne mezereum*   95
   Juniper *Juniperus communis*   96
   Lords and ladies *Arum maculatum*   98
   Some minor fruits   100
   Introduced and cultivated plants   102

2 THE FRUIT-EATERS   109
   Blackbird *Turdus merula*   111
   Song Thrush *Turdus philomelos*   120
   Mistle Thrush *Turdus viscivorus*   125
   Fieldfare *Turdus pilaris*   133

6 Contents

    Redwing *Turdus iliacus* 138
    Robin *Erithacus rubecula* 142
    Blackcap *Sylvia atricapilla* 147
    Other warblers and flycatchers 154
    Starling *Sturnus vulgaris* 158
    The Corvidae 163
    Occasional or scarce fruit-eaters 166
    Waxwing *Bombycilla garrulus* 168
    Woodpigeon *Columba palumbus* 171
    Seed-predators and pulp-predators 175
    The finches 176
    The sparrows 182
    The tits 183

3 INTERPRETATION 188
    Physical characteristics of fruits: size and seed burden 189
    The colour and taste of fruits 194
    Competition between plants for dispersers 198
    Adaptation of fruiting seasons and fruit quality to changing bird populations 202
    Seed-predators and the plants' defences 206
    Physical limitations to fruit-eating 210
    Adaptations for a fruit diet 214
    Fruit preferences 219
    Foraging strategies of fruit-eaters 227
    Fruit as a component of the total diet 231
    Time and energy budgets for fruit-eaters 235
    The question of coevolution 239

    Appendices
    1 Plants with fleshy fruits native to England and Wales 245
    2 Design components of native bird-dispersed fruits, southern England 247
    3 Design components of some introduced and cultivar bird-dispersed fruits, southern England 249
    4 Nutritive values of some bird-dispersed fruits, southern England 250
    5 Monthly totals of timed watches at wild fruits, southern England 1980–85 251

References 253

Index 261

List of Figures

1 Seasonal availability of wild fruits for dispersers. Vale of Aylesbury area, 1980–85   19
2 Records of birds feeding on hollies of different status with respect to defence by Mistle Thrushes   23
3 The feeding rate of Mistle Thrushes at defended mistletoe in winter   38
4 Records of dispersers eating haws   46
5 Numbers of different species of introduced/ornamental fruits available per month, compared to native fruits   103
6 Mean meal sizes of Robins for 15 different kinds of fruits   145
7 Diameters of bird-dispersed fruits, southern England, and gape-widths of the main dispersers   190
8 Relationship between seed number and individual seed weight in 36 native bird-dispersed fruits   191
9 Relationship between mean seed number and seed burden in sample ivy berries   192
10 Relationship between fruit size and seed burden in samples of sloe and wild cherry   193
11 Fruiting seasons of 18 species of *Miconia* in Trinidad   200
12 Seasons of ripening of 40 European fruit species in southern England   203

List of Tables

1 Number of species, and growth forms of plants with fleshy fruits native to England and Wales in the study area   18
2 Holly: monthly summary of feeding records   23
3 Feeding rates at defended and undefended hollies, and at hollies being overwhelmed   25
4 Months of stripping of fruit crop of 32 holly trees   26
5 Yew: monthly summary of feeding records   29
6 Ivy: monthly summary of feeding records   33
7 Mistletoe: monthly summary of feeding records   37
8 Dog rose: monthly summary of feeding records   42
9 Haw: monthly summary of feeding records   47
10 Rowan: monthly summary of feeding records   50
11 Whitebeam: monthly summary of feeding records   53
12 Wild cherry: monthly summary of feeding records   55
13 Bird cherry: monthly summary of feeding records   57
14 Sloe: monthly summary of feeding records   58
15 Blackberry: monthly summary of feeding records   61
16 Wild raspberry: monthly summary of feeding records   63
17 Crab-apple: monthly summary of feeding records   65
18 Elder: monthly summary of feeding records   66
19 Wayfaring tree: monthly summary of feeding records   69
20 Guelder rose: monthly summary of feeding records   71
21 Common and perfoliate honeysuckles: monthly summary of feeding records, dispersers   73
22 Common and perfoliate honeysuckles: monthly summary of feeding records, seed-predators   73
23 Dogwood: monthly summary of feeding records   75
24 Privet: monthly summary of feeding records   77
25 Rate of feeding on privet fruit by dispersers in different months   78
26 Buckthorn: monthly summary of feeding records   80
27 Spindle: monthly summary of feeding records   82
28 Wild currant: monthly summary of feeding records   84
29 Woody nightshade: monthly summary of feeding records   87
30 White bryony: monthly summary of feeding records   90
31 Black bryony: monthly summary of feeding records   93
32 Juniper: monthly summary of feeding records   96
33 Lords and ladies: monthly summary of feeding records   99
34 Introduced/ornamental plants: summary of feeding records, all dispersers   103
35 Blackbird: monthly summary of feeding records   112
36 Blackbird: incidence of aerial sallying to pluck fruit   113
37 Blackbird: success rate according to method of taking fruit   113
38 Records of Blackbirds being chased from fruits   115
39 Blackbird: meal sizes for different fruits   116
40 Records of Blackbirds feeding fruit to young   117
41 Song Thrush: monthly summary of feeding records   121
42 Song Thrush: incidence of aerial sallying to pick fruit   122
43 Song Thrush: success rates according to method of taking fruit   122

List of Tables

44 Song Thrush: meal sizes for different fruits   122
45 Mistle Thrush: monthly summary of feeding records   126
46 Mistle Thrush: success rates according to method of taking fruit   126
47 Mistle Thrush: incidence of aerial sallying to pluck fruit   127
48 Mistle Thrush: meal sizes for different fruits   130
49 Fieldfare: monthly summary of feeding records   134
50 Fieldfare: meal sizes for different fruits   135
51 Redwing: monthly summary of feeding records   139
52 Redwing: success rates according to method of taking fruit   139
53 Redwing: records of sallying to pluck fruit   140
54 Redwing: meal sizes for different fruits   140
55 Robin: monthly summary of feeding records   143
56 Robin: meal sizes for different fruits   144
57 Blackcap: monthly summary of feeding records   148
58 Blackcap: meal sizes for different fruits   150
59 Success and meal sizes of Garden Warblers and Blackcaps feeding on bird cherry   155
60 Starling: monthly summary of feeding records   159
61 Starling: success rates for taking different fruits, compared with rates for thrushes   160
62 Starling: meal sizes for different fruits   161
63 Magpie: monthly summary of feeding records   164
64 Jay: monthly summary of feeding records   164
65 Carrion Crow: monthly summary of feeding records   165
66 Woodpigeon: monthly summary of feeding records   173
67 Bullfinch: monthly summary of feeding records   177
68 Greenfinch: monthly summary of feeding records   179
69 Chaffinch and Brambling: monthly summary of feeding records   180
70 Great Tit: monthly summary of feeding records   184
71 Blue Tit: monthly summary of feeding records   184
72 Marsh Tit: monthly summary of feeding records   186
73 Seed burden and water content of pulp of native fruits ripening at different seasons, southern England   204
74 Lipid content of dry pulp of native fruits ripening at different seasons, southern England   205
75 Impact of seed-predators on some fruits   207
76 Wing-loadings of thrushes and other fruit-eaters   212
77 Incidence of flight-sallies by thrushes feeding on ivy berries   213
78 Records of five thrushes feeding on some important fruits   220
79 Fruit preferences shown in choice situations by Blackbirds   221
80 Fruit preferences shown in choice situations by Song Thrushes   221
81 Fruit preferences shown in choice situations by Mistle Thrushes, Redwings and Fieldfares   222
82 Fruit preferences shown in choice situations by Starlings   223
83 Estimated energy intake by Mistle Thrushes feeding on defended hollies and haws in very cold weather   237

Dedicated to Alexander F. Skutch, the only ornithologist we know whose research technique has been to sit and watch; also to BBONT, our local naturalists' trust, on whose reserves we sat.

Introduction

BIRDS AND FRUITS – THE NATURE OF THE INTERACTION

Most small fleshy fruits are adapted for dispersal by birds. The relationship between plant and bird is mutually advantageous: the plant invests some of its resources in coating the seed with a nutritious pulp which it offers to the bird, and in exchange the bird eats the fruit, digests the pulp and ejects the seed, either in its faeces or by regurgitation, or in special cases by some other means. If the interaction is successful, from the plant's point of view, some at least of the seeds are dropped away from the parent plant in places where they can germinate and new plants may establish themselves.

No botanist or ornithologist would doubt the truth of this very general statement. However, the further implications and the intricacies of particular interactions are far from being fully understood. The relationship is certainly a case of 'coevolution': that is to say, plants and birds (at least the more specialised frugivorous birds) have been mutually modified in the course of their evolution. Thus the size, form and nutritive content of fruits of different plants have become adapted for dispersal by certain kinds of birds, and the plants make their fruits conspicuous in various ways and physically accessible to birds (and perhaps inaccessible to other animals). Similarly the bills, digestive systems and other characteristics of frugivorous birds have become adapted for the efficient plucking, swallowing and digesting of fruits.

But mutualistic systems of this sort are vulnerable to cheating. Some other organism may break into the system and obtain the rewards without fulfilling its part of the bargain. Thus the coevolved bird/fruit system may be exploited by birds that eat and digest the seed (either with its pulp, or actually discarding the pulp), or by birds that eat the pulp but do not disperse the seed. The former may be called 'seed-predators' and the latter 'pulp-predators', while the birds that participate legitimately in the system may be called 'legitimate frugivores' or, more simply, 'dispersers'. Failure to recognise these distinctions has much reduced the value of some of the literature on the eating of fruit by birds (*eg* Turček 1961). In many handbooks birds such as thrushes (which are dispersers) and parrots (which are seed-predators) are indiscriminately described as fruit-eaters.

Much research has been devoted to the ecology of fruit-eating in the last 15 years, most of it in the tropics where the relationships between birds and plants are most highly developed. An important exception is the work carried out in southern Spain by Dr C.M. Herrera and Dr P. Jordano. We shall frequently have occasion to refer to it. There is now an extensive literature on this aspect of bird ecology, but most of it is in the ecological journals or other specialised publications, and little has filtered down into more general publications on natural history. We ourselves began to take an interest in the subject in Trinidad, and later did further field work in South and Central America, where we studied some of the most striking cases of very specialised fruit-eating birds that are absolutely dependent on a year-round supply of nutritious fruits. Tropical studies will probably continue to be at the forefront of research into bird-plant coevolution. We hope to show, however, that even in England, with our comparatively impoverished avifauna and flora, a study of the interactions between birds and fruits can reveal unsuspected facts and enhance our understanding of natural history.

THE STUDY AREA

From the beginning of 1980 to the end of the 1984–5 winter we carried out systematic observations on birds and fruits in a fairly typical area of southern England, in Buckinghamshire and adjacent parts of Hertfordshire. The area includes part of the Vale of Aylesbury and some of the hills that border the Vale, especially the northern flanks of the Chilterns from Ivinghoe to Princes Risborough. In the north it includes some of the old oak woodland of Bernwood Forest between Quainton and the Claydons, and extends east to the Brickhills. The Vale of Aylesbury is mainly agricultural country on heavy clay soils consisting of medium-sized fields with hedges, with a few small woods and thickets. The Chilterns with underlying chalk are very different, with woods and open or bushy downland, as well as arable and pasture fields. This is the part of our study area richest in plants that provide fruit for birds. To the east, from the Brickhills to Leighton Buzzard, a belt of Lower Greensand produces a rather different kind of country, agriculturally poor and well wooded, with much planted pine and larch in addition to the native woodland dominated by oak and birch.

All the detailed, systematic observations which we summarise were made in this area; but we have made more casual observations elsewhere in Britain and continental Europe. Where these are mentioned, the locality is stated. We have made as complete a survey as possible of the earlier British literature on birds and fruits, and of the European literature.

METHODS OF STUDY

Throughout our observations we tried to keep some sort of balance between following a planned programme and maintaining flexibility. Professional ecological research (especially if it is dependent on financial grants) now usually demands a set plan of observation or experiment, designed so that it will produce statistically valid answers to certain specific questions. If, however, one is carrying out a broad survey, and one is not certain what it will reveal – as was the case when we began the study – flexibility is essential. We found that the pattern of fruit availability changes constantly; one year differs from another, and unusual seasonal changes produce new situations. Inevitably one finds new areas of interest and new plants worth watching. If one is working strictly to a pre-set plan one may produce answers to the questions asked, if all goes well; but one may miss much that is of equal or even greater interest, because the questions initially asked were based on a limited understanding of the complexity of the situation.

In any study of the feeding habits of birds various methods may be used; each method will have advantages and disadvantages, and some of them will be subject to bias. The most obvious method of studying fruit-eating is, of course, direct observation of a bird eating a fruit; but in order to avoid bias one needs to adopt

Introduction 13

a system of observation. Casual records are not to be recommended, as the more conspicuous birds or fruits will always be over-recorded and the less conspicuous under-recorded, and one will probably miss seeing some fruits being eaten altogether. Our principal method has been to carry out timed watches at fruit sources that could be watched with a fair probability of being able to see all the birds visiting them.

TIMED WATCHES

The vast majority of our observations were made during timed watches at fruit-bearing plants (see Appendix 5). Each visit by an individual bird was scored as one record (a 'feeding visit') if the bird was seen to eat at least one fruit. We found that a feeding visit is a natural unit to use in recording fruit-eating. In most circumstances, unless it is disturbed, a bird that is eating fruit visits a fruit source, takes a 'meal' — that is, apparently the amount of fruit that it can accommodate in its stomach — and then leaves. It does not normally return for its next meal for ten or more minutes; but as a bird may sometimes for some reason not follow this usual routine, and instead takes smaller feeds at short intervals, we arbitrarily decided that a second record could be made for a bird only if separated by at least five minutes from a previous record. In the case of spatially separated trees and shrubs, if a bird moved to another plant, and fed at it, a second record was made; but only one record was made if a bird fed at a compact clump consisting of several plants of the same kind. (When watching large flocks of winter thrushes feeding on farmland, we sometimes used an alternative method of recording by scanning the hedges where they were feeding, recording the number of birds actually plucking fruit, then allowing five minutes to elapse before the next scan. This accounted for only 2.6% of records, mostly of birds feeding on haws.) The feeding visits thus recorded form the basis of our tabulated data.

As far as possible, we recorded the following for each feeding visit: age and sex of bird; time of day; height above ground; number of attempts to take fruit; number of fruits actually taken; method of taking (or trying to take) fruit; interactions with other birds; any other points of interest. It was often impossible, of course, to record all of this information, especially if many birds were visiting the fruit under observation.

We aimed to accumulate watches totalling at least ten hours for each of the main months of availability of all fruits except some of minor importance. Overall, our timed watches have totalled 1,673 hours, equivalent to about one working year. In the course of these watches we have occasionally been objects of curiosity, if not suspicion, to passers-by or householders who have seen us staring through binoculars apparently at nothing or, even worse, at an upstairs window of someone else's house. The most interesting holly trees often grow close to houses.

RECORDING DEPLETION OF FRUIT CROPS

A more indirect method of studying fruit-eating is to count some, or even all, the fruit on a marked plant, or part of a plant, and then count the number remaining at intervals until all the fruit has gone. This method has been widely used by others, even to the exclusion of making direct observations (*eg* Thompson & Willson 1979). We used the method on occasion, especially for fruits where for some reason we found it difficult to see the fruit-eaters' visits. This sometimes alerted us to the fact that the fruit was being taken, so that we were able to watch at the right time. We have not made much other use of these counts, as in most cases we have enough direct observations. The indirect method has two major disadvantages. First, one does not know who has taken the fruit, nor with what success. Second, there is liable to be a bias in selecting the fruits to be counted. Without special equipment large shrubs and trees are too high to reach, except the lower branches. Often birds begin to strip a tree from the top. Or they may begin on the better protected parts of a tree or shrub, leaving the outer, more exposed fruit until last, at whatever level they are feeding.

EXAMINATION OF GUT CONTENTS

We obtained only a very little information, additional to that obtained from direct observation, from the examination of faeces or regurgitated seeds, but this method has been widely used by others. In particular Herrera and Jordano in southern Spain have obtained most of their data by trapping birds and analysing the faecal or regurgitated matter produced while they were held in bags, or flushing out the contents of the digestive tract with a weak salt solution (*eg* Herrera 1981a, Jordano 1987). The method has the advantage of producing data that are not subject to biases arising from difficulties of direct observation; disadvantages are that it gives no information about the birds' methods of taking fruit, their success rate, or behavioural interactions between different individuals and species.

NUTRITIVE QUALITY OF FRUITS

To find out how good a food fruit is, how it compares with animal food and how one fruit compares with another, some sort of analysis must be made of a fruit's composition. We carried out a standard, simple analysis of all the main kinds of native fruits in our study area, and of many of the introduced fruits (Appendices 2 and 3). In most cases, several analyses were made of each kind of fruit, from samples taken from different plants. For each sample, means were obtained for the following, generally based on 10 or more fruits:
1. Dimensions (diameter and length if ovoid, diameter only if more or less globular).
2. Weight of whole fresh fruit.
3. Total weight of seed(s) ('seed burden').
4. Number of seeds per fruit.
5. Individual seed weight.
6. Water content of pulp, *ie* edible part of fruit whatever its botanical status.

Two further statistics were derived from these:
(a) Seed burden as a percentage of the weight of the whole fruit. This gives a measure of the amount of 'ballast' which a bird must ingest if it swallows the fruit whole – ballast that is useless for the bird and is quickly got rid of, but from the plant's point of view the vital part of the fruit.
(b) Amount of dry, mainly nutritive matter, obtained by the bird, expressed as a percentage of the total fruit weight (derived from the figures for seed burden and water content of pulp). This is a useful measure of the profitability of eating a fruit, and different kinds of fruits can easily be compared with one another. It has been used by Herrera in his studies of fruit-eating in southern Spain, and we use the term coined by him, 'relative yield'. The dry matter is, as stated above, mainly but not entirely nutritive, as there is always some indigestible fibre and a little mineral matter (which may or may not be nutritionally valuable).

The whole complex of measurements of size, shape, weight, seed weight and water content of pulp, *ie* all those components affecting a bird's ability to swallow a fruit and the fruit's value as a food, apart from its chemical composition, may

Introduction 15

be called its 'design components' – another term coined by Herrera (1981b). Obviously, for a full understanding of the ecology of fruit-eating the chemical composition of fruit pulp is also of fundamental importance (the 'chemical components'). For many of the fruits in our study area we have been able to obtain analyses of the chemical constituents of the dry pulp, and for some others data are available from published sources (Appendix 4). These analyses give figures for the total amounts of carbohydrates, fats and proteins. They add significantly to what can be found out from an analysis of design components only, but they are still very incomplete. For instance, proteins contain many different amino acids, and they are not all of equal importance in a bird's diet. Similarly, fats are composed of different fatty acids, and a balanced diet must contain enough of all the essential ones. It will be a long time before a complete analysis is available of the nutritive value of any wild fruit to its disperser; in the meantime, only a very rudimentary understanding is possible.

PHYSICAL LIMITATIONS, AND ADAPTATIONS, OF THE FRUIT-EATERS

An important question, which one cannot avoid asking even if a full answer cannot be given, is the extent to which the various bird species that eat fruit are adapted to a fruit diet, structurally or physiologically. And connected with this, how do their structural characteristics limit the range of fruit that they can eat, and their methods of obtaining the fruit? As regards physiology, we have made no personal investigations and have had to rely on the little published information that is available. As regards structure, in order to supplement the field observations we have collected data on various structural components which most obviously affect a bird's ability to reach, pluck and swallow fruits of different kinds. The main ones are: the bird's overall size as indicated by weight; wing-length; wing-area (which, combined with weight), gives wing-loading; bill-length and bill-width ('gape'); and leg length. Analysis of these structural characters, incomplete though they are, throws some light on the efficiency of different bird species in exploiting different kinds of fruits, but does not necessarily indicate whether they are to be regarded as adaptations for fruit-eating. This question is discussed separately, on page 214.

'FRUITS' AND 'BERRIES' – A MATTER OF TERMINOLOGY

At the outset, we must deal with a possible source of confusion. Botanists use these two terms in a different sense from the general public. For a botanist, 'the term *fruit* ... is used to denote that part of the plant in which the seeds are found. It consists essentially of the ripe ovary, but it may also include other floral parts which are connected with the ovary. The name *fruit*, used in this sense, includes much more than the popular term *fruit*; it embraces not only what are generally known as fruits but also some vegetables, and even dry, inedible structures' (Brown 1935). And similarly for berries, which botanically are 'fleshy, indehiscent, few-seeded to many-seeded fruits ... The word *berry*, like the word *fruit* or the word *nut*, has thus very different botanical and popular meanings. The date is a one-seeded berry ... Tomatoes, grapes and bananas are berries, while oranges, cucumbers, melons, and apples are examples of special classes of berries' (Brown 1935). It would be confusing for the reader to maintain the strict botanical usage in this book; thus we would not be able to write about 'elderberries' (which are drupes), and we would continually have to qualify the word 'fruit'. And so for convenience we use the two words in their popular sense: in this book 'fruits' are fleshy fruits, unless otherwise qualified, and 'berries' are small fleshy fruits of the kinds with which we are mainly concerned; thus the two words are largely

16 Introduction

interchangeable. The correct botanical terms for all the fruits with which we deal are given in Appendix 1.

Similarly we use the word 'seed' in a general sense, as the part of the fruit that is dispersed by birds, whether it be the unprotected seed (*ie* the seeds of berries) or the seed enclosed in a hard endocarp (*ie* seeds of drupes). We have generally avoided using the word 'stone' for the latter, as it might cause confusion for some readers.

THE ARRANGEMENT OF THE BOOK

The main part of the book is factual: it gives an account of what we observed, and of what others have observed. It can be treated as a straightforward natural history of fruit-eating by birds in a small area of southern England, with some comparison with other parts of Britain and Europe. Following this main part are a number of chapters of what may best be called interpretation. In these chapters we try to go beyond the facts, or rather, to interpret them in a wider ecological context and to see them as the outcome of evolutionary processes moulding both birds and plants. This is an area of rapidly developing research, and we do not hope to do more than open the eyes of any readers, whether ornithologists or botanists or general naturalists, who are unfamiliar with the issues that are discussed.

In presenting our observations in the main part of the book we have had to cope with a problem of presentation: almost everything we describe involves an interaction, and can be seen from two points of view, that of the plant or that of the bird. When dealing with a plant we must either repeat a certain amount that is said under the birds that feed on it, or give an incomplete account with page references to relevant sections elsewhere in the book; and similarly when dealing with a bird. We have in fact compromised, by repeating a certain amount but referring to other sections when something that needs to be said concerns both bird and plant but is clearly more relevant to one or the other. Thus we hope that the sections on plants can be read by a botanist without too much distracting cross-reference to the birds, and the sections on birds without having to refer continually to the plants. We hope, of course, that most readers will be interested in both the birds and the plants.

Acknowledgements

For help in the preparation of this book we are very grateful to: Dr C. M. Herrera and Dr P. Jordano (Estación Biológica de Doñana, Sevilla, Spain) for undertaking chemical analyses of some fruits and for freely placing their data, some unpublished, at our disposal; Dr E. W. Stiles (Rutgers University, Piscataway, New Jersey) for undertaking chemical analyses of many samples of fruit pulp; and Derek Goodwin, for sending us his recent observations of fruit-eating together with much other stimulating correspondence.

A few passages have been drawn, with slight modification, from previously published papers (Snow 1971, Snow & Snow 1984, Snow & Snow 1986); we are grateful to the British Ornithologists' Union and Dr W. Junk, Publishers, Dordrecht, for permission to re-use these passages.

1: THE FRUITS

At least 39 different species of plants with fleshy fruits are native in our study area. There are in addition a further three species, found mainly in gardens, that may not occur naturally in the area but are native elsewhere in southern England, and one (bird cherry) that occurs in an apparently wild state in our area but has a more northerly range in Britain and has probably been introduced. We treat all of these as wild fruits; their main characteristics are summarised in Appendices 2 and 4.

These 43 species comprise about two-thirds of all the species of plants with fleshy fruits that are native to England and Wales, if a few species that are very rare or local are left out of consideration. The 43 species are listed in Appendix 1, and Table 1 shows a break-down into different growth-forms. It will be seen that our area contains nearly all the trees and climbers on the list for England and Wales, but only 16 of the 28 species of shrubs. The main species in the latter group that are missing are the small shrubs of heathland and moorland, such as bilberry and cranberry.

There is also, in gardens in the study area, a considerable number of exotic plants and cultivars with fleshy bird-dispersed fruits. We have made observations on the more important of them, but have made no attempt to find and watch all of them. The fruit characters of 16 of them, including all those for which we had significant numbers of feeding records, are summarised in Appendix 3.

In the following sections we discuss, species by species, the wild fruits, dealing with their seasons, the birds that feed on them, and other points of interest. After dealing with our own observations we discuss, more briefly, published observations on the fruits from other parts of Britain and from continental Europe, concentrating on points that amplify or contrast with our observations, or are of

TABLE 1 *Number of species, and growth-forms, of plants with fleshy fruits native to England and Wales, and occurring in the study area*

	Number of species occurring in	
	England and Wales	study area
Trees	9	9
Small trees/shrubs	9	6
Shrubs	28	16
Climbers	6	5
Herbs	11	6
Epiphytes	1	1
Totals	64	43

Notes. Data for England and Wales based on Clapham, Tutin & Warburg (1973). Aggregates of very closely related forms (*eg* in *Rosa*) are treated as single species. A few very rare or local species are omitted (*Cotoneaster integerrimus, Polygonatum odoratum, Asparagus officinalis*).

interest in other ways. Our observations on introduced and cultivated fruits are dealt with in less detail in a final section.

The seasonal pattern of fruiting

When we began systematic observation in our study area we had had some experience of the humid tropics, where fruit is available the year round. We did not expect the same situation in southern England, but in fact we found that here too wild fruits of some kind are available to birds for the whole year. There are, of course, marked seasonal fluctuations in the variety and quantity of fruits available. We may begin by looking at seasonal changes in the variety of fruits, without taking into account their abundance. It is reasonable to do so, because the wild fruits that are present in our study area today are much the same as would have been present when the vegetation was in its natural state; but their relative abundance must have been drastically altered by forest-clearing, cultivation and urbanization. Thus the relative abundance of different fruits now is largely artificial, and liable to rapid alteration; but there is no reason to think that their seasons have been altered.

The 'fruit year' begins, not in January, but with the ripening of the first summer fruits. The earliest of these, the wild cherry, currant and strawberry, ripen in June. Of these three, only the wild cherry is an important source of food for

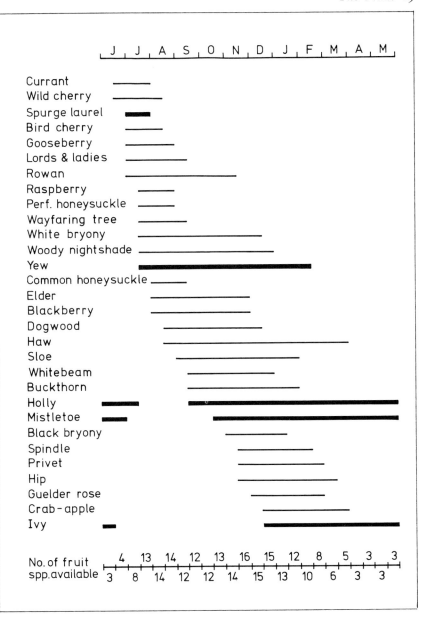

Figure 1 Seasonal availability of wild fruits for dispersers, Vale of Aylesbury area, 1980–85.

The figure is based on observations of fruit being eaten, except for gooseberry for which observations of ripening and depletion of fruits have been used. The season for hip refers to *Rosa canina* only. Evergreen plants are distinguished by a heavy line.

birds, the other two being local as wild plants in our area. July sees the ripening of several more, especially rowan, wayfaring tree, woody nightshade and lords and ladies. The earliest yews ripen at the end of July. More fruits ripen in August, much the most important being the elder, the bulk of the yews, and the earliest haws and blackberries. In September come whitebeam, dogwood, and the bulk of the haws and blackberries. The earliest hollies ripen at the end of September, but most in October, when a locally common small tree, the buckthorn, and mistletoe are added to the list. Privet and spindle are the main fruits starting to ripen in November, also the earliest hips of the dog rose (*Rosa canina*, the less common sweet-briar and field rose being much earlier). Most dog rose hips ripen later, in December or January, and it is in late December that the earliest ivy fruits ripen. Different strains of ivy ripen over a prolonged period, from late December to April. Some of the autumn- and winter-ripening fruits remain on the plants – for various reasons, which we deal with later in the accounts of the fruits concerned – until the spring, so that, in some years at least, in addition to ivy a certain amount of haw, holly and mistletoe fruit remains until February or March, and some hollies and mistletoes retain fruit until much later, in extreme cases until June or July, so that they overlap the beginning of the next fruit year.

The fruiting seasons are shown graphically in Fig. 1. This figure illustrates more clearly than words the succession of fruits throughout the year, and in particular it shows that three common evergreen plants – yew, holly and ivy – between them maintain an almost year-round fruit supply. The importance of the evergreen habit in enabling a plant to maintain a long fruit season is discussed later.

Although there is some fruit in every month, there is a much greater variety in some months than others. From late July through to the end of January, 12–16 different kinds of fruit are available in each half-month period; from April to June only three or four are available. Not all parts of our study area, however, enjoy a year-round supply of fruit. Thus ivy, the most important spring fruit, is locally absent from some of the higher parts of the Chilterns, and mistletoe is very local. Holly, the only other fruit that spans the early spring period of fruit dearth, is not available everywhere because only certain holly trees retain their fruit after midwinter, as discussed later. Thus locally there may be a complete absence of fruit from March to May.

Fig. 1 shows the availability of ripe fruit. It does not follow that each fruit is taken at anything like a uniform rate throughout the period when it is available. Some fruits are preferred to others, and a less popular one may be neglected for some time after it has ripened if more popular alternatives are available. Or a fruit may begin to be eaten and then, when a preferred fruit ripens, may be neglected until the more popular fruit is finished. Thus woody nightshade fruits are eaten a lot in July and August, very little during the period of great fruit abundance in September and October, and then are finished off in November and December when feeding conditions become harder.

There are minor local differences in fruiting seasons even within our study area. For some fruits the season is a little later in the higher parts of the Chilterns than on lower ground. There are also annual differences: ripening times have been later in some years than in others, probably as a consequence of the weather prevailing at the time when the fruits were developing. Our data are insufficiently detailed for a proper examination of such variations, but they do not affect the main seasonal pattern.

Essentially the same pattern presumably prevails over much of southern England, but we have been unable to find comprehensive data from elsewhere for comparison. Sorensen's (1981) data for a number of fruits near Oxford indicate a similar seasonal pattern, as would be expected, except for sloe. It is only from the extreme south of Europe, in southwestern Spain, where Herrera has carried out very detailed studies of all aspects of fruit-eating by birds, that we have comprehensive data for comparison with our study area. The seasonal differences in fruit availability between southern Spain and southern England are discussed on p. 203.

We did not attempt to assess seasonal changes in the *amount*, as opposed to the variety, of fruit available. But some of the fruits shown in Figure 1 are not abundant anywhere in our area; others are locally abundant, but absent from most parts. Overall, the quantity of fruit available is determined mainly by the seasons of the more abundant, widespread species which grow in semi-open as well as in more wooded country, especially ivy, hip, haw, sloe, blackberry and elder. These are available mainly from September to February, the two most abundant being largely finished by November (elder) and December (haw). Many of the fruits in our study area which appear to be of minor importance, such as lords and ladies, spurge laurel, currant, raspberry and gooseberry, are plants of the woodland shrub and ground layer, and all fruit in summer. Before the advent of man, when southern England was forest, these plants presumably played a more important role in providing fruit for birds. Moreover they provide it at a time when fruit is often in high demand.

Holly *Ilex aquifolium*

Holly is a common under-storey tree in well-drained woodland in our study area, but absent from woodland on heavy clay soils. It is also locally abundant as a hedgerow tree, and is common in gardens. It flowers mainly in the second half of May and early June; male and female flowers are borne on separate plants, so that only about half of the trees bear fruit. The well known red berries (technically drupes) average about 9 mm in diameter, small enough for all the main fruit-eating birds to swallow whole. They contain a toxic substance, ilicin, which can cause vomiting, diarrhoea and drowsiness in man, but it seems not to be known whether the toxin is present in all parts of the fruit or concentrated in the seed (Cooper & Johnson 1984).

Of all the bird-fruits in our area, holly berries have the longest season. The earliest ripen in the second half of September (our earliest record of their being eaten was on 23 September), and a small number of trees in some years retain fruit and continue to be fed on until late July. August is the only month for which we have no records, but it is quite likely that a few very late fruits are eaten in this month, too, as we have seen occasional trees retaining a few of the previous year's fruits in August when the next crop was beginning to ripen.

There are two reasons for this very long season. First, holly berries are long-lasting; they do not go bad, do not all fall to the ground, and they are resistant to extreme cold. Second, many holly trees are defended by Mistle Thrushes (single birds or pairs), which prevent other birds from feeding on the fruit and so conserve them as a long-term food supply which can last through the winter and on into the following spring and summer, when the Mistle Thrushes stop defending them and the fruit becomes available again to other species. These two reasons for the holly's long season are not independent of one another. Undoubtedly it is the long-lasting qualities of the fruit that make holly especially suitable for long-term defence. The defence of fruit by Mistle Thrushes is a subject in itself, and is dealt with more fully in another section.

The resistance of holly berries to deterioration over a period of many months, including spells of intense cold which damage other kinds of fruit, is presumably linked to the evergreen habit, though we have been unable to find botanical studies dealing with this aspect of fruit biology. It seems likely that an active, if reduced, metabolism through the winter enables the plant to maintain the water content, and perhaps the other constituents, of the fruit. At all events, holly berries lose little moisture during the long period that they remain on the tree: our measurements of the water content of the pulp of samples collected in different months gave the following figures: October–December 71–75%; January–March 66–72%; April–June 62–67%.

Our monthly feeding records for the eight bird species we have seen feeding on holly berries are shown in Table 2, and the seasonal pattern is broken down into its main components in Figure 2. The differently shaded areas in this figure show that the overall pattern is made up of a complex of records from holly trees of different 'status' with regard to defence, or lack of defence, by Mistle Thrushes.

(1) *Undefended hollies.* Only a few hollies in any area are defended. Observations of birds feeding at undefended hollies account for the great majority of our October and November records, and nearly half of the December and January

TABLE 2 *Holly: monthly summary of feeding records*

	Sep	Oct	Nov	Dec	Jan	Feb	Mar	Apr	May	Jun	Jul	Totals
Blackbird		59	72	170	49	64	61	16	27	76	18	612
Mistle Thrush	1	8	14	59	135	63	39	23	21	5	2	370
Song Thrush		5	5	23	9		1	2	1	9	2	57
Redwing		1	11	326	204	32	8					582
Fieldfare				56	3	25						84
Robin		4	2	2	2	5						15
Blackcap			1		30	13	1					45
Woodpigeon			4	2	12	2	9	3				32
All species	1	77	109	638	444	204	119	44	49	90	22	1797

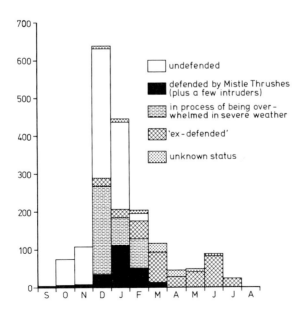

Figure 2 Records of birds feeding on hollies of different status with respect to defence by Mistle Thrushes.

records (most of the latter being in the first half of the month). In the years when there was a very cold spell in December, all undefended hollies were completely stripped by the end of the month; in the years when the early winter was mild, some undefended hollies retained fruit until well into January. It is broadly true to say that the defence of hollies by Mistle Thrushes is the reason why there are always some holly berries available for Christmas decorations.

(2) *Defended hollies.* Mistle Thrushes in possession of defended hollies feed at them rather little in October and November; in these months, while actively preventing other birds from feeding at their hollies, they themselves feed more on fruits from undefended plants nearby, if any are available (Table 3). By December they are feeding mainly at their defended hollies, and throughout the hardest part of the winter may be completely dependent on them during very severe spells, especially if there is deep snow cover. Defence of fruit wanes in early spring, when Mistle Thrushes may switch to feeding on ivy, and also are active in territorial defence and beginning to breed. They may continue to feed occasionally at the hollies which they had defended, especially when they have young to feed, but after the end of March we had no records of their driving other birds away. All records of birds feeding at hollies that had retained some fruit as a result of having been defended by Mistle Thrushes have been placed in a third category.

(3) *Ex-defended hollies.* Most of the March records shown in Fig. 2 are in this category, and all of those from April onwards (with the exception of a small number that have been classified as 'unknown' because the previous history of the trees concerned had not been observed, but which almost certainly come into the 'ex-defended' category). The small number of earlier records (December–February) refer to hollies that for some reason had been abandoned prematurely by the Mistle Thrushes, in some cases possibly as a result of the death of the defending bird.

(4) *Hollies in the process of being overwhelmed.* The fourth category of records is of particular interest. During periods of intense cold, especially if a deep snow cover makes ground-feeding impossible, Mistle Thrushes defending hollies (and other kinds of fruit) may come under such intense pressure from other birds, especially Redwings and Fieldfares, that they are eventually overwhelmed. When this happens there is a short period of intense feeding activity, as described elsewhere (p. 130), and the tree is soon stripped, usually within 48 hours of the Mistle Thrushes losing control. We had five such periods during the five years of our study (in December 1981, January 1982, February 1983, January and February 1985); they account for a large number of our December records, and for smaller numbers of our records in January and February.

Table 3, showing the rates of feeding at hollies in these four categories, provides quantitative support for the general statements made above. It can be seen that the feeding rate (number of bird-visits per hour) at undefended hollies is about ten times the rate at defended hollies. The rate at defended hollies is low in October and November, when, as already mentioned, Mistle Thrushes feed preferentially on other fruits nearby. The feeding rate increases in December at defended and undefended hollies, and it is easy to see how it is that the latter, with a rate of about 25 visits per hour (hence losing some 200 fruits per hour, as about eight fruits are commonly taken per visit), are generally stripped in this month or in early January. Hollies in the process of being overwhelmed in severe weather show rates two or three times as high. Hollies that have been successfully defended by Mistle Thrushes and have retained fruit into the spring show a low rate of feeding visits in April, when ivy, a preferred fruit, is often available near

TABLE 3 *Feeding rates at defended and undefended hollies, and at hollies being overwhelmed*

	Defended		Undefended/ ex-defended		Being overwhelmed	
	Hrs of obs	Visits per hr	Hrs of obs	Visits per hr	Hrs of obs	Visits per hr
October	9.4	0.8	9.2	6.8	–	–
November	15.9	0.6	12.1	6.4	–	–
December	16.6	2.4	12.2	24.5	2.3	79.0
January	35.0	3.3	6.4	26.8	1.3	50.4
February	20.7	3.4	–	–	2.1	46.2
March	–	–	10.6	7.7	–	–
April	–	–	30.8	1.0	–	–
May	–	–	20.8	2.1	–	–
June	–	–	29.0	2.9	–	–
July	–	–	3.8	5.9	–	–

Notes. 'Ex-defended' hollies (figures for March–July) are hollies that had been defended through the winter by Mistle Thrushes and so retained fruit (see text). The comparatively high rate of feeding visits in March (7.7) was influenced by data from one year when other food was apparently scarce, the rate for other years being 0.5 per hour.

by, general feeding conditions are good, and few young birds are yet being fed. There is an increase from May to July, when increasing numbers of young birds are being fed, some hungry juveniles are already independent, and drought tends to reduce the availability of earthworms.

The very wide range of dates when individual hollies were finally stripped of their fruit crop is shown in Table 4. One very early-ripening holly was stripped in late October in 1982; in the other years it was defended. Probably its fruit ripened even earlier than usual in 1982, or the Mistle Thrushes were late in starting to defend it. At the other extreme, some hollies that retained a large fruit crop after being successfully defended by Mistle Thrushes were not stripped until the end of the following July. Those that were overwhelmed by invading birds in very cold weather were mainly stripped in December and February, but this simply reflects the fact that the most intense cold spells were in those months during the period of our study.

To be suitable for defence, a holly tree should be of moderate size, standing free of other trees but preferably with other perches (trees, wires, etc) near by from which a look-out can be kept. A sparse foliage is more suitable than a thick, bushy growth, since if a tree is too large and thick a Mistle Thrush may be unable to see intruders coming from all directions, or not see them feeding if they manage to fly in undetected. Many of the defended hollies in man-made habitats were adjacent to grassland suitable for foraging, from which the defended tree was in view.

It is clear that holly trees in the course of their lives pass through a stage during which they may be suitable for defence by Mistle Thrushes. Before this stage they do not provide enough fruits, and after it they may be too large to be effectively defended. Periodic lopping and trimming of garden hollies by man

TABLE 4 *Months of stripping of fruit crop of 32 holly trees, Vale of Aylesbury area 1980–86*

	Undefended trees	Trees initially defended, then abandoned for unknown reasons	Trees overwhelmed in very cold weather	Trees successfully defended by Mistle Thrushes
October	1	–	–	–
November	1	–	–	–
December	10	–	12	–
January	9	7	3	5
February	–	3	4	4
March	–	–	1	6
April	–	–	–	7
May	–	–	–	8
June	–	–	–	12
July	–	–	–	5

Note. For some trees records were complete over all winters; others were recorded over fewer winters, depending on opportunities for observation.

probably prolongs the period for which they are suitable for defence. Obviously, the transition from being suitable for defence to being unsuitable must be gradual, probably lasting a number of years. We had several cases of attempts to defend rather large hollies which were abandoned apparently because even in average winter weather (*ie*, when the pressure from other birds was not especially great) the Mistle Thrush could not deal effectively with intruders. These were presumably hollies that had got beyond the defendable stage. One of them, with hindsight, we know was formerly defendable, as some years ago it used to retain its berries until spring.

We did not attempt to count the number of fruits that ripened on all the holly trees that were defended. Estimates made on a few trees indicated that defended hollies started with a fruit crop of the order of a few tens of thousands of berries. A crop of such a size can easily meet a pair of Mistle Thrushes' needs throughout the winter; but if it is being stripped at rates of 200 fruits or more per hour (see above), the crop does not last long.

It will be obvious that the Mistle Thrush's defence of hollies has a big effect on the opportunities for other birds to feed on the berries. The four other thrushes all take holly fruits regularly, but with distinctive seasonal differences (Table 2). Fieldfares take holly only in severe weather, when alternative fruits, especially haws, are scarce or non-existent; almost all our records are from periods when defended hollies were being overwhelmed. Redwings show a much more extended seasonal pattern, with single birds and small groups regularly feeding at hollies when the weather is mild, at any time from October to March. Song Thrushes seem not to favour holly berries if alternative food is available; the lack of records in February, for instance, and the small number in January probably indicate that snails are a preferred winter food. Blackbirds take holly berries at all times; for them the seasonal pattern of records is much the same as that for all dispersers combined (Fig. 2).

We have records for two smaller birds taking holly berries, Robin and Blackcap. The records for Robin, though few, come from several hollies and all years, so that it seems likely that the seasonal pattern shown in Table 2 is valid, and that Robins do not feed regularly on holly berries after late winter. But the irregular distribution of our Blackcap records is probably largely a matter of chance. Most of them come from one area, where in one winter at least four individuals were feeding in a loose association.

Woodpigeons in our area seem to feed on holly berries rather irregularly and only in small numbers, probably when other food is scarce (p. 173). They are evidently dispersers of holly and not seed-predators, as the seeds appear undamaged in their faeces. Thus of a sample of 16 holly seeds in a Woodpigeon's droppings 15 were intact; the one that was broken was probably unripe, being noticeably paler than the others.

It must surely be significant that, having watched fruit-laden hollies for over 280 hours and recorded nearly 1,800 feeding visits by other birds, we have no record of any of the crow family taking holly fruit or showing any interest in it. We cannot suggest any good reason why they should not do so, except that the seeds may be toxic for them; but, as noted below, from Continental Europe there are records for Magpie and Jay.

HOLLIES ELSEWHERE

The only quantitative information about holly-berry eating that we can find from other areas in Britain is that given by Hartley (1954). From a broad survey of fruit-eating in England and Wales, records indicated that holly berries were eaten in two distinct seasons, October–December and May–June – a seasonal pattern more or less corresponding to what we have found. He thought that different holly trees might vary in their season of ripening or in their edibility, and quoted Turner (1935), who wrote: 'It seems to me that of all fruits holly berries are the least sought after by birds. In exceptionally hard winters, Fieldfares and Redwings will soon strip a tree, but normally the berries are left...' Ignorance of the Mistle Thrush's defence of holly trees has been responsible for this rather common belief. Much longer ago, however, at least one observant naturalist recognised the true situation. William Borlase in *The natural history of Cornwall* (1758) described how the Mistle Thrush in winter 'feeds upon holly-berries, each bird taking possession of his tree, keeping constant to it, as long as there is fruit, and driving away all other birds' (see also p. 131).

In the exceptionally cold spell of December 1981, Derek Goodwin told us of exactly the same scenes in Kent as we had been witnessing, of Mistle Thrush hollies being overwhelmed by flocks of Redwings and Fieldfares. Almost certainly, what we were seeing in our study area was typical of what was happening to holly trees over much of England.

Holly has a mainly southern and western distribution in continental Europe. In a survey of fruit-eating by birds in the Dresden area, at about the latitude of Salisbury but with a continental climate and much colder winters than anywhere in England, Creutz (1953) recorded holly berries being eaten by Blackbirds and Song Thrushes, but he noted that holly is not much planted and does not regularly fruit in his area, and so presumably does not maintain itself as a native plant. It is presumably for this reason that there are rather few records from central Europe of the eating of holly berries by birds. Among them is a record for the Magpie (Maas 1950).

In the Cantabrian Mountains of northern Spain holly is locally abundant and is ecologically of great importance, providing food for birds and in one way or another being responsible for the winter survival of the majority of vertebrates (Guitian 1984). As a provider of cover it takes the place of conifers, which are absent. Large flocks of Redwings feed on the holly berries, and smaller numbers of Blackbirds and Mistle Thrushes. Guitian does not mention whether Mistle Thrushes defend hollies, but a lone holly covered with berries, which we saw in these mountains in mid May, suggests that some hollies are defended. He records Jays as eating the holly berries, the only record that we have found for this species, and mentions Bullfinches as seed-predators, collecting holly seeds from the snow below the trees or extracting them from the fruit on the tree. Our lack of records of Bullfinches showing any interest in holly berries in our area, as also of any Jay records, suggests that a more detailed comparison between England and Spain would be of interest.

Ilex is a cosmopolitan genus and there are records of the fruits of some tropical and subtropical species being eaten by birds in southeast Asia (Ridley 1930), South Africa (Frost 1980), Malawi (F. Dowsett-Lemaire, in litt.), Trinidad (Snow & Snow 1971) and Jamaica (Lack 1976). Several species occur in North America, but there does not seem to have been any thorough study of their fruits being eaten by birds. Petrides (1942) noted that *Ilex opaca* fruits in late winter in Maryland, and had records of their berries being eaten by birds in March. Stiles (1980) tabulated the seasons of dispersal of the five species of *Ilex* (including *I. opaca*) occurring in the eastern deciduous forests of North America, all of them being from autumn until either winter or spring. In late April in Connecticut one of us (DWS) saw fruits on two garden hollies (*I. opaca*) which had been fed on through the winter, and presumably guarded, by a resident American Robin (Dr T.H.G. Aitken, pers. comm.), which suggests an interesting project for garden birdwatchers in New England.

Yew *Taxus baccata*

As a native tree, yew is apparently confined to hilly country in Britain (Bentham & Hooker 1937). In our study area it occurs in a natural state only in woodland and scrub on the Chilterns; elsewhere it is common in gardens, parkland and churchyards. Most of our observations on it were made in these man-made habitats.

The red, translucent fruits are berry-like. We call them berries for convenience, but botanically they are arillate fruits, the naked seed being half immersed in a fleshy cup-like aril. Yew is well known to be a poisonous plant. According to Cooper & Johnson (1984), all parts of the plant are poisonous except the flesh of the fruit; but as discussed below, we think it is likely that there is a significant difference in the amount of poison in the seed-coat and the rest of the seed.

In our area the first yew berries ripen on some early-fruiting trees at the end of July. On each tree the fruits ripen asynchronously over a period of a

TABLE 5 Yew: monthly summary of feeding records

	Jul	Aug	Sep	Oct	Nov	Dec	Jan	Feb	Total
Dispersers									
Blackbird	1	95	150	135	85	11	3		480
Song Thrush	6	55	120	109	124	53	43		510
Mistle Thrush	2	26	68	44	54	28	10	1	233
Redwing				34	19	1	1		55
Ring Ouzel			3						3
Robin		2	9	5	4	1	5		26
Blackcap	1	3							4
Starling		32	230	445	127	3			837
All dispersers	10	213	580	772	413	96	62	1	2148
Seed predators									
Greenfinch			15	42	31	87	32	4	211
Great Tit				1		6			7
									218

few weeks, so that small green fruits as well as ripe ones are to be found in the first weeks of the fruiting period. Uneaten fruits persist in good condition for many weeks.

Our earliest record of yew fruit being eaten was on 28 July. The fruit is fed on increasingly through August and September, reaching a peak in October (Table 5). Many trees are stripped in the course of November, and by mid December the more easily accessible fruits have usually been eaten, those that remain being mostly the ones hanging below the branches. Those trees that still have a lot of fruit are mainly ones that have been defended by Mistle Thrushes. Only in one year (1985) did we have a substantial amount of fruit remaining into January. At least one tree had been defended by Mistle Thrushes, but others had not. Several kinds of fruit had persisted unusually late in the early winter of 1984–5, and it seemed that with other fruits available the pressure on the yew crop had been lighter than usual.

Song Thrushes, Mistle Thrushes, Blackbirds and Starlings are the main dispersers of yew fruit in our area. Of these, undoubtedly the Song Thrush is the fondest of yew; the 510 Song Thrush records (Table 5) represent nearly one quarter of the total for that species. (Possible reasons for the preference are discussed on p. 222.) For Mistle Thrush and Starling, yew records make up respectively 16% and 18% of the totals of feeding records, and for Blackbirds only 7%. For three other species – Redwing, Robin and Blackcap – yew records comprise only 1–4% of the totals. We have never recorded Fieldfares feeding on yew berries, but they have been recorded eating yew berries elsewhere (see below). In our study area, during the main period of the yew crop, Fieldfares feed almost exclusively on haws, mainly in semi-open country, and only come into the more wooded habitats where yews are found in severe weather, by which time the easily accessible yew fruits are gone.

The yews in our area have one important seed-predator, the Greenfinch, one less important one, the Great Tit, and probably two occasional ones, Blue and Marsh Tit. Greenfinches are almost always to be seen on and around fruiting yew trees. When they visit a tree to feed they may remain in it for many minutes on end, dealing with the fruits at a rate of one every minute or two, and so on average they must account for more berries per visit than the much larger thrushes, which take on average 8–10 per visit. Our 207 records for Greenfinches thus indicate that, as seed-predators, they may destroy a fraction of the yew crop comparable to that eaten by one of the main dispersers of the fruit. They also begin to take the fruit before it is ripe.

A Greenfinch dealing with a yew fruit first mandibulates it until the pulp falls away, then rotates the seed in the bill, stripping off the brown seed coat before eating the rest of the seed. This suggests that the poison which yew seeds are known to contain is concentrated in the seed coat. It further suggests a possible reason why yew fruits are avoided by another seed-predator, the Woodpigeon, which swallows fruits whole and must then grind up the seed-coat with the seed; and might also account for the fact that the crows – or at least the larger ones, as Creutz (1952; see below) had records for the Jay – do not eat yew berries. Though they are 'legitimate' fruit-eaters, the larger crows might nevertheless subject yew seeds, which are not at all hard, to sufficiently harsh treatment in the stomach to break up the seed coat.

By comparison with Greenfinches the tits are relatively unimportant predators. Some Great Tits pluck whole yew fruits, discard the pulp and eat the seeds, holding them in the feet and hammering them to break them up; and we have seen Marsh Tits do the same, though not in our study area. Further details are given on p. 183. Chaffinches, too, feed on yew seeds, but we do not include them as seed-predators as they take only fragments of endosperm remaining from those opened by Greenfinches.

YEWS ELSEWHERE

Hartley (1954), in his survey of wild fruits in the diet of British thrushes, found that the Song Thrush was the main eater of yew berries in two areas of southern England, not only when calculated as a percentage of the total number of records for each species but in absolute number of records (as we also found). He had only a single Redwing record and, like us, none for the Fieldfare. Probably the poor showing of the two winter thrushes at yew in his areas, as in ours, was due to the fact that most of the observations were in man-made habitats. In his account of the ancient yew wood at Kingley Vale in West Sussex, Williamson (1978) mentions that the berry crop lasts from August to mid-December (just as in our area) and draws as many as 2,000 Redwings and Fieldfares from about mid-October onwards. Though he does not make clear the relative amount that they feed on the yews and haws that are common in the area, the implication is that they feed extensively on both. He mentions that some yews are guarded by resident Mistle Thrushes, which drive off intruding birds and conserve the fruit for use later in the winter.

For continental Europe, the only detailed observations seem to be those made by Creutz (1952, 1953) in the Pillnitz area, near Dresden. In his area yew fruits were eaten mainly in September and October by Song Thrush, Mistle Thrush, Blackbird, Fieldfare, Robin and Jay; Greenfinch, Hawfinch, Nuthatch, Great Tit, and Green and Greater Spotted Woodpeckers were seed-predators. He noted that about 10–12 berries constituted a 'meal' for the thrushes (within the range observed by us, but a little higher than our average meal-sizes), and mentioned that after eating the fruit birds apparently need to drink frequently and to wipe their bills, evidently to get rid of the rather glutinous pulp. Of the seed-predators, he reported that Great Tits hold the seed against a branch and hammer it, as a few of our Great Tits did, while the Nuthatch and woodpeckers take the fruit to a neighbouring or occasionally a more distant tree, wedge it in a crack in the bark, and open the seed.

Creutz stressed that he never saw Starlings eating yew fruit, although they were often to be seen near yew trees. In this connection it may be significant that two detailed feeding studies of Starlings, in Hungary (Szijj 1956–7) and Czechoslovakia (Havlin & Folk 1965), make no mention of yew fruits being eaten, nor is the Starling listed as feeding on yew fruits by Turček (1961), who worked in Czechoslovakia and drew on much central European literature. Heymer (1966), however, gave a record of Starlings eating yew berries in Saxony. The relationship between Starlings and yew fruit in central Europe needs further study.

Ivy *Hedera helix*

Ivy is abundant over most of our study area, especially on heavy clay soils in low-lying areas where many poplars and willows bordering the fields have thickly ivied trunks. It is sparser at high levels in the Chilterns, and not very abundant on the poor and well-drained soils of the Lower Greensand.

Ivy is unique among European fruit-bearing plants in that it flowers in autumn. In our area there is great variability in the flowering and fruiting seasons of different plants. Some plants observed over three or four seasons flowered at the end of August and had ripe fruit by January or the end of December. Others near by in very similar sites flowered in October and the fruit did not ripen until May. Between these two extremes most ivies flowered in September and ripened their fruit in March or April. It seems that different genetic strains are involved. Individual plants may also ripen their fruit over a period of a few weeks. The flowering stems terminate in a main inflorescence, but usually smaller lateral inflorescences are produced; these flower later, and may either produce later-ripening fruit or small fruit that abort before ripening.

Ripe fruits have bluish black or greenish black skins and purple pulp, and contain 1–4 seeds. They average 8–9 mm in diameter (Appendix 2) and so are available to all except the smallest fruit-eating birds. They are among the most nutritious of British wild fruits, having an especially high fat content. Ivy contains poisons which are apparently concentrated in the fruits (Cooper & Johnson 1984), but whether in the seeds or in the pulp seems not to be known. The plant is moderately poisonous to mammals if eaten in quantity. A report that ivy seeds eaten by poultry in France caused their death (Mahe-Quinio et al 1975, quoted by Cooper & Johnson) suggests that the poison may be concentrated in the seeds, which on theoretical grounds seems most likely (p. 209). The seeds, even when mature, are unusually soft, a fact that may be significant in determining which birds exploit the fruit.

Ivy fruits ripen mainly at a time of year when the weather is gradually becoming milder, but excessive cold may affect the earlier fruit crop when it is ripening so that it withers and aborts. We recorded such effects in 1981 and 1982. We have never found that fruit in good condition has been left uneaten; the crop is usually finished some time in May or early June, but early fruiting plants are stripped much sooner.

Dispersers. Our records cover nearly six months of the year, from late December to early June, with a peak (37% of all records) in April (Table 6). In addition to many records for all the resident and wintering thrushes we have several records for a Ring Ouzel on spring passage in the third week of April 1981. Most of our Fieldfare records were obtained in the same period, when an unusually large crop of fruit was available on ivy-grown trees in an area of pasture with hedges. The intensity of feeding on ivy fruit then was the greatest that we have ever seen. It was made even more spectacular by a pair of Mistle Thrushes, which had a nest on a bough of the best of the ivied trees and constantly attacked any bird that tried to feed near their nest, including at times the Ring Ouzel.

In addition to the thrushes, Robins, Blackcaps and Starlings feed a great deal on ivy fruit. Our Blackcap records are mainly in April and May, when fruiting ivy attracts newly arrived migrants; the earlier records, in January and February, were from the sparse population of wintering Blackcaps.

We had no records for corvids, and no indication of any of them showing an interest in ivy fruits. In view of the unusual softness of the seeds, which probably contain the highest concentration of the fruit's poisons, we suggest that crows would be in danger of damaging the seeds and releasing the poisons during digestion of the fruit, and may avoid them for this reason; but there are records of Carrion Crows eating ivy berries on the Continent (see below).

Seed-predators. Woodpigeons are the only seed-predators of ivy in our area. They begin to take the fruit while it is still unripe, and continue to take unripe and ripe fruit throughout its season (Table 6). The total of feeding records for Woodpigeons is rather less than a quarter of that for dispersers; but since

TABLE 6 *Ivy: monthly summary of feeding records*

	Dec	Jan	Feb	Mar	Apr	May	Jun	Totals
Dispersers								
Blackbird	18	180	137	86	243	203		867
Song Thrush	4	46	76	36	138	78	1	379
Mistle Thrush		21	26	25	26	21		119
Redwing		8	7	123	30			168
Fieldfare			1	1	101	7		110
Ring Ouzel					12			12
Robin	4	6	11	17	26	12		76
Blackcap		5	6		82	32	1	126
Starling		55	8	73	138	5		279
All dispersers	26	321	272	361	796	358	2	2136
Seed-predator								
Woodpigeon								
unripe fruit	17	179	43	15	2			256
ripe fruit		22	67	38	10	3		140
not differentiated		75	5					80

Note. Woodpigeons may also act as dispersers of ripe fruit (see text).

Woodpigeons take about 30 times as many fruits per feed as the thrushes it is clear that they must destroy a major part of the ivy seed crop. On the basis of our records we calculate that 78% of the seed crop is destroyed by Woodpigeons while it is still unripe (p. 207). To this must be added a percentage destroyed when ripe fruit is eaten. As discussed below, there is evidence that Woodpigeons eating ripe ivy berries may sometimes act as dispersers as well as seed-predators.

Ivy seeds, being soft, should be easily dealt with by Bullfinches and other small seed-predators. We have no records of any finch or tit showing interest in ivy fruit (nor are there any records in the literature), and it seems probable that the toxicity of the seed, to which Woodpigeons must be immune, is an effective defence against predation by finches and tits.

IVY ELSEWHERE

Ivy has an essentially southern and western distribution in Europe, extending to 60°30′N in Norway but only occurring sparsely north of the Alps in central Europe, where winters are severe. Even in southern England distribution approaches its climatic limit on high ground. In our area it becomes sparse on top of the Chilterns, and Macleod (1983) found that in the Cotswolds plants growing above 265 m had smaller and less abundant fruit, which ripened later, than plants at lower altitudes (below 80 m).

In her Cotswold study area Macleod recorded most of the same disperser species that we recorded. Surprisingly, she concluded that the Woodpigeon is a potential disperser, as ivy seeds obtained from droppings germinated. Woodpigeon droppings that we have examined, from birds which had evidently been feeding on ivy berries (droppings stained deep purple at a time when no other fruit of this colour was available), contained no trace of intact seeds. This point needs further investigation; it may be that the treatment of ivy seeds depends on what other food Woodpigeons are eating and how much grit the gizzard contains, as discussed on page 173.

Other British records of birds eating ivy berries do not add significantly to what we found. It is clear, however, that the period when ivy fruit is available is not the same all over the country; in more northerly districts the season may be somewhat later. Thus in 1983 we saw birds feeding on a heavy crop of ivy berries in Dovedale, Derbyshire, on 21 May, when the ivy fruit in our study area had all been eaten.

In view of its distribution on the Continent it is not surprising that records of ivy fruit being eaten by birds in northern Europe are few. Schuster (1930), who in addition to his own observations attempted as complete as possible a coverage of the German literature, could cite only a handful of references to ivy fruit being eaten, by Capercaillie, Woodpigeon, Blackbird, Mistle Thrush, Blackcap and Waxwing, the Blackbird being the only species for which he had several references. In the spate of notes and papers from Germany stimulated by Schuster's review, only Creutz (1953), in the Dresden area, recorded ivy fruit being eaten, again by Blackbirds. In Switzerland, Müller-Schneider (1949) wrote that he had seen Blackbirds eagerly eating ivy fruit near Biel in April.

In northern France, Heim de Balsac (1928a) noted that unripe ivy berries are much eaten by Woodpigeons in winter, and in spring the few ripe berries that are left are eaten by other birds and mammals (the latter surprising, in view of the fruit's toxicity). Further south, in France and the Iberian peninsula, the only dated record that we have found is by Jollet (1984), who found that Carrion Crows eat ivy fruit in April and May in the Limoges area of southwestern France. This is also the only record we have found of any of the crows eating ivy berries. In the first half of May 1986, in the mountains south of Grenoble in southeastern France, we found ivy plants loaded with ripe fruit that apparently had not been touched, which suggests that it is taken much later than in our study area. Also, in northern Spain in the foothills of the Picos de Europa, we found ivy plants with plenty of ripe fruit between 14–24 May 1987, on which Blackcaps, Blackbirds, Song Thrushes and Robins (in that order of abundance) were feeding; at one large plant we recorded a rate of 80 feeding visits per hour. In southern Germany, Berthold (1984) found that Blackcaps fed their young on ivy berries between 13 May and 17 June in the unusually cold and dry spring of 1984, when other food was apparently scarce, which also suggests a later season than in our study area (see p. 203). It would be of interest to make a comparative study of the phenology of ivy in Britain and on the Continent, and of the dates when the fruit is dispersed. Further south still, in the Mediterranean area, an early season seems likely; we have seen wintering Robins eating ivy berries in Malta in late December.

Mistletoe *Viscum album*

Mistletoes are very specialised plants. In addition to their parasitic habit, one of their main specialisations concerns the way in which their seeds are dispersed to lodge on the branches of trees in sites suitable for germination. Birds are intimately involved in this process. In the tropics, where the greatest diversity of mistletoe species is to be found, birds of several families have become specialised, structurally and physiologically, as eaters of mistletoe fruits and dispersers of their seeds. In Britain, where only one species of mistletoe occurs, no bird is, as far as known, specialised for feeding on mistletoe berries. The fruits, however, are beautifully adapted for dispersal by birds. In particular, the single seed is surrounded by a layer of extremely sticky jelly which resists digestion in the bird's gut, so that it adheres readily to tree branches when the bird gets rid of it, whether by defaecation or by wiping it off the bill. The fruits are quite small (mean diameter 8.3 mm; Appendix 2), so that they can be swallowed by even small frugivores, and the

36 *The Fruits*

pulp, in spite of its watery appearance, has a high lipid content and so is very nutritious (Appendix 4).

In our study area mistletoe is a local plant. A few clumps grow, or grew, on lime trees in Tring, but many more are to be found at Halton, not far from Wendover; these too are mainly on lime trees. In our own village, Wingrave, there are a few small plants on apple and other garden trees, some and perhaps all propagated by man. We made regular observations at the only large group of mistletoes in Tring over two winters (until the lime tree was cut down to make way for new housing) and similarly over three winters at some of the Halton clumps. All these mistletoes were fed upon almost exclusively by Mistle Thrushes, which defended the larger groups of plants from early November throughout the winter and continued to feed at them, if any fruit remained, until the following spring or summer. The seasonal distribution of our records is shown in Table 7.

The association between Mistle Thrushes and mistletoes is of course well known, and has given the bird its scientific and English names, and its name in several other European languages. It may be loosely called a mistletoe 'specialist'; but it is really a specialist at defending a fruit supply – a point discussed further on p. 127. The fact that we hardly saw any other birds feeding on mistletoe berries does not mean that the berries are acceptable only to Mistle Thrushes, or that Mistle Thrushes are in some way specially adapted to feed on them. It results from three facts, valid for our study area but not necessarily elsewhere: first, that Mistle Thrushes defend the plants against other birds; second, that mistletoes are not common, and hence there are enough Mistle Thrushes to defend all sizeable clumps (unlike hollies in some parts of our study area, which are too numerous for all of them to be defended by Mistle Thrushes); and third, that the siting of the mistletoes in discrete groups or clumps of plants, mainly high in lime trees and clear of other vegetation, makes them easy for Mistle Thrushes to defend effectively (unlike many hollies, whose size and position make effective defence impossible).

In late autumn, when they first ripened, we had evidence that other species of birds were keen to take mistletoe berries. Thus in 12.1 hours of observation in late October and November we had 12 records of Mistle Thrushes chasing other thrushes (Blackbirds, Song Thrushes and Redwings) from their defended mistletoes, as well as a few records of their chasing away other Mistle Thrushes. In December, in 7.7 hours of observation we recorded only one chase, of an intruding Mistle Thrush, and from January onwards, in over 32 hours of observation we had no records of attacks on intruding thrushes or other potential eaters of the berries. In all months we had a few records of perfunctory attacks on occasional Woodpigeons and small passerines that happened to perch near the mistletoe, and we had two records of a passing Blackbird being briefly chased. It seems that the initial period of active defence in October and November was effective in discouraging other thrushes from attempting to get at the fruit.

Our records of other species eating mistletoe berries were few, and some were the result of special circumstances. In October and November, during the period of active defence against other thrushes, we had records of two Blackbirds and a Redwing managing to snatch a quick feed before being chased off. A small group of low-growing mistletoes in a Wingrave garden, which were evidently unsuited for effective defence and were only intermittently defended and fed at by a Mistle Thrush, was seen to be fed at by two Blackbirds and a Song Thrush in January and February; and in March and April, when the Mistle Thrushes' defence was waning, we had records at Tring of two Robins and a Blackcap taking mistletoe fruit. Finally, in January a Robin took a fruit from a mistletoe plant which we

TABLE 7 *Mistletoe: monthly summary of feeding records*

	Oct	Nov	Dec	Jan	Feb	Mar	Apr	May	Jun	Totals
Mistle Thrush	2	30	46	57	70	18	16	6	6	251
Blackbird	1	1			2					4
Song Thrush				1		1				2
Redwing			1							1
Robin				1		1	1			3
Blackcap						1				1
All species	3	32	46	59	72	21	17	6	6	262

had brought into our garden.

This last observation was of special interest. A large part of the largest clump of mistletoes at Halton had been cut down and removed around Christmas time, severely disrupting our observations; but for some reason a few good-sized bunches, covered with fruit, had been left lying on the ground at the foot of the tree. On 13 January we took these home, tied them to our two apple trees in conspicuous positions and awaited results. The nearest other mistletoes (those mentioned above) were about 350 m away. For 14 days no bird showed any interest in the mistletoe berries, although the weather at the time was very severe and several hungry Blackbirds, Song Thrushes and one Fieldfare were feeding on rotten apples and other food that we had put out just below the mistletoes. Then on the morning of 27 January a pair of Mistle Thrushes flew in and almost at once began to feed on the mistletoe berries. Within a few minutes one of our garden Robins also flew up and took a berry, and was at once driven off by one of the Mistle Thrushes. We saw no other bird attempt to take berries, but the Mistle Thrushes chased off Blackbirds which occasionally came close. The pair of Mistle Thrushes, which had been defending a cotoneaster about 120 m away, continued to inhabit our garden until they had practically stripped the mistletoe 19 days later.

The lack of interest shown by our garden Blackbirds and Song Thrushes must have been due to their failing to recognise the mistletoe berries as potential food, and this failure was probably attributable to their rarity in our area and perhaps their white colour and unusual habit of growth. Similarly, it seems likely that it was the sight of the Mistle Thrushes eating the fruit that alerted the Robin, a very versatile frugivore, to the fact that they represented food. Though we do not know the extent of the area over which the two Mistle Thrushes had previously foraged, they are wide-ranging birds and it is quite likely that they, but not our local Blackbirds and Song Thrushes, had visited the other Wingrave mistletoes and so recognised the berries as food. An alternative, less likely explanation is that Mistle Thrushes have an innate recognition of mistletoe berries which other thrushes lack.

In the very severe spells of winter weather when many of the hollies defended by Mistle Thrushes were overwhelmed by other thrushes and stripped (p. 130), none of the defended mistletoes lost their fruit. In all cases the berries lasted until they had been finished by the Mistle Thrushes, as early as February in the case of two of the smaller clumps which had not fruited very prolifically, but not until spring or summer in other cases.

As would be expected (and as we also found for hawthorns defended by Mistle Thrushes), the rate of depletion of the mistletoe berry crop depends to some

38 The Fruits

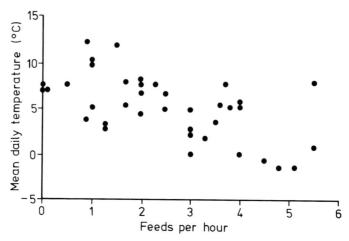

Figure 3   The feeding rate of Mistle Thrushes at defended mistletoes in winter (November–February) in relation to mean daily temperature; results of 39 watches, each of at least 60 bird-minutes. ($r = -0.34$; $P < 0.05$.)

extent on the severity of the weather. Fig. 3 shows that the hourly rate of Mistle Thrush feeding bouts varies inversely with the mean daily air temperature: when mean temperatures are around freezing point, a feeding rate of about five bouts per hour is usual. In mild weather (mean temperatures of 4°C or higher), rates of about one feed per hour, or even less, are usual. The correlation is impaired to an unknown extent by the fact that the number of berries taken by a Mistle Thrush in the course of a feeding bout (the 'meal size') is highly variable. In only a minority of cases was it possible to see exactly how many berries were taken, so that this variable cannot be taken into account. For instance, the anomalous point in Figure 3 (feeding rate 5.5 per hour, temperature 8°C) refers to a watch during which the sizes of most of the feeds were recorded and were smaller than average, with a mean of 4.9 berries per feed compared with an overall mean of 7.7. Our highest recorded meal sizes were 23, 19 and 19. Clearly, if fewer berries are taken during a bout the bird is likely to compensate with a higher rate of bouts per hour. The reason for the high variability is probably that Mistle Thrushes spend a great part of the day perched beside their mistletoes (whereas those defending hollies spend much time perched on vantage points further away), so that it is easy for them to take occasional small feeds.

Mistle Thrushes defaecate the seeds of mistletoe berries, and in our area must be easily the most important disperser of the plant. Blue Tits are almost certainly mistletoe seed predators. They regularly search the twigs and branches around mistletoe clumps, and though we had no proof it seemed most likely that they were looking for mistletoe seeds adhering to the bark. As mentioned below, there is definite evidence from France that Blue Tits feed regularly on mistletoe seeds.

MISTLETOE ELSEWHERE

We have found little of significance published on the consumption of mistletoe berries elsewhere in Britain, except for notes by Hardy (1969, 1978) from north-western England and North Wales. He pointed out (1969) that mistletoe is scarce in northern England, except where planted by man on low-level apple boughs; high-level, bird-sown plants are scarce north of Shropshire and the lower Dee area of Cheshire. At Loton Park, Shropshire, he noted that Mistle Thrushes fed on mistletoes less frequently than Song Thrushes, Blackbirds and Redwings. We wonder what is the basis for his statement that 'the fruit is unpalatably unripe until February and March'; it suggests that he may have been misled by the fact that feeding rates are often low at Mistle Thrush-defended plants during the winter. In the later paper (Hardy 1978) on winter foods of Blackcaps in Britain, he recorded mistletoe berries eaten in a garden in North Wales from late January to mid-February. A female and two male Blackcaps fed on them and were aggressive to one another.

The situation in two areas in France was investigated in the 1920s by two of the outstanding French ornithologists of that time, H. Heim de Balsac and N. Mayaud (Heim de Balsac 1928b, Heim de Balsac & Mayaud 1930, Mayaud 1928). They were primarily interested in the method of propagation of mistletoe seeds by birds. In Lorraine, in eastern France, Heim de Balsac found that Mistle Thrushes begin to eat the fruit as early as mid-September, when it is still not quite ripe (not full-sized, greenish, and with a rather tough skin). From October onwards they spend the greater part of the day at their mistletoe clumps, driving off any other bird that approaches. Heim de Balsac commented especially on the fact that he saw no other species of thrush feeding on the fruit, and supposed that they do not like mistletoe berries ('... dédaignent systématiquement le Gui'), rather surprisingly ignoring the more likely explanation that they are prevented from feeding on them. In the second area, in Anjou in west-central France, Mayaud similarly recorded Mistle Thrushes, and no other thrush species, feeding on mistletoe berries from midwinter through to early March.

In both areas the only other species that Heim de Balsac and Mayaud saw feeding on mistletoe fruit was the Blackcap. In Anjou Blackcaps were found to subsist, apparently almost exclusively, on mistletoe berries through the most severe winters, in particular the exceptional winter of 1928–29. Where mistletoes were abundant the local density of Blackcaps was very high, each bird defending a group of plants against other Blackcaps. Mayaud collected ten of them for analysis of their gut contents and found that the total number of birds remained about the same; when a bird was killed its place was taken by another. Clearly there was a tightly organized system of individual feeding territories. In Lorraine the situation was very different. There, Blackcaps were found to be wholly migratory and only ate mistletoe fruit on their return to the area, in March and April, by which time it seems (though this is not definitely stated) that Mistle thrushes were no longer defending them.

Heim de Balsac & Mayaud described in detail how Blackcaps deal with the berries. They separate the seed from the skin and pulp, taking the fruit and placing it on a branch, then squeezing out the seed. The skin and pulp, still retaining its spherical form, is then swallowed and the seed is left adhering to the branch on which they perform the operation, which is rarely a branch of the mistletoe itself but usually of the host tree. Heim de Balsac & Mayaud therefore considered that Blackcaps were the main dispersers of mistletoe in their areas, as Mistle Thrushes defaecate the seeds haphazardly and only a small proportion of them land in sites where they may be able to germinate.

Mayaud (1928) at first thought that Blue Tits also ate mistletoe berries, but more careful observation (Heim de Balsac & Mayaud 1930) showed that their frequent visits to mistletoe clumps and the branches and twigs near by were for two purposes: to forage for insects within the mistletoe clumps, and to take mistletoe seeds adhering to the twigs and branches. The seeds may be pecked at in situ or may be taken to another perch, held between the toes, and opened; in either case the kernel is extracted and the seed case left.

In the Dresden area of East Germany, where the winters are much more severe than in eastern France, Mistle Thrushes are apparently totally migratory. Creutz (1952, 1953) reported an interesting observation of mistletoe berry-eating at what is nearly the eastern limit of the plant's range in central Europe. In early March, Mistle Thrushes (how many is not stated), soon after their return to the area, attempted to defend a large clump of mistletoe plants against a flock of about 150 Waxwings which had been feeding on the fruit all the winter. By this time only remnants of the fruit crop were left. The Waxwings went from plant to plant in groups of up to 20 or more birds, and were persistently attacked by the Mistle Thrushes. Their attacks effectively broke up the Waxwing parties, but the Waxwings, evidently very hungry, did not move away far but kept on returning and succeeded in taking the fruit.

These observations from France and Germany raise a number of intriguing questions. Mistletoe fruit apparently allows Mistle Thrushes to remain through the winter in Lorraine, where winters are moderately severe, but not, it seems, near Dresden where winters are much more severe. Where is the eastern limit set; or do they remain in some winters but not others, depending on the severity of the weather or the abundance of the mistletoe fruit crop? Similarly, what are the geographical limits of the area where Blackcaps over-winter at mistletoe clumps? And how do these two species interact in areas (if there are any) where both of them are dependent on mistletoes for winter survival?

Dog rose *Rosa canina*

Dog rose is very common in our study area, occurring almost everywhere in hedges and bushy places, and along woodland edges. It also grows within woodland, sometimes climbing high up the trees. It flowers in June. The hips redden in September, but remain hard and firmly attached to their stalks until late November or December. They then soften, and can be plucked by birds. There is no abscission layer at the junction between fruit and stalk, but when the hip is pulled it tears away at the base of the pulp. If not eaten, the hips remain on the plant in good condition for several weeks, but in late February and March many dry up and eventually blacken, still attached to the plant; a few remain in good condition until well into March.

The spindle-shaped hips of dog rose are larger than most other fruits in our study area, with an average diameter (the critical dimension for swallowing) of nearly 13 mm and lengths usually of 20–24 mm (Appendix 2). Sizes are rather variable, even on the same bush, but most fruits are larger than the smaller thrushes can swallow, and all are beyond the size range possible for Robin and warblers. The elongated seeds are hairy and form a tightly grouped mass in the middle of the fruit; their number is very variable, and often a fruit contains many undeveloped seeds as well as fully developed ones. The 'relative yield' of hips is high, mainly because the water content of the pulp is low (Appendix 2), but they contain a high concentration of tannins (Herrera 1984b), which may prevent birds from digesting them efficiently (Herrera 1982). As mentioned below, there is some evidence that birds do not extract nutriment from hips efficiently.

In several ways the fruits of *Rosa* spp. are untypical of bird-dispersed fruits, and tend to be avoided by dispersers if other fruits are available (p. 224).

Dispersers. In our study area the hips of dog rose are eaten almost exclusively by the three larger thrushes – Blackbird and Fieldfare, and to a lesser extent Mistle

42 The Fruits

TABLE 8 *Dog rose: monthly summary of feeding records*

	Nov	Dec	Jan	Feb	Mar	Totals
Dispersers						
Blackbird	7	158	224	200	14	603
Song Thrush		1	4	2		7
Mistle Thrush		1	6	5		12
Redwing			4			4
Fieldfare		96	73	262		431
Robin		1				1
Blackcap			2			2
Woodpigeon		9	6			15
All dispersers	7	266	319	469	14	1075
Others						
Greenfinch		11	4	7		22
Blue Tit		2	5			7

Note. All the November records were in the second half of the month, and all the March records in the first half.

Thrush – which together account for 99% of all our records for dispersers (Table 8). We have a few records of Song Thrushes and Redwings taking hips, but their firmness of attachment and size make it hard for these small thrushes to pluck and swallow them. Our single record for a Robin was of one that picked a whole fruit and flew off to eat it under a bush; and our records for the Blackcap were of birds picking out pieces of pulp from hips in situ in very cold weather.

Our records of dispersers eating hips extend from late November to the first half of March, with a peak from late December to early February. Bushes which are sparsely distributed along hedgerows may be stripped by mid December, but in areas where dog rose is abundant, especially on some Chiltern slopes, masses of hips remain until late winter. In cold weather, Blackbirds and Fieldfares concentrate on the hip crops on these slopes, and individual Fieldfares may take possession of discrete clumps, defending them for several days against other thrushes. Mistle Thrushes and Blackbirds may also temporarily defend concentrations of hips at times when other fruits are scarce.

In very cold weather, when flocks of fieldfares are feeding on hips, the snow-covered ground may be littered with their red, sausage-shaped droppings, containing seeds, skin and apparently also a good deal of undigested pulp. From distant observations we have suspected that the Fieldfares sometimes eat these droppings, and W. D. Campbell (*in litt*) has suspected the same, also unfortunately without being able to be certain. Re-cycling of the fruit material would make sense if, as seems likely, hips are not digested very efficiently on first passage through a bird's gut.

Woodpigeons are probably regular eaters of hips, but we have records only for periods of severe weather in two winters. Examination of a large number of hip seeds from Woodpigeon droppings showed that they were intact (p. 171); hence this is one of the plants of which Woodpigeons are dispersers, not seed-predators.

Others

Greenfinches are regular seed-predators of *Rosa canina*, as they are of other roses. Blue Tits are occasional pulp-predators, pecking small pieces of pulp from hips in situ.

DOG ROSE ELSEWHERE

The little that has been published on the eating of dog rose hips elsewhere in Britain is in good agreement with our observations. In a wooded area near Oxford, Sorensen (1981) found that they were eaten from early December, but mainly in January and February; she recorded only Blackbirds and Woodpigeons. Hartley (1954), in a survey of fruit-eating by thrushes, had records for Blackbirds only. Murton (1965) recorded hips in the diet of Woodpigeons in East Anglia. Campbell (1981) described a case of a Fieldfare guarding a supply of hips in a briar tangle in the very cold spell of December 1981, in Oxfordshire, just as we found on the Chiltern slopes near Tring in the same winter.

The period when the hips of *Rosa canina* are eaten in continental Europe seems to be much the same as in England, even in southern Spain, where the hips ripen from mid to late September but are eaten, mainly by Blackbirds, in November-February (Herrera 1984a). These southern Spanish hips are much smaller than English hips (mean diameter 9.2 mm, mean length 15.0 mm) and only slightly larger than the haws that occur with them in the same habitat. Herrera (1984b) studied the effect of the haws, which are preferred by birds, on the dispersal of hip seeds.

In Germany, Schneider (1957) mentions that in snowy weather Blackbirds and Fieldfares compete for the little that is left of the hip crop. Waxwings take hips when they invade central Europe (*eg* Schüz 1933), and were reported doing so in mid March in England during the 1950/51 Waxwing invasion (Gibb & Gibb 1951. Greenfinches are evidently regular seed-predators (Heymer 1966, Schlegel & Schlegel 1965). However, in some areas or in some winters on the Continent, hip crops may be untouched. In the French Alps at heights of 1,000–1,500 m, in the first half of May 1986, we found that virtually the whole of an abundant crop of hips of *Rosa canina*, a very common plant in the area, remained on the plants, still red but completely dried up. Presumably by the time the crop had ripened the local Blackbirds, certainly the commonest of the potential dispersers, had moved down to lower levels in the valleys and when they returned the hips were inedible.

Other roses

There are two other wild roses in our study area, the field rose *Rosa arvensis* and the sweet-briar *Rosa rubiginosa* agg. (The 'agg.' indicates that sweet-briar may be subdivided into a number of closely related species, as is also the case for the dog rose.) Both are much less common than the dog rose and have a different season of ripening. We had only a single record of a disperser taking the fruit (of field rose), but we saw seed-predators feeding at them or had evidence of their activities.

Field rose. The field rose is much commoner than the sweet-briar in our study area, both in the Chilterns and in association with the ancient woodlands on the clays north of the Vale of Aylesbury; but it is much less common than the ubiquitous dog rose, and we saw no signs that it was successfully invading new scrubland or hedges, as dog rose does. It flowers in the first three weeks of July, later than the dog rose, but its fruit ripens earlier, from the end of September to early December. The fate of the fruit had to be assessed mainly by examining the bushes. Rodent seed-predators, probably field mice (*Sylvaemus sylvaticus* or *S. flavicollis*), bite through the pedicel a few millimetres below the fruit, leaving characteristic tooth marks, and then transport it to some safe place below or near the bush to eat the seeds. Greenfinches open the fruit and eat the seeds on the bush, leaving the skin and pulp in place. The scar where a hip has been recently plucked by a disperser is also recognisable.

We spent eight hours watching field roses with ripe fruit from October to December and had only one record of a disperser taking them, a Blackbird in early December. We examined many bushes and saw no signs that fruit had been plucked by a disperser. In an area where yew, the Greenfinch's most favoured seed, was abundant, we saw no Greenfinches feeding on hips of field rose until December (when the yew fruit crop is mainly finished). In an area where yew was scarce, examination of field rose bushes gave evidence of Greenfinch seed-predation from the end of October. Depredation by mice starts earlier, by mid-October.

Sweet-briar. We found only one area in the Chilterns where the sweet-briar grew and we monitered the progress of the flowering and fruiting of adjacent sweet-briars and dog roses there in 1980–82. The sweet-briar flowered approximately ten days later than the dog rose, but ripened its fruit earlier, in late September and October. In three hours spent watching ripe sweet-briar hips we saw no birds at them, although during one watch (in November) a Blackbird and a Robin fed on black bryony near by.

In 1980 and 1981 only 4–9% of the fruit crop was left by mid November. Judging from the piles of fruit and seed fragments under the bushes, most of the seed crop had been eaten by mice. In 1982 we had no evidence of mouse activity, but much of the fruit crop rotted on the plants or fell to the ground. Although we did not observe Greenfinches eating the seeds, in mid December we found typical evidence of their seed-predation on a few remaining fruits.

Our evidence indicates that the dog rose enjoys much better seed dispersal than the two other roses in our area, and that this advantage is related to the time of ripening of the fruit. By ripening earlier, the field rose and sweet-briar expose their seeds to predation by rodents and Greenfinches for longer, and lose a great part of their fruit crop in this way; and they are also in competition with haws, which are preferred by dispersers to hips. The dog rose fruit crop, ripening later, comes mainly after the end of the haw season, so that although it is not a preferred fruit it is available at a time when conditions may be severe and other fruit is often scarce. Although its seeds are also taken by Greenfinches and mice, we had no evidence of these seed-predators taking them when unripe. So in December local seed-predators and large flocks of wintering thrushes simultaneously turn to this food source, and the seeds are largely dispersed. We suggest that its much more successful dispersal is one of the reasons, and perhaps the main one, why the dog rose is so much more abundant in our area than the other two roses.

Haws *Crataegus* spp.

Because it is abundant, being the main constituent of hedgerows, hawthorn is one of the most important producers of bird-fruits in our area, and indeed over much of Britain. In our study area the hawthorns are almost entirely *Crataegus monogyna*, the single-seeded species with deeply dissected leaves; but the 'midland hawthorn' *C. laevigata*, two-seeded and with less dissected leaves, also occurs, both in woods and locally in hedges, and there are in addition plants intermediate between the two. It would be interesting to make a detailed study of the ecology of fruit-dispersal in the two hawthorns – *C. laevigata* is said to fruit earlier (Bentham & Hooker 1937) – but we became aware of the presence of small numbers of *laevigata* in our study area too late to note any evidence of a difference in their fruiting seasons, and have had to combine our records for all wild hawthorns. From the bird's point of view *C. monogyna* should be a better fruit to eat, as the single seed makes up a smaller proportion of the total fruit weight (Appendix 2).

Haws vary greatly in size. In the samples that we measured, summarised in Appendix 2, the mean diameter was 9–14 mm (most being 9–10 mm) and mean weight 0.41–1.16 g. Song Thrushes, Redwings and Robins are unable to swallow the largest haws, but such fruits are infrequent. The mealy pulp is moderately nutritious (details in Appendix 4). Botanically the fruit is a drupe, the stone (seed with encasing endocarp) being very hard and largely immune to all the seed-predators that occur regularly in our study area.

The seasonal pattern. In our area hawthorn flowers in May and early June, and the fruits begin to redden in August. We have a few records of their being eaten by Blackbirds in the second half of August (earliest date, 22 August), but few plants seem to have fully ripe fruit before September, when Blackbirds begin to eat them increasingly. In October, with the arrival of the wintering flocks of Fieldfares and Redwings, haw-eating increases dramatically (Fig. 4). The peak comes in November, and in a typical year the bulk of the haw crop has been

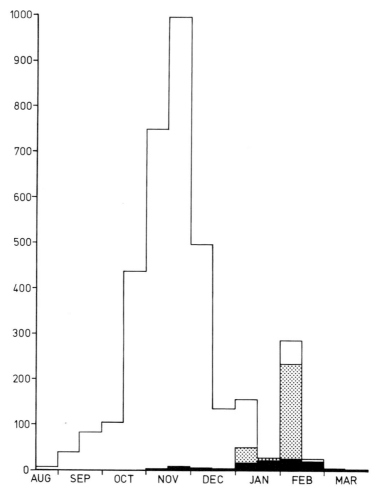

Figure 4 Records of dispersers eating haws, by half-month periods. Open parts of histograms: undefended haws. Shaded parts: haws defended by Mistle Thrushes. Black parts: records from defended haws in process of being overwhelmed by other thrushes during severe weather (see text).

eaten in the low-lying parts of our study area by the end of November or early December. More or less sparse fruit remains on the higher slopes of the Chilterns until January. In the 1984/85 winter, when all autumn-ripening fruit ripened late and persisted later than usual, a considerable quantity of haws lasted until January in all areas. If the winter is not too severe, uneaten haws last in good condition to the following spring, but prolonged spells of very cold weather may destroy the greater part or all of the remaining fruit crop, leaving the fruit blackened and inedible. Our records are summarised in Table 9.

This broad seasonal pattern is modified, to a small but ecologically significant extent, by the fact that a few of the hawthorns are protected as a long-term food supply by Mistle Thrushes, and in consequence retain their fruit long after the others have been stripped. The long-term defence of fruit by Mistle Thrushes is dealt with elsewhere (pp. 127–131); here we mention only those points that are especially relevant to hawthorn.

Long-term defence of haws. We have observations on three hawthorns that were protected over two or more successive years by pairs of Mistle Thrushes and were evidently especially suitable for defence, being small trees clear of surrounding vegetation but with higher perches near by from which the defending birds could keep watch. Other hawthorns, some of which were clearly less suitable for defence, were protected for one year only, some of them by single birds. (Another, which appeared very suitable for defence and was protected by a pair, was not found until the last year of observation.) In several cases, the protection extended to one or two small hawthorns adjacent to the main one. Defended hawthorns may be very obvious in late winter, remaining crimson with fruit when all the surrounding plants have been stripped. Indeed, we learned that the mere presence of a hawthorn in such a state is a sure sign that it has been, and is still being, protected by Mistle Thrushes. This may not be obvious at a casual glance, especially if the weather is mild and the Mistle Thrushes are feeding a little distance away or simply standing guard in a neighbouring tree. As mentioned elsewhere, the Mistle Thrushes' actual feeding visits to their protected fruit supply may be quite infrequent as long as alternative food is available, while the neighbouring Blackbirds and other thrushes have learnt to keep clear of it, so that one may get the misleading impression that the fruit has been overlooked or is for some reason unattractive.

We recorded five instances of Fieldfares protecting hawthorns, but none of them for more than a few days. They mainly occurred during spells of cold weather, and all involved single birds. Blackbirds, which regularly protect small, defendable fruit supplies within their territories, may protect garden hawthorns;

TABLE 9 *Haw: monthly summary of feeding records*

	Aug	Sep	Oct	Nov	Dec	Jan	Feb	Mar	Totals
Dispersers									
Blackbird	8	118	344	695	262	64	64	4	1559
Song Thrush		4	36	48	15	6	4		113
Mistle Thrush			15	33	24	28	47		147
Redwing				63	310	118	45		536
Fieldfare				40	504	201	48	197	990
Robin					12	7			19
Starling			1	45	146	6			198
Woodpigeon			2	24	60	4	6		96
All dispersers	8	125	567	1808	637	191	318	4	3658
Pulp-predator									
Blue Tit				6	20	2	1	17	46

Note. Woodpigeons are also to a small extent seed-predators (see text).

we have never had a case of persistent defence of a hawthorn by a Blackbird in open country, but they may occasionally do so.

Hawthorns are not as suitable for long-term defence as hollies, mainly because, as mentioned above, the fruit is ruined by periods of very cold weather. The haws eventually blacken, even if they have remained edible through the cold period, and the whole of the remaining crop is inedible. Nevertheless, the fruit may last for long enough to tide the owning Mistle Thrushes over the worst part of the winter. The history of one hawthorn that was kept under observation over five successive winters well illustrates what may occur. Over the first three winters it was successfully protected, the fruit being spoilt by early February in one year after an exceptionally cold January, but in the two other years remaining in good condition until March, when the Mistle Thrushes abandoned their defence. In the fourth year the haw crop was very poor and the tree was undefended. In the fifth year it was defended as usual by a pair of Mistle Thrushes, but was overwhelmed and stripped during a very severe spell in January when one of the defended hollies under observation was also overwhelmed by a flock of Redwings and Fieldfares.

Seed- and pulp-predators. Woodpigeons frequently eat haws and their droppings collected from beneath roosts have contained large numbers of haw seeds, of which some were broken but the great majority were intact. Hence Woodpigeons must be mainly dispersers of haws, and we include them with the dispersers in Table 9. Blue Tits are pulp-predators; we have many records of them pecking small pieces of pulp from haws in situ, leaving them with yellow spots and patches where the underlying pulp is exposed. The Hawfinch is the only finch seed-predator that can tackle haws (as its name implies) but it is rare or absent in our study area. According to Mountfort (1957), Hawfinches eat haws but they are not a chief item of their diet; they take them over a very long season, from August to April, but mainly from October to January.

HAWS ELSEWHERE

We have found no published observations on the eating of haws by birds in Britain that add significantly to what we recorded, except for seed-predation by Hawfinches, mentioned above. On the continent of Europe, haws are a far less important fruit for birds than in Britain, as hawthorns are generally much sparser. Hawthorn hedges, which provide the main food of our migrant thrush flocks in autumn and early winter, are a distinctively British phenomenon. Thus in their accounts of more or less systematic observation of fruit-eating in different parts of Germany, neither Maas (1950), Creutz (1953) nor Schlegel & Schlegel (1965) make any mention of haws. At Bad Aibling, southeast of Munich, Zedler (1954) recorded haws as a supplementary food taken by Blackbirds in winter. In Saxony, Heymer (1966) recorded haws (*C. oxyacantha* = *laevigata*) as eaten in late December and early January. Among other more casual references we have come across none that suggests that haws are more than a rather minor fruit for wintering birds on the Continent.

In southern Europe, hawthorn *C. monogyna* is a mainly montane plant. Herrera (1984b) studied it at 1,280–1,450 m in southern Spain. There the fruit ripens in mid or late September, a little later than in our study area, and is dispersed mainly by Blackbirds in November–February, also rather later than in our area.

Rowan *Sorbus aucuparia*

In the northeast of our study area, Rowan is a common tree, in woods on the Lower Greensand; on the Chiltern slopes it is sparsely distributed, elsewhere it occurs only as a planted tree (including cultivated varieties), mainly in gardens and along suburban streets. It flowers in May and early June, the earliest fruits ripen about mid July, and by August the whole fruit-crop is ripe. Like whitebeam, rowan fruits tend to dry up and turn brown if not eaten, but there is a good deal of variation in this. Some trees kept under observation have had practically all their remaining fruit shrivelled and bad by mid September, others have stayed in good condition until October or even early November. Probably the lasting quality of the fruit is dependent, at least in part, on weather conditions.

The ripe fruit is orange (reddish in some cultivated varieties) and contains a variable number of rather soft, small seeds (1–4 full-sized seeds in our samples, plus some very small and undeveloped). The 'seed burden' (seed as a percentage of the total fruit weight) is unusually low, averaging 3.4% (Appendix 2). A mean diameter of about 9 mm allows the fruits to be swallowed whole by all except the smallest frugivores. Warblers may be able to swallow the smaller fruits whole but cannot swallow full-sized ones.

Dispersers. Blackbirds are the main dispersers of rowan fruit in our area, accounting for 78% of the total records (Table 10). Song Thrushes take the fruit comparatively little, perhaps because they strongly prefer elder and yew fruits which are available at the same time as rowan. Starlings, too, seldom take rowan fruit, probably for the same reason. Robins usually sally to pluck a fruit, then take it to a perch before swallowing it – their usual procedure with fruits that are at the upper size limit. Our few records of warblers eating rowan fruit were of birds taking fragments of fruit shredded by Bullfinches.

Our earliest record of rowan fruit being eaten was on 15 July, when a Blackbird was seen to visit an early-ripening tree and take a single fruit, which did not look quite ripe. By the end of July the fruit is being increasingly taken, and August is the peak month. Some time in September most trees are either stripped or, if they have fruit left, it has gone bad; but some trees retain fruit in good condition in October. In busy suburban areas where the pressure on fruit is less and some rowan cultivars may have fruit that is less attractive to birds, the fruit may occasionally last into early November.

50 The Fruits

TABLE 10 *Rowan: monthly summary of feeding records*

	Jul	Aug	Sep	Oct	Nov	Totals
Dispersers						
Blackbird	56	373	70		6	505
Song Thrush		20	2			22
Mistle Thrush	5	21	4	2		32
Robin		7				7
Blackcap	1	7	4			12
Garden Warbler		1				1
Lesser Whitethroat		3				3
Starling		30	13	20		63
Magpie		2	1			3
Jay		1				1
All dispersers	62	465	94	22	6	649
Seed-predators						
Bullfinch	18	66	21			105
Blue Tit			2			2

Note. All but one of the July records were in the second half of the month, and all November records were in the first half. Blue Tits may be pulp- rather than seed-predators (see text). All warbler records are of birds scavenging fragments of pulp after the seeds had been removed by seed-predators.

Seed-predators. Bullfinches are major seed-predators of rowan, the number of records exceeding that for all of the dispersers except Blackbird. Some rowans have a large fraction of their total seed crop destroyed by Bullfinches, which leave unmistakable signs of their work, shredded fragments of skin and pulp from which the seeds have been extracted. In one area in 1981 more than 90% of the fruit crop was attacked by Bullfinches. Blue Tits, the only other potential seed-predators recorded, were seen pecking at the fragments of Bullfinch-shredded fruits but we are not sure whether it was to take the pulp or a few seeds that may have been left by the Bullfinches.

ROWAN ELSEWHERE

Rowan is well known to be one of the most important bird fruits in northern Europe, as indicated by its German name *Vogelbeer* (bird-berry). Its nutritive value has been exhaustively analysed by Pulliainen (1978). The size of the rowan crop in Fennoscandia affects the migration of Fieldfares and the eruptive movements of Waxwings. In years when there is a good crop, part of the Fieldfare population remains in Finland after the normal time of migration (Tyrväinen 1975). Similarly, Waxwings stay in the north all winter when rowan fruit is available (Siivonen 1941), and rowan fruit is much the most important food of Waxwings wintering in Germany (Schüz 1933). Tyrväinen mentions that the critical level of the decreasing fruit supply, triggering emigration, is higher for the Fieldfare than for the more specialized Waxwing.

In a highland glen in Inverness, northern Scotland, Swann (1983) has found that Redwings pass through rapidly on autumn migration, with the peak passage usually 10–24 October. Individual birds remain for less than 24 hours, during which they feed mainly on rowan fruits. The annual totals of birds trapped in different years were correlated with the size of the rowan crop.

There are many general references to rowan fruit being eaten in central Europe. The most thorough observations seem to be those made by Blaschke (1976) who kept records over 14 years of birds eating the fruit of 15 rowan trees growing in a residential area in northern Germany, making almost daily observations from the ripening of the fruit in July until it had been eaten. Starlings were the most important species; in some years flocks of them more or less stripped the trees in a few days in October or November. Next were Waxwings, which fed on the fruit from early or mid October until late November or early December. Next in order of importance were Blackbirds, Fieldfares and Redwings; the Fieldfares usually, and Redwings always, as single birds. Six species were irregularly or rarely recorded, four or five of them presumably as seed-predators (this is not made clear): Crossbill (in summer, one year only), Bullfinch, Greenfinch, Brambling, Greater Spotted Woodpecker and Blue Tit. Bullfinches, the main seed-predator of rowan in our study area, were irregular in winter in Blaschke's records (being perhaps scarce in the area); and Greenfinches (which we never recorded) were very few compared with the numbers present at feeding places near by. At least in some years, some fruit remained until February. This indicates a striking difference from the seasonal availability of rowan in our area, where the fruit, if not eaten, goes bad by the late autumn. In Scandinavia, too, the observations cited above indicate that rowan fruit may persist in good condition through the winter.

Whitebeam *Sorbus aria*

In our study area, whitebeam grows only on the Chiltern slopes, where it is a locally common tree along wood edges and in bushy places, and less common within closed woodland. It flowers in May or June, and the fruit ripens in September. For reasons that we do not understand, its fruiting is very irregular from year to year. 1980 was a year of massive fruiting; after the leaves had fallen, in October, the loaded whitebeams stood out as orange patches on the flanks of the Chilterns. In 1981 there was a complete failure; few inflorescences were produced and those that there were failed to set fruit. In 1982 most trees failed; the few large old trees that did produce fruit seemed not to ripen it properly. A random sample of 80 fruits picked up below three of these trees had the long fruit stalks still firmly attached and were unpluckable at the abscission layer. In 1983 and 1984, whitebeam fruit was everywhere abundant, but in 1985 the fruit crop failed completely, and in 1986 almost completely. In 1987 there was a heavy crop.

With a mean diameter of 11 mm, whitebeam fruits are considerably larger than those of the related rowan and well above the upper size limit for smaller frugivores such as Robin and Blackcap to swallow whole. Like rowan fruits they have a very small 'seed burden' (mean 7.1%; Appendix 2), but the pulp is less watery than the pulp of rowan fruit so that, measured by their 'relative yield', they are among the most rewarding for a fruit-eater in our study area. They tend, however, to turn brown and then dry up after about six weeks on the tree, if not eaten. The fruit of some trees kept under observation had gone bad within a month of ripening. Possibly weather conditions affect the length of time that the fruit remains in good condition.

Dispersers. The seasons when the fruit was eaten differed by a month in the two years of heaviest fruiting: second half of September to second half of November in 1980, and second half of October to end of December in 1984. The latter was a generally late year for autumn-ripening fruits. All the thrushes were recorded taking the fruit (Table 11), Fieldfares least of all as they feed very little on the Chiltern slopes at the time when whitebeam fruit is available. The combined thrush records constituted 88% of the total. Carrion Crows, Magpies and Jays together made up nearly 10% of the total, a higher corvid percentage than we recorded for any other fruit; and they were certainly under-represented, for the reasons discussed elsewhere (p. 163). We had only a few Starling records. Possibly the size of the fruit makes it unattractive to Starlings, which for their body size cannot swallow fruits as large as thrushes can (p. 160). In addition, at the time when whitebeam is fruiting, Starlings on the Chiltern slopes tend to concentrate on dogwood and ignore other fruits. (Outside our study area we have seen Starlings feed in small flocks on the very similar fruits of the introduced Swedish whitebeam.) Whitebeam fruit is certainly unsuitable for Robins and Blackcaps, for which we have only two and three records respectively. Blackcaps were recorded pecking fragments of pulp remaining from fruits attacked by Bullfinches. Robins were twice seen sallying to pick at fruit.

Seed-predators. Whitebeam seeds are rather soft, and are much exploited by seed-predators (Table 11). Woodpigeons swallow the fruits whole. Bullfinches and Greenfinches bite into the fruit, extracting the seeds which they then mandibulate, apparently de-husking them before swallowing them. The skin of the fruit and some pulp is left on the tree and is a characteristic sign of finch predation. They may remain in a tree for 5–10 minutes at a stretch, destroying many fruits in the course of a visit, and so account for a greater part of the fruit crop than is indicated by the figures in Table 11. Greenfinches, however, tend to concentrate on brown, dried-up fruit, when brown and orange fruits are both available, and so largely destroy seeds that would not be dispersed by legitimate fruit-eaters. We have not seen Bullfinches doing this, but only attacking fruit in good condition. Chaffinches were recorded opening the fruits on the trees and eating the seeds, and also picking seeds from fruits already damaged by Bullfinches. Bramblings were also recorded opening the fruits on the trees and eating the seeds.

In the mild autumn of 1987, when whitebeams produced a heavy fruit crop, seed-predators, apparently mainly Greenfinches, had destroyed approximately 80% of the fruit on two whitebeam trees by the end of October.

Blue Tits seem to be both seed- and pulp-predators of whitebeam. They peck at fruits in situ, apparently taking pulp and seeds; and they also detach whole fruits, taking them to another perch, holding them under the foot and pecking into them. Blue Tits feeding at dried-up fruit in situ in January were presumably also taking the seeds. Great Tits are seed-predators: the two records are of birds

TABLE 11 *Whitebeam: monthly summary of feeding records*

	Sep	Oct	Nov	Dec	Totals
Dispersers					
Blackbird	33	57	123	44	257
Song Thrush	16	16	4		36
Mistle Thrush	41	9	14	5	69
Redwing		1	147	22	170
Fieldfare			3		3
Robin	1		1		2
Blackcap	3				3
Starling			11		11
Carrion Crow		12	5	7	24
Jay	1	5	1		7
Magpie		15	10	2	27
All dispersers	95	115	319	80	609
Seed-predators					
Woodpigeon		18			18
Bullfinch	4	24	43	26	97
Greenfinch	2	1	39	11	53
Chaffinch	3	6	42	4	55
Brambling		21			21
Great Tit			2		2
Seed/pulp-predator					
Blue Tit		20	5	7	32

Note. Two January records of Blue Tits feeding on dried-up fruits are omitted.

plucking fruits and taking them to another perch, where they held them underfoot and extracted the seeds.

WHITEBEAM ELSEWHERE

We have found only one reference of any significance to whitebeam fruits being eaten by birds elsewhere in Britain. Oakes (1942) published some observations on whitebeam in Lancashire (misidentified as *Sorbus domestica* in his paper, later corrected; *ibid* p. 182). He recorded Blackbird, Song Thrush and Mistle Thrush taking the fruit, and Greenfinch, Chaffinch, Great Tit and Blue Tit as seed-predators. In particular, he was impressed by the quantity of seeds eaten by the Greenfinches, noting that up to 60 of them fed in a single tree 'for weeks at a time'. He did not specify the exact dates of his observations.

On the continent of Europe whitebeam has a southern distribution, from Spain east to the Carpathians. We have found no reference to its fruit being eaten by birds, except for Turček's (1961) list of 11 species, ranging from Capercaillie to Garden Warbler, presumably mostly from eastern Europe but unsupported by any details.

Wild service tree *Sorbus torminalis*

The wild service tree is very local in our study area. We did not know of any until we were almost at the end of our systematic field work, when we found some trees in one of the old oak woods in the north of the area. Our records for them are very incomplete.

The fruits are unusual in being brightest in colour while still unripe; on ripening they change from orange to brown. In 1985 and 1986 their fruit crop failed almost entirely, but in 1987 they fruited prolifically. The fruits ripened late: on 5 October they were unripe, but on 4 November there was abundant ripe fruit, some of which had already been attacked by seed-predators. In $6\frac{1}{2}$ hours of observations, covering early and late November and early December, we saw Bullfinches taking seeds on every occasion, and recorded a total of 19 feeding visits. On 27 November we saw a Redwing and two Song Thrushes feeding or attempting to feed on the fruit (it was not certain whether the Song Thrushes were successful in their attempts as little intact fruit remained); these were our only records of dispersers at the fruit. By that date a great part of the fruit crop had been attacked by seed-predators, mainly Bullfinches; the shredded skins and pulp had been left attached and most were dry and blackened. On 8 December we had three records of Marsh Tits taking the seeds. On two occasions they were plucking fruits and holding them under the foot to extract the seeds; on one occasion a bird extracted a seed from a fruit in situ.

At the time when these wild service trees had ripe fruit the much more abundant hawthorns, as well as a number of other plants, were providing alternative sources of fruit, which were probably safer to exploit than the service trees' fruit 3–15 m up on terminal twigs of the leafless branches. We do not know whether the very poor dispersal that these trees were evidently achieving is typical; if it is, it suggests that in the semi-open man-made habitats of southern England their November fruiting season is a disadvantage, and that poor dispersal, through unsuccessful competition with other plants, may be a reason for their being rather scarce and local. Furthermore, the impact of seed-predators must, other things being equal, be more severe for them than for the other two *Sorbus* species, as wild service tree fruits are single-seeded whereas rowan and whitebeam average about 2 seeds per fruit (Appendix 2). Thus, in order to take the same number of seeds, a seed-predator will have to destroy twice as many wild service tree fruits.

We have found no published references to the fruit of wild service tree being eaten by birds in Britain; for the Continent, Turček (1961) lists 14 species, dispersers and seed-predators, but without details.

Wild cherry *Prunus avium*

The wild cherry is a common woodland tree throughout our study area. It flowers mainly in the second half of April and first half of May, and the fruit ripens in late June or early July. The period when ripe fruit is available is short: nearly all our records, totalling over 500, were between 17 June and the end of July (Table 12). By some time in the second half of July practically all the trees are stripped.

TABLE 12 *Wild cherry: monthly summary of feeding records*

	2nd half June	July	1st half August	Totals
Blackbird	110	278	6	394
Song Thrush	8	40		48
Mistle Thrush	8	32		40
Blackcap	2	1		3
Garden Warbler		1		1
Starling	46	21		67
Magpie		3	1	4
Jay		7		7
Carrion Crow		8		8
Woodpigeon	1	8		9
All species	175	399	7	581

The red fruits, with an average diameter of nearly 12 mm, are larger than most British wild fruits and above the upper size limit for the smaller frugivores. The thrushes can swallow them whole; their main difficulty is in plucking them. When the fruits are only just ripe, they often have to pull at many of them before one comes away; and often the cherry has its stalk attached and is then dropped by the bird. Many such cherries, with the bird's beak marks, may be found on the ground beneath the trees. Starlings have great difficulty in swallowing the fruit whole. Usually they pick out bits of the flesh in situ, eventually leaving a more or less clean cherry stone hanging on its stalk; sometimes the whole fruit comes away, and the Starling then drops it. Our few records for Blackcap and Garden Warbler were of birds picking bits of pulp from the fruit in situ. Robins almost certainly could not swallow the fruit whole, and probably cannot pluck them. We have no records of them attempting to feed, like the warblers, on cherries in situ.

Woodpigeons are the only seed-predators that we have seen eating cherries; but in fact this is one of the fruits for which they act as dispersers (see below; also p. 172). Hawfinches, the only seed-predators adapted to tackle cherry stones, are either absent from our area or so rare or local that we have never seen one during our watches at fruit.

WILD CHERRY ELSEWHERE

We have found no references to birds eating wild cherries in Britain that add significantly to what we found. In Czechoslovakia, Turček (1968) made a study of the dispersal of *Prunus avium* by birds, in woodland at 600–800 m above sea level. He concluded that Blackbirds and Song Thrushes were the two most important dispersers, and had records also for Jay and Blackcap. Blackcaps in his area were evidently able to swallow the cherries whole; possibly they are smaller in the mountains than on our lowland trees. Naumann (1897–1905) mentions that Blackcaps may swallow cherries whole; but that this may be hazardous for them is indicated by his reference to a Blackcap found dead with a cherry stone stuck in its throat. Turček found most seedlings within 50 m of the parent tree. He kept the four observed dispersers in captivity, and some other birds that are known to eat wild cherries, and recorded the time taken to regurgitate the seeds (see p. 217); for Blackbird, Song Thrush and Blackcap, the times were all between 10 and 30 minutes.

Turček did not mention Woodpigeons, but in the Erz Mountains of Germany, at about 600 m, Schlegel & Schlegel (1965) found that they ate the fruits of wild cherries and that, when they were feeding on them, their droppings consisted mainly of cherry stones, which were then taken and broken open by Hawfinches.

Cultivated cherries. Planted varieties of *Prunus avium* in gardens and suburbs in our study area have larger fruits which ripen earlier than the wild cherry; but we did not watch them much and our records, like those for wild cherry, are mainly in the first half of July. On the Continent, where cherries are grown commercially on a large scale, there are many literature references to their being eaten by birds; Starlings in particular are considered a pest (p. 161). In southern Germany and Austria, flocks of Black-headed Gulls pluck cherries from the trees in flight (Gewalt 1986). This is apparently a recently developed habit, as it is hardly mentioned in the earlier literature.

Bird cherry *Prunus padus*

Bird cherry is mainly a northern and western plant in Britain, not occurring naturally in the wild state south of Gloucester, Derby and East Yorkshire (Clapham, Tutin & Warburg 1973). Cultivated varieties are quite common in gardens in our area, but a few trees apparently of wild type – though presumably introduced – occur in one place, and most of our observations have been made on these. They flower in May and the fruit ripens in July. The cultivated varieties flower at the same time; they fruit much less prolifically, and those that we have watched have slightly larger fruit which ripens a little later than the wild type. What follows refers only to the plants of wild type.

The cherries, borne on long drooping racemes, turn black when they are ripe. They are small, averaging only 7–8 mm in diameter, and can be swallowed easily by Robins and warblers. They are much sought after, and the trees are soon stripped. Ripe fruit is thus available over a very short period, almost all our feeding records being in the second half of July (Table 13). Thrushes account for

TABLE 13 *Bird cherry: monthly summary of feeding records*

	1st half July	2nd half July	1st half August	Totals
Blackbird	13	74	2	89
Song Thrush	6	15		21
Mistle Thrush	2	26		28
Robin		11		11
Blackcap	1	8		9
Garden Warbler	2	3		5
Magpie		1	1	2
All species	24	138	3	165

83% of the total; most of the remaining 17% are records of Robins and warblers. We have no records of Starlings taking the fruit; the reason is undoubtedly that, when the bird cherry fruit has been available, the local Starlings have been feeding in flocks on acres of cultivated strawberries on a neighbouring fruit farm. We have no records of seed- or pulp-predators.

BIRD CHERRY ELSEWHERE

Bird cherry is an important source of fruit for birds in the north of Britain but we have found no account of any detailed observations. Campbell (1984) mentions that bird cherries and rowans in the Rannoch area in the Central Highlands of Scotland swarm with winter thrushes in autumn, before they move south into the rest of Britain. We have a casual record of a Robin taking bird cherry fruit in the Lake District on 8 September. Clearly the bird cherry fruit season is a good deal later in the north of Britain than in our study area.

On the Continent bird cherry is widespread in northern and central Europe, with a montane distribution in the south of the range. Schuster (1920) and Möhring (1957) both give July as the month when the fruit is eaten in Germany. In southern Finland, Siivonen (1939) recorded Song Thrushes taking the fruit in early August. In the Giant Mountains of Czechoslovakia, at about 600 m above sea level, we found bird cherries with ripe fruit in the second half of August 1980. Blackcaps and Garden Warblers were feeding on the fruit in numbers; we often saw several together in the same tree. Blackbirds and Song Thrushes in smaller numbers were also coming to take the fruit. By the end of the month, when our visit ended, the crop was almost finished.

Where they are present, Hawfinches are important seed-predators of bird cherry. In the Giant Mountains, in August 1980, family parties of Hawfinches were spending much time in the bird cherry trees, plucking the fruit, discarding the flesh and cracking the stones. In England, Mountfort (1957) recorded Hawfinches feeding on bird cherry seeds in the months August to October, and occasionally in December and January. These late records must surely have been of birds taking seeds lying on the ground.

58 The Fruits

Blackthorn *Prunus spinosa*

Blackthorn is an abundant shrub or small tree in our study area, especially in hedgerows and thickets. As we are mainly concerned with its fruit, the sloe, we refer to it hereafter in the text and in the tables by the name of its fruit. It flowers in March or April, depending on the warmth of the spring weather, and the earliest fruits ripen in September. Ripe fruits are variable in size but most are 12–15 mm in diameter, and have a bluish bloom which wears off, leaving the fruit shiny black (see p. 196, where the significance of the bloom for fruit-eating birds is discussed). At first very sour to the human taste, the sloes become a little less sour in the course of the winter. Some wither and shrivel after a few weeks, but most remain on the bushes in good condition until late in the winter.

In addition to typical sloes of the wild type, with fruit diameters of 11–16 mm and very spiny twigs, other varieties occur in parts of our study area, less spiny and with a more open growth form and larger fruits up to 23 mm in diameter, far too large for thrushes to swallow. These are presumably forms of 'bullace' *P. domestica*, of garden origin. Nearly all our records have been obtained from sloes of the wild type. Even wild-type sloes are close to the upper limit of size for the smaller thrushes to swallow. Blackbirds manage the larger ones with difficulty, often taking them to the ground to deal with them. Song Thrushes take more sloes than Blackbirds, but often have difficulty with them, sometimes swallowing them only at the second or third attempt and occasionally rejecting one that is too large. In freezing weather, when the sloes could be harder to pluck, Song Thrushes sometimes peck pieces of pulp from the fruit in situ. Starlings, too, have difficulty in swallowing sloes and will also attack the pulp; they seem to be inefficient at eating them and, taking account of the large Starling population, we had very few records of their eating sloes. All our records for Robins are of birds picking out pieces of pulp.

Our records of sloes being eaten extend from early September to the end of January (Table 14). Of the total of 781 records, 91% were in the months November–January, the time when most of the bushes are stripped. It might seem that the sloe crop is not ready to be eaten until November, as Sorensen (1981) supposed, but some sloes are certainly ripe much earlier than this, even

TABLE 14   *Sloe: monthly summary of feeding records*

	Sep	Oct	Nov	Dec	Jan	Totals
Blackbird	54	3	38	89	84	268
Song Thrush		2	130	102	76	310
Mistle Thrush		13	14	20	6	53
Redwing		1	15	14	30	60
Fieldfare			8	3	22	33
Robin			1	12	8	21
Starling			17		1	18
Carrion Crow		1	12	4		17
Magpie			1			
All species	54	20	236	244	227	781

before the haws. We have records of sloes being eaten in September in two successive years and in three different areas. (We did not appreciate this possibility and had not watched sloes as early as September in the earlier years of our study.) All these early records were for Blackbirds. In one area Blackbirds were regularly coming from some distance to take the sloes, flying across an open field to converge on a few blackthorns surrounded by hawthorns loaded with red but probably not fully ripe fruit.

After September we recorded comparatively little sloe-eating in October, followed by the main November–January peak. Song Thrushes contribute very largely to this peak, the number of records being almost equal to the number for all the other thrushes combined. This preference is discussed on p. 225.

The most likely explanation for the paucity of sloe records in October is that the new haw crop is fully ripe and provides an alternative fruit source that is preferred by Blackbirds, Redwings and Fieldfares, while the wintering Song Thrushes, which prefer sloes, may not have arrived. When the haw crop is beginning to be exhausted, usually some time in November, sloes are again taken in quantity. In cold weather they may then become a most important source of food, especially in open agricultural country with hedges where haw and sloe are the only abundant winter fruits.

SLOE ELSEWHERE

We have found very little published about the eating of sloes by birds in Britain. Hartley (1954) gives only a single record, for the Blackbird. Sorensen (1981), who studied fruit-eating in a wood near Oxford over one winter, found that sloes were not taken until late January, in a cold spell, and speculated that they may be unpalatable until their taste is altered by a hard frost; but, as mentioned above, this is clearly not the case.

Blackthorn is widespread on the continent of Europe, but not much seems to have been recorded about the time when sloes are eaten. In northern and central Europe it is evidently a winter fruit, as in Britain – taken by Starlings in November in Czechoslovakia (Havlin & Folk 1965); by Blackbird, Fieldfare and Magpie in Saxony on 21 December (Heymer 1966). It extends to southern Spain, where it flowers in February and the fruits ripen in summer (Herrera 1982); but it is apparently not abundant and Herrera gives no records of its being eaten by birds.

The Hawfinch is the only species known to be a seed-predator of sloe. According to Mountfort (1957), they usually take the seeds of fruit that has fallen to the ground.

60 The Fruits

Blackberry *Rubus fruticosus*

The bramble is abundant in our area, as it is almost everywhere in England. It begins flowering in June and the fruit ripens from early August onwards, until some time in October when cold weather prevents further ripening of the late fruits. Ripe fruit, if not eaten, soon grows a mould and eventually dries up on the plant.

Ripe blackberries are very variable in size, most being in the range 12–15 mm in diameter and large ones up to 17 mm. The latter are too large to be easily swallowed by thrushes, and all blackberries of normal size are above the range for Robins and warblers. But, being composite fruits, they offer birds an alternative and can either be swallowed whole, or individual drupelets may be picked out and swallowed.

Dispersers. We have records of eight species eating blackberries, from early August to late November (Table 15). The most notable aspect of the seasonal pattern of blackberry-eating in our area is that it is double-peaked, with a first peak in early September and a second in late October and early November. This is not apparent from the monthly groupings shown in Table 15, but grouping by half-month periods gives the following totals:

August 1	6	October 1	32
August 2	3	October 2	210
September 1	135	November 1	107
September 2	27	November 2	1

The first peak is mainly of Blackbirds, Robins and warblers eating blackberries in bushy places such as thickets and wood edges. The second peak is almost entirely of Starlings feeding in more open country. In late October flocks of Starlings, having finished the elderberry crop, turn to the blackberries growing in field hedges and along roadsides, and strip all the ripe fruit remaining in good condition. The relative size of the two peaks shown by our records is not necessarily a valid indication of the amount of feeding on blackberries in different habitats, as flocks of Starlings are much easier to see than Blackbirds and warblers unobtrusively feeding among thicker vegetation; moreover from an observation point in open country long stretches of hedge can be monitored.

Blackberries are among the few fruits in our area which, from the point of view of a fruit-eating bird, are normally superabundant, at least in many of the man-altered habitats where they grow. Being watery, with a thin skin, they are not long-lasting and large numbers of them go mouldy and then dry up. For example, counts in late October in an area where Blackbirds, Robins and warblers had been feeding on blackberries earlier, showed that of 291 fruits 60% had dried up uneaten.

Seed-predators. Bullfinches and Greenfinches are seed-predators of blackberry in our area (Table 15), but they were outnumbered by Blue Tits which regularly take seeds from blackberries, though we cannot exclude the possibility that some records were of birds probing the fruit for grubs.

BLACKBERRIES ELSEWHERE

We have found no published observations of blackberry-eating by birds in Britain that add significantly to our findings, but some additional species are

TABLE 15 *Blackberry: monthly summary of feeding records*

	Aug	Sep	Oct	Nov	Totals
Dispersers					
Blackbird	6	121	10	1	138
Song Thrush		1	4		5
Robin	3	22	1		26
Blackcap		5			5
Garden Warbler		2			2
Common Whitethroat		2			2
Lesser Whitethroat		3			3
Starling		1	227	107	335
Moorhen		5			5
All dispersers	9	162	242	108	521
Seed-predators					
Blue Tit	4	24	2		30
Bullfinch	1	5	1		7
Greenfinch			1	4	5

Note. Not included in the table is a record of a Bullfinch taking seeds from a dried-up fruit in February.

recorded, for example Ring Ouzel (Olney 1966), Magpie (Holyoak 1968), and Woodpigeon (Murton 1965), the first two certainly dispersers and the last perhaps an occasional disperser as well as a seed-predator.

Easily the most thorough study of blackberry-eating by birds in continental Europe is that made by Jordano (1982) in southern Spain. Although nearly 1,500 km further south than our area the season of ripening of blackberries in Jordano's area, at 500 m, is almost exactly the same. Jordano recorded 20 passerine species feeding on blackberries, most being autumn passage migrants that peaked in abundance between mid September and early October. Blackberries are probably one of the most important sources of easily obtained energy for migrant passerines passing through the Iberian peninsula in autumn, as was also found by C.J. Mead (pers. comm.) in northwest Portugal (see also p. 155). On the basis of counts of large samples, Jordano found that only 7.2% of ripe fruits dried up uneaten.

Elsewhere in continental Europe records of blackberry-eating are predominantly casual ones, the few dated records indicating much the same seasonal pattern as in our area. For example Möhring (1957) recorded Blackcaps eating blackberries in August and September in Germany. Szijj (1956/57) reported that, in Hungary, Starlings eat blackberries in late summer and autumn, earlier than in our area, presumably because the elderberry crop in Hungary ripens about a month earlier than in southern England and is finished earlier.

Wild raspberry *Rubus idaeus*

Wild raspberry is locally abundant in undergrowth in open woodland and bushy places in the Chilterns, and locally present but not abundant elsewhere in our study area. It flowers in June and July and the fruit ripens in the second half of July and early August. Ripe fruit is soon taken, all being gone by the end of August. The 'polydrupes' are variable in size but tend to be small, often consisting of only a few drupelets.

We have not been very successful in watching birds eat wild raspberries, mainly because in places where they are abundant enough to be worth watching most of the fruits are well hidden, and birds do not expose themselves when eating fruit unless they have to. All but one of our 15 records, involving four species, were in the second half of July and the first half of August (Table 16). Blackcaps

TABLE 16 *Wild raspberry: monthly summary of feeding records*

	Jul	Aug	Totals
Blackbird	3	2	5
Robin	3		3
Blackcap	2	4	6
Magpie	1		1

Note. All July records were in the second half of the month, and all but one of the August records in the first half of the month.

pick single drupelets, or two or three drupelets together, from a perch; Robins snatch single drupelets in rapid flight sallies. Blackbirds have usually been too well hidden for the details to be seen, but they presumably pluck whole fruits. We have no evidence of seed-predators attacking wild raspberries.

WILD RASPBERRY ELSEWHERE

We have found no published records of birds eating wild raspberries in Britain. (We have disregarded statements that 'raspberries' are eaten, *eg* in Witherby *et al* (1938), as they probably refer to cultivated raspberries, for which evidence is all too easily obtained.) There are a number of references to wild raspberries being eaten on the continent of Europe, mostly by *Sylvia* warblers. Möhring's (1957) records for Blackcaps in Germany (July–September) cover a longer period than ours. There are two records of warblers feeding their young on wild raspberries, Lesser Whitethroat (Wetzel 1878) and Garden Warbler (Schlegel & Schlegel 1965). Turček's (1961) record of the Willow Warbler eating raspberries is one of the few references to fruit-eating by this species.

64 The Fruits

Crab-apple *Malus sylvestris*

Crab-apples are common small trees in old hedges and along woodland edges in our study area. They flower in May and the apples are ripe by September or October. They begin to fall from October onwards, but there is much variation in the time of fruit-fall, perhaps related to weather or other local conditions. The small yellow apples, some 25–30 mm in diameter, remain in good condition on the ground for a long time (unlike many cultivated apples).

Crab-apples are almost certainly not adapted primarily for bird-dispersal. They are far too big for European frugivorous birds to swallow (Fig. 7). Unlike bird-adapted fruits, which mostly have an abscission layer between fruit and stalk, when it is ripe the crab-apple's stalk remains firmly attached. The size of crab-apples, and the fact that they fall to the ground, both suggest dispersal by mammals. In our area they are rapidly removed by the introduced muntjac in places where these deer occur. In more natural conditions in Europe it seems probable that wild pigs and native deer are the chief dispersers.

Birds do, nevertheless, eat crab-apples on the ground. We have no evidence that they do so in autumn or early winter, as they do cultivated apples in gardens and orchards; all our records are in late winter. They extend from late December to March, 78% of the total being in February (Table 17). We have records for all the thrushes (the Blackbird, as usual, predominating), Robin, Starling and Carrion Crow. Of these the Carrion Crow and occasionally the Blackbird are potential dispersers; the others peck out the pulp without moving the apple appreciably. We have seen Carrion Crows take fallen crab-apples and fly off with them for at least 150 m, presumably to eat or possibly store them, and have six records of Blackbirds moving whole apples 5–10 m.

Chaffinches and Great Tits are seed-predators of crab-apples. We have records of both species extracting seeds from fallen apples which had been opened up by Blackbirds. We have also seen a Marsh Tit extracting seeds and flying off with them into woodland near by, doubtless to store them.

CRAB-APPLES ELSEWHERE

We have found no reference to crab-apples being eaten by birds in Britain, except for the following interesting note by Campbell (1980). In snowy weather in December, on the banks of the Evenlode, 'A marsh tit appeared, with a marble-sized yellow crab apple suspended from its bill by a long thin stem and, after some hesitation, dived between my legs into a buttress-root cavity; on looking into the cranny later, I discovered a cache of about a dozen crabs piled in a little pyramid.' This Marsh Tit had evidently developed a different storing technique from the bird mentioned above.

The crab-apple is widespread in continental Europe, but there seems to be almost no information on the eating of the fruit by birds. Schuster (1930), who combed the literature up to the time of his writing, gave six references under the heading 'wild or domestic apple' (translated), but all refer certainly or probably to apples in orchards.

TABLE 17 *Crab-apple: monthly summary of feeding records*

	Dec	Jan	Feb	Mar	Totals
Blackbird	3	21	100	7	131
Song Thrush		2	11		13
Mistle Thrush		1	12		13
Redwing			1		1
Fieldfare			7		7
Robin			13		13
Starling		3	1		4
Carrion Crow		6	6		12
All dispersers	3	33	151	7	194
Seed-predators					
Chaffinch		1	16		17
Great Tit			3		3
Marsh Tit			2		2

Note. Omitted from the table is a record of a Chaffinch taking seeds in April.

Elder *Sambucus nigra*

Elder is almost ubiquitous in our area. The small, shiny black berries, borne profusely on spreading cymes, are attractive to almost all frugivorous birds, including some that are not usually thought of as fruit-eaters. Consequently it is one of the most important fruit-bearing plants; furthermore, it has an impact on other plants as the availability of elderberries influences the extent to which other fruits are eaten.

In our area the elder flowers in June; the elderberries begin to ripen in the first half of August and are available in quantity from the second half of August until some time in October. Our earliest record of the fruit being eaten was on 4 August. In the course of September and early October, Starlings, numerically

66 The Fruits

TABLE 18 *Elder: monthly summary of feeding records*

	Aug	Sep	Oct	Nov	Totals
Dispersers					
Blackbird	35	107	22	5	169
Song Thrush	31	123	28	5	187
Mistle Thrush	1	3			4
Redwing			1		1
Fieldfare			1		1
Robin	22	61	12	5	100
Blackcap	39	51	18		108
Garden Warbler	3	12			15
Common Whitethroat		3			3
Lesser Whitethroat	5	7			12
Starling	66	478	172	6	722
Spotted Flycatcher	2	2			4
Carrion Crow			1		1
Jay		2	1		3
Magpie	5	9	3	2	19
Moorhen	1				1
All dispersers	210	858	259	23	1350
Seed-predators					
Woodpigeon	5	47	3		55
Collared Dove	2				2
Bullfinch	2	14	1	12	29
Great Tit			1		1
Blut Tit	6	99	20	1	126
Marsh Tit	2				2

Notes. The earliest August record for a disperser was on 4 August; 80% of August records were in the the second half of the month. Only one of the November records was in the second half of the month. Omitted from the table is a record of a Woodpigeon taking unripe elderberries in July. Woodpigeons, though listed here under seed-predators, may also be dispersers of elder, as we have found intact seeds in droppings collected below their roosts.

easily the most important of the eaters of elder fruit (Table 18), strip most of the elder bushes in and around towns and villages, working each bush from the top, so that by mid October it is common to see bushes with only a fringe of fruit on the lowest branches. By the end of October all elders have usually been stripped except those growing under the canopy of woodland or in other shady places, where the fruit is late in ripening and, in addition, Starlings do not usually feed. Some fruit may persist in such places until late November and may eventually dry up uneaten.

As described elsewhere, the elder fruit crop affects the impact of frugivores on two other fruits that are available at the same time, woody nightshade (p. 87) and wayfaring tree (p. 68). These ripen before the elder and are usually fed on intensively at first, then, as the elder crop becomes available, are neglected. Later, when the haws ripen, which Blackbirds prefer to elder, the elder in turn may be neglected in areas where Starlings do not feed and, as mentioned above, may remain uneaten.

We have records of 17 species of dispersers eating elderberries, a greater number of species than for any other fruit, and records of six species of seed-predators. In addition to the nutritive quality of the fruit, which is quite high in spite of the pulp being rather watery (see Appendix 4), its popularity is undoubtedly due in large part to the fact that it ripens early and is then the only abundant fruit source in many places. It is also physically easy to exploit: the berries are readily plucked, and all the fruit-eaters can swallow them whole. Starlings, which find it hard or impossible to pluck some kinds of fruit (p. 159), have no difficulty with elder. Among the less usual fruit-eaters we have four records of Spotted Flycatchers eating elderberries, and one of a Moorhen climbing about in a canal-side elder to take the fruit.

Of the six seed-predators the Woodpigeon is certainly the most important (but see footnote to Table 18). We have more feeding records for Blue Tits (126) than for Woodpigeons (55), but the pigeons eat a very large number of berries at each visit, Blue Tits only a few. Even so, Woodpigeons probably have a negligible effect on the huge elderberry crop; they feed mainly on the exposed tops of large elders, in undisturbed areas. In addition to the seed-predators, House Sparrows spoil a certain amount of elder fruit by plucking the berries, squeezing out the juice, and dropping the skin and seeds.

ELDERS ELSEWHERE

As would be expected, there are numerous references in the ornithological literature to the eating of elderberries by birds, and only some of special interest are mentioned here. Most observations refer to the widespread black-fruited *Sambucus nigra*, the only species that occurs in our study area. There are far fewer references to birds eating the fruits of the two other European species, the red-fruited *S. racemosa*, a northern or montane elder not native to Britain, and the dwarf elder *S. ebulus*, a plant widespread on the Continent but local in Britain.

Black elder *S. nigra*. The list of birds recorded as eating its fruit on the continent of Europe includes almost every possible species, but almost certainly the fruit is most important as an energy source for migrating *Sylvia* warblers. Schmidt (1965) studied an area in Hungary thickly grown with elder bushes. He compared the number of birds present with the estimated number of fruits available through the late summer and autumn. In this area elder fruits usually ripen in mid July, nearly a month earlier than in our study area. He found a good correlation between the number of fruits and the number of frugivorous birds present, especially *Sylvia* warblers and nightingales which were on passage and clearly being attracted to the concentration of fruit. Zedler (1954) also mentions that in Germany elderberries are the main food of migrating *Sylvia* warblers, and this applies too to migrants on the North Sea islands. Thus elderberries are a main food of migrating Garden Warblers on Heligoland (Banzhaf 1932, Kroll 1972), and of Common Whitethroats on Hiddensee, off the German Baltic coast (Emmrich 1973-74).

Red elder *S. racemosa*. There seem to be few detailed records of red elder fruits being eaten, though a large number of bird species is listed by Turček (1961). Ringleben (1949) mentioned that in the Erz Mountains of Germany red elder is called *Rotkehlchenbaum* (Robin's tree), which indicates that it must be common knowledge that Robins feed on the berries. They ripen earlier than the fruits of

68 *The Fruits*

black elder, and all dated records that we have found are in July and August. In the Bad Nauheim area, north of Frankfurt, Schuster (1920) found that red elder fruit ripened in the second half of July and was then immediately fed on by Black and Common Redstarts, who seemed to feed almost exclusively on the fruit all day; also smaller numbers of Lesser Whitethroats and Garden Warblers, occasionally Blackbirds and once a Song Thrush, these last two much preferring the fruit of *Prunus virginiana* near by (an introduced plant). On Heligoland, where Banzhaf found that Garden Warblers fed mainly on black elder fruit, he noted that red elder was also present and had abundant fruit but he never saw it being eaten. Presumably black elders are preferred to red, but one would like to have more evidence on this.

In the Giant Mountains of Czechoslovakia, in the second half of August 1980, at 500–700 m, we found red elders with abundant ripe fruit and saw many Blackcaps and Garden Warblers, and a few Black Redstarts, Blackbirds and Song Thrushes feeding on it. This was above the upper altitudinal limit of black elder; at 450–500 m we saw a few *S. nigra* bushes, still with only small, green fruit.

Dwarf elder *S. ebulus*. The only dated record that we have found is by Möhring (1957), who reported that in central Germany dwarf elder fruits are eaten by Blackcaps in July and August. Turček (1961) listed 13 species recorded eating dwarf elder fruits, including the Wood Warbler, but gave no details.

Wayfaring tree *Viburnum lantana*

Wayfaring tree is a locally common shrub in our study area, found mainly on the Chilterns or along the Chiltern base. It flowers in June; the earliest fruits ripen in late July and most of the crop in August, the times of ripening of individual plants varying by up to three weeks, depending largely on the plant's exposure to sunlight. This is the most striking case in our area of a two-colour fruit display (p. 195). As the fruits begin to ripen they turn from green to bright red and then, finally, when fully ripe they change rapidly to black. Ripe fruits are thin-skinned and very juicy, oval in shape, and contain a single flattened seed. They are in good condition for a very short time and after about two weeks begin to dry up and shrivel.

For a period in late summer wayfaring tree fruit is a favourite food for thrushes and warblers. The season when they are taken is a very short one in our study area; our extreme dates are 26 July and 10 September, 86% of our records being in August (Table 19). In the first week of August we recorded, in the course of one watch, five species of frugivores feeding at a rate of 17 visits per hour at a single bush, a rate comparable to recorded visits to fruit sources in hard weather in winter. Later in August the visiting rates declined dramatically, coincident with the ripening of the much more abundant, and even more highly preferred,

TABLE 19 *Wayfaring tree: summary of feeding records, dispersers*

	2nd half July	August	1st half Sept	Totals
Blackbird	3	33	4	40
Song Thrush		6		6
Robin		21	7	28
Blackcap		23	2	25
Garden Warbler	1	2		3
Lesser Whitethroat		7		7
All dispersers	4	92	13	109

Note. For seed-predators, see text.

elderberry crop. If, due to such factors as the shadiness of the site (late ripening) or exposure to predators (isolated bushes in the open), the fruit has not been taken by mid- to late August, removal is then slow, and the fruit begins to dry up and is wasted.

While it is in good condition the fruit is hardly attacked at all by seed-predators; we have a single record of a young Bullfinch taking the seeds in August. We also have two records of Bullfinches taking seeds from dried-up fruit in January; they may do so regularly in winter, but we did not watch wayfaring trees systematically at that time of year.

WAYFARING TREE ELSEWHERE

We have found no significant published observations on the eating of wayfaring tree fruit by birds elsewhere in Britain. For the continent of Europe, the only information of significance comes from Germany. Schuster (1920) recorded that the fruit ripens in August and that Blackcaps, and probably also Blackbirds and Song Thrushes, eat it. Zedler (1954) makes the surprising statement that the fruits, as soon as they are *red*, are preferred to all other available fruits by Blackbirds, Bullfinches (presumably as seed-predators, though he makes no distinction) and apparently also Robins. He goes on to say that the fruits soon blacken, as sap is withdrawn, and in this state are especially sought after by Bullfinches, and no longer by Blackbirds. One can only suppose that Zedler did not examine the fruits very carefully or watch feeding birds at close quarters. Schneider (1957) had single records of Blackbird and Fieldfare (the latter presumably resident in his area) taking wayfaring tree fruit. He mentioned that most fruit remains on the bushes and dries up, and concluded that it is not much liked by birds, as we might have concluded if we had observed only at a few sites where this happened. This is a good example of how limited observations in particular habitats may give a misleading impression.

Guelder rose *Viburnum opulus*

Guelder rose is a sparsely distributed shrub on the Chiltern slopes, and very local in the flat country at the foot of the Chilterns. Elsewhere in our study area a very similar species, *V. sargentii*, is found in some gardens. These garden plants seem identical in their fruits and fruiting season to the wild plants, and their fruits are taken at the same time. We include them here, but the great majority of our records come from wild plants.

The showy panicles of white flowers come out in June and July. The fruits redden in September and look ripe from October onwards, but they remain firmly attached to their stalks and are unpluckable by birds until the end of November. In December the fruits become easy to pluck and from then onwards are taken by birds. Ripe fruits are bright red and almost translucent, and contain a single flattened seed. With a diameter of 8–9 mm, they are potentially available to all dispersers. The pulp has a rather disagreeable, sickly scent.

Dispersers. In our area guelder rose is very distinctly a winter fruit, taken by birds in the coldest months of the year, December to February (Table 20). In over ten hours of observation at bushes with ripe-looking fruit before December we had no record of a bird taking, or trying to take, the fruit. Song Thrushes are easily the most important dispersers in our study area, accounting for 81% of all records. In three places we suspected that guelder rose bushes might have been protected by Mistle Thrushes, at least for part of the winter (see below), but never had satisfactory evidence, and in fact obtained only four feeding records for Mistle Thrushes.

Seed-predators. Bullfinches are important seed-predators of guelder rose. The number of records is almost half of the total for all dispersers combined (Table 20); and as Bullfinches often remain for a long time in a bush and destroy many more fruits than a disperser takes at one visit, their impact is greater than the number of records suggests. Some guelder rose bushes that we monitored had the greater part of their fruit crop attacked by Bullfinches, which leave fragments of skin attached to the stalks as evidence of their activity.

GUELDER ROSE ELSEWHERE

We have found few references in the literature to guelder rose fruit being eaten by birds elsewhere in Britain. They are included with records from continental Europe, below. Sorensen (1981), in her survey of fruit-eating by birds in woodland near Oxford, had no records. She showed, by feeding artificial 'dough-fruits' flavoured with the juice of guelder rose and other fruits to caged Blackbirds and Song Thrushes (p. 197), that the taste of guelder rose fruit came low in their order of preference, and concluded that thrushes actively avoid them. Her lack of records, however, was the outcome of only five hours of observation, and if these were not carried out at the critical time in midwinter negative results were to be expected.

Guelder rose is widespread in central Europe and there are several references to its fruit being eaten by birds. From these it is evident that the fruit ripens much earlier than in our study area. Zedler's observations (1954) from Bad Aibling, near Munich, are the most detailed. He noted that guelder rose, especially in parks and gardens, keeps its fruit longer than any other native tree or shrub. He

TABLE 20 *Guelder rose: monthly summary of feeding records*

	Dec	Jan	Feb	Totals
Dispersers				
Blackbird	11	3		14
Song Thrush	19	76		95
Mistle Thrush	1	1	2	4
Redwing		1		1
Robin	2	1		3
Blackcap	1			1
All dispersers	34	82	2	118
Seed-predators				
Bullfinch	26	27	4	57

saw Blackbirds eating the berries only in hard weather in midwinter. Much of the fruit was not eaten; it slowly shrivelled and fell to the ground, where the seeds were eaten by mice. In field hedges, however, in the autumn migration season Song Thrushes and, especially, Fieldfares descended on the bushes in flocks. This distinction between the time when guelder rose fruit is eaten in open country and in gardens is supported by other authors. Bettmann (1953) reported that flocks of Fieldfares near Bonn descend on guelder rose bushes in the hedges and strip them; he did not give dates, but stressed that this happens when the weather is quite mild. Bodenstein (1953) reported that Blackbirds feed on guelder rose fruit in winter, in gardens, but that otherwise it is hardly touched by birds. Turček (1961) lists three *Sylvia* warblers and the Spotted Flycatcher among the species recorded eating guelder rose fruit, and these can only be autumn records. Waxwings are evidently fond of guelder rose fruit; there are several German records of them taking guelder rose berries in winter during invasion years, and they have been seen doing so in Norfolk (Ellis 1985).

Heim de Balsac (1928), discussing the eating of mistletoe berries by birds in Meurthe-et-Moselle, in France, noted that they were eaten by Mistle Thrushes but by few other birds (probably because they are protected by the Mistle Thrushes, though he did not realise this). He mentioned that Mistle Thrushes also exploit another fruit that is 'dédaigné' by other birds, guelder rose. This suggests that Mistle Thrushes may protect guelder rose bushes in France, as we suspect that they occasionally do in our study area.

Bullfinches are recorded as seed-predators of guelder rose in Germany (Thein 1954), and Hawfinches also are occasional seed-predators, both in England (Mountfort 1957) and in Germany (Heymer 1966). Oakes (1942) noted that Marsh Tits in Lancashire are fond of guelder rose fruit, but gave no details; presumably they are seed-predators.

Honeysuckles *Lonicera* spp.

The common honeysuckle *Lonicera periclymenon* is abundant locally in woodland in our study area, especially in oak woods. We have seen the very similar perfoliate honeysuckle *L. caprifolium* only in gardens. Although probably not a native plant, perfoliate honeysuckle is naturalised and widespread in southern England, and we include it here rather than among the introduced and cultivated fruits because comparison with the common honeysuckle is interesting. For some reason the perfoliate honeysuckle apparently fails to propagate itself in our area. Both honeysuckles grow in our garden, and are certainly adapted for bird-dispersal. *L. periclymenon* produces seedlings in abundance, but we have never seen a seedling of *L. caprifolium*.

L. caprifolium is the earlier of the two species, both in flowering and in fruiting; it flowers in early summer, and the orange fruits ripen in the second half of July. *L. periclymenon* flowers in mid and late summer and the red fruits begin to ripen in the first half of August, but most not until late August. Their fruits are very similar, except in colour; those of *caprifolium* are on average smaller, but both are variable in size. In water content and nutritive composition of the pulp they are similar, but as the fruits of *caprifolium* have a smaller seed burden their 'relative yield' is a little higher (Appendix 2). Thus weight for weight *caprifolium* fruits should be a little more profitable for a frugivore than *periclymenon* fruits.

This slight difference does not, however, seem adequate to account for the fact that we have regularly recorded far more feeding visits to *caprifolium* than to *periclymenon*, the hourly rates for visits by dispersers being respectively 4.3 and 0.9. The reason seems to lie in the different ripening times of the two species. *L. caprifolium*, ripening in the second half of July, is one of the early fruits, coming before the ubiquitous elder and locally abundant rowan have ripe fruit; it exploits a period when fruit availability is rather low and, for thrushes, the availability of animal food may also be low, especially if the weather is dry. The fruits of *caprifolium* in our area are thus quickly eaten; they have one of the shortest seasons of any fruit, almost all our records being between 26 July and 13 August. By contrast, the fruits of *periclymenon* are a bit too late, ripening at a time when other abundant fruits are becoming available; they are much less sought after, and some fruit may remain on the plants until October or even early November.

We have records of six species of dispersers taking the fruits of *caprifolium* (Table 21, bottom). Blackbird records outnumber all the others combined, the majority of them being juveniles. For *periclymenon* we have records for only four species (Table 21, top), the Robin being for some reason much more prominent than it is in our records for *caprifolium*.

TABLE 21 *Common and perfoliate honeysuckles: monthly summary of feeding records, dispersers*

	2nd half July	1st half Aug	2nd half Aug	1st half Sep	Totals
Common honeysuckle					
Blackbird		9	10	1	20
Song Thrush		2			2
Robin			16	2	18
Starling			1		1
All dispersers		11	27	3	41
Perfoliate honeysuckle					
Blackbird	36	42			78
Song Thrush	5	6			11
Robin		5	1		6
Blackcap	14	24			38
Lesser Whitethroat		9			9
Starling	9	5			14
All dispersers	64	91	1		156

TABLE 22 *Common and perfoliate honeysuckles: monthly summary of feeding records, seed-predators*

	Jul	Aug	Sep	Oct	Nov	Totals
Common honeysuckle						
Bullfinch		40	6	6		52
Blue Tit			1		7	8
Marsh Tit			6	3		9
Perfoliate honeysuckle						
Bullfinch	21	44				65
Blue Tit		(3)				(3)

Note. All the Blue Tit records for common honeysuckle were of birds taking seeds from dried-up fruits. The Blue Tit records for perfoliate honeysuckle were of birds picking pieces of pulp, and perhaps also taking seeds.

Both of the honeysuckles are exploited by a major seed-predator, the Bullfinch, and wild honeysuckle fruits in woodland are exploited also by Marsh Tits and Blue Tits (Table 22). Hourly rates of visits by seed-predators to wild and perfoliate honeysuckle are very similar, 1.2 and 1.4 respectively, suggesting that they find little to choose between the two fruits. For the wild honeysuckle the hourly rate of feeding visits by seed-predators is actually higher than the rate of visits by legitimate frugivores (0.9) – for no other fruit have we found this to be so. Bullfinches, the main seed-predators, may remain feeding for minutes on end, destroying more fruit at a visit than the number eaten by a legitimate frugivore, and it is clear that the greater part of the fruit crop is destroyed and not dispersed.

HONEYSUCKLES ELSEWHERE

We have found no published records from Britain that add significantly to what is written above. Leach (1981) lists *Lonicera* spp. as major fruits eaten by wintering Blackcaps in Britain; presumably cultivated honeysuckles, but this is not stated. On the continent of Europe most information is available for the widespread *L. xylosteum* (which also occurs in southern England but not in our study area) and the Mediterranean species *L. implexa*. *L. xylosteum* ripens in July in Germany (Möhring 1957) and is taken by birds in July and early August (Schuster 1920). Blackbirds and Blackcaps are recorded feeding its fruits to their young (Schuster *loc cit*; Wahn 1950). Schuster considered that in general it is not greatly sought after; he noted that in mid August most fruit was over-ripe, after which it dried up and fell off. From this it seems that its seasonal peak of dispersal is the same as that of perfoliate honeysuckle in our area, and it is probable that it gets the same advantage from early ripening, and later loses in competition with other kinds of fruit as they become available.

The Mediterranean *L. implexa* has a very different fruiting season. It ripens in September in southern Spain, is at its peak in October, and a little fruit persists into November (Herrera 1984a). Robins and Blackcaps are recorded as eating it (Herrera *loc cit*; Jordano & Herrera 1981). This is one of many examples (discussed further on p. 202) of fruits ripening later in the Mediterranean area than farther north in Europe, not earlier as might be expected. Tutman (1962) records that a species of *Lonicera*, unfortunately not specifically identified, is eaten by Blackbirds in January at Dubrovnik, Jugoslavia.

In broad-leaved woodlands of the eastern USA, Stiles (1980) lists five species of *Lonicera*, all of which he classifies as 'summer small-seeded' fruits, with periods of dispersal beginning in June, July or August and lasting until September or October – essentially similar to our two *Lonicera* species but apparently with somewhat longer seasons.

Dogwood *Cornus sanguinea*

Dogwood is a very common shrub along the Chiltern escarpment. In places the young plants are so abundant, invading open ground, that they have to be controlled if the more specialised chalk flora is to survive. In the flatter country of the Vale of Aylesbury along the base of the Chilterns, dogwood is locally common in hedgerows, but as most of it is regularly trimmed it usually succeeds in fruiting only in a few neglected bushy places. Although most abundant on or near the chalk, dogwood is also locally present on the Lower Greensand hills in the north of our area. It flowers in June or early July; the first fruit ripens in late August but most of it in September. The fruit is a drupe, with a hard, two-celled stone, which we refer to for convenience as the 'seed' (p. 16). Ripe fruits are black and globular, a little over 7 mm in diameter.

TABLE 23   *Dogwood: monthly summary of feeding records*

	Aug	Sep	Oct	Nov	Dec	Jan	Totals
Blackbird	1	2	34	25	10		72
Song Thrush		4	9	29	16		58
Mistle Thrush			1				1
Redwing			31	51	19		101
Robin	1	22	8	13	7	2	53
Blackcap		6	2	1			9
Starling		2	1408	540	20	6	1976
Magpie		3	7	3			13
Carrion Crow				2			2
Green Woodpecker			3	1			4
All species	2	39	1502	665	72	8	2289

Notes. The earliest record was on 28 August. All January records were in the first half of the month. Omitted from the table is one of a Song Thrush taking dried-up fruits in very cold weather in early February.

Our records of dogwood fruit being eaten by birds extend from the end of August to early January (Table 23). For the first few weeks the fruit crop is depleted slowly; most of our August and September records (59% of the total) are of Robins, the rest being Blackcaps, Blackbirds, Song Thrushes and Starlings. These were birds coming singly or in small groups to the fruit. In nearly 21 hours of watching in these two months, we recorded 1.6 feeding visits per hour. Then, in October, the rate of depletion of the dogwood crop increases dramatically as flocks of Starlings, having finished the elder fruit crop, descend on the dogwood: in October, in nearly 28 hours of observation at dogwood, we recorded a rate of 51 feeding visits per hour. In places where it occurs, dogwood continues to be the main fruit eaten by Starlings until it is finished. They begin on the dogwood growing at low levels, along the base or on the lower slopes of the Chilterns, later moving up to take the fruit at the top of the escarpment, where it usually lasts until the end of November. In the winter of 1984/85, following an autumn when most fruits were unusually late, dogwood fruit persisted until early January on the Chilterns, by which time most of that that remained had been blasted by the exceptional cold and was dry and withered on the plant.

The Starling accounts for 86% of all our dogwood records, followed by Redwing, Blackbird, Song Thrush and Robin (in that order), the only other species that contribute significantly to the total. We have only one Mistle Thrush record and none of the Fieldfare, but several records of Magpies and four of Green Woodpeckers, this being the only kind of fruit that we saw taken by Green Woodpeckers. We have no evidence that any seed predators exploit dogwood fruit in our area, the seed being probably too hard for them to tackle.

The behaviour of Starlings feeding on dogwood fruit is spectacular. In October, when they turn to the dogwood, one may see flocks of up to 150 birds wheeling about over the Chiltern slopes, swooping down as if to land but again and again climbing and circling until they eventually decide that it is safe, whereupon they land precipitately in the low dogwood clumps and feed hard for a few seconds before taking wing again.

Not surprisingly, our observations involving a choice between dogwood and haw, the other main fruit available at the same time and in the same places as dogwood, show that Starlings overwhelmingly prefer dogwood to haw (1,300:0; see also p. 158). Song Thrushes and Redwings also prefer dogwood; only the Blackbird shows a definite preference for haw where both are available. Almost certainly it is the ease with which the smaller thrushes can pluck and swallow dogwood fruit that determines their preference.

DOGWOOD ELSEWHERE

We can find no significant published observations on the eating of dogwood fruit by birds elsewhere in Britain. On the Continent of Europe several species of small birds, mainly warblers, are recorded eating dogwood fruit. Blondel (1969) mentions that in southern France dogwood fruit ripens in the autumn (exact dates not given) and that passage migrants gorge on them. During a visit to southeastern France in October 1984, in the limestone mountains of the Vercors, south of Grenoble, we saw flocks of Song Thrushes, good numbers of Blackcaps and some Blackbirds feeding on dogwood, which at that time was the main fruit available in open areas at 1,000–1,500 m.

Privet *Ligustrum vulgare*

In our study area privet occurs as a wild plant mainly on the Chiltern slopes, in bushy places, and locally in the clay country to the north. (We ignore the evergreen *L. ovalifolium* from Japan, which is common in gardens but is usually clipped too frequently to be able to fruit.) It flowers late, mainly in July. The green fruits begin to turn black in late September or early October, and most are fully black by the end of October. When ripe, the shiny black berries are unusually variable in size, depending on how many seeds they contain. The full complement is four, but small fruits contain only one or two. Of a sample of 50 fruits, only 14% contained the full complement, most containing two or three seeds. The full-sized fruits are not very large (about 8 mm diameter) and are easily plucked and swallowed by all fruit-eating birds.

Although the fruits appear fully ripe by the end of October, they are not taken by dispersers until about a month later; at least, not in any quantity. That a small number may be eaten in late October was indicated by a 5% depletion of a sample of 460 marked fruits in the last ten days of October. Our earliest observations of privet fruit being eaten by a disperser are in the second half of November (Table 24). The overall rate of consumption of privet fruits is low in November and December, rising markedly in January (Table 25). Individual bushes may be stripped at any time from December to February, depending on year and locality. There seems little doubt that this seasonal pattern is because other fruits, if available, are generally preferred to privet. Privet fruits are not very nutritious (Appendix 4), and their terminal position on the branches of bushes that are themselves often in exposed sites is a further disadvantage in competition with other kinds of fruits growing in better protected sites. Table 25 summarises the

TABLE 24 *Privet: monthly summary of feeding records*

	Oct	Nov	Dec	Jan	Feb	Totals
Dispersers						
Blackbird		13	13	56	28	110
Song Thrush		1	1	10		12
Mistle Thrush				3	2	5
Redwing			3	4		7
Fieldfare		1		1	1	3
Robin			3	7	9	19
Blackcap			5	3		8
Magpie		1				1
All dispersers		16	25	84	40	165
Seed-predators						
Bullfinch	9	17	9	11	2	48
Woodpigeon		2	2			4
Pulp-predators						
Blue Tit	3			1	5	9
Marsh Tit				1		1

TABLE 25 *Rate of feeding on privet fruit by dispersers in different months*

	hrs obs	Mean no. other fruits available	Rate of feeding on privet (visits/hr)	Rate of feeding on other fruits (visits/hr)	Feeds on privet as % of total feeds
October	4.9	2.6(1–3)	0	9.1	0
November	9.8	2.7(1–4)	1.2	11.2	10
December	9.8	2.4(1–4)	1.6	14.5	10
January	15.3	2.0(1–3)	5.0	5.7	47
February	5.3	1.7(1–2)	7.1	9.4	43

Note. All observations were made in the same area, on the Chiltern slopes above Great Kimble.

results of watches made on privet fruit, simultaneously with other kinds of fruit, in one restricted area in the Chilterns. When up to four other kinds of fruits were available in the immediate neighbourhood, only 10% of feeding records were at privet (November and December), but when fewer alternative fruits were available (January and February), privet records made up around 45% of the total. The difference in the number of other fruits available is much under-estimated in the table, since it was not practicable to keep more than about four kinds of fruit under observation at the same time. In November and December, six or more other kinds of fruit were usually available within a short distance of the observation sites, but in January and February the number was usually only two or three, and they were also much less abundant.

Blackbirds are the main eaters of privet fruit in our area, accounting for 70% of records. Robins score next highest, with 10%; and Blackcap records, considering the small size of the wintering population, are more numerous than might be expected (Table 24). Almost certainly, the relatively high number of records for these two species is because they can pluck and swallow privet berries more easily than those of most of the alternative fruits available.

Bullfinches are the only seed-predators that we have seen regularly attacking privet fruits. We have records from October to February, the total number (48) being second only to our Blackbird records and nearly equalling the total for all other dispersers. Locally Bullfinches may destroy a good proportion of the privet fruit crop. Thus by 18 December a solitary privet bush monitored on Ivinghoe Beacon had no intact fruits left. Some had been taken by 'legitimate' fruit-eaters, but the many dry fragments of tattered skin and pulp showed that a high proportion had had their seeds extracted by Bullfinches. In areas where privet is plentiful, however, the effect of seed-predation by Bullfinches seems to be minor.

In addition to Bullfinches, Woodpigeons are occasional eaters of privet fruit and we include them under seed-predators, though they may be occasional dispersers as well (p. 172). Blue Tits and Marsh Tits are occasional pulp-predators.

PRIVET ELSEWHERE

We have seen no published references to privet fruit in Britain that add significantly to our findings, except for a record of Waxwings eating the fruits in mid March during the Waxwing invasion of 1949–50 (Gibb & Gibb 1951), a late date that indicates that the crop is not always stripped by the end of February. On the Continent, privet occurs in southern, western and central Europe, but it is only from Germany and Hungary that significant observations have been reported. There are several records of Waxwings eating privet fruits but, as in our area, the Bullfinch is in the main, perhaps the only regular, seed-predator. The few dated records indicate that the fruit may be eaten very late in the winter and in the following spring, supporting the general statement by Creutz (1953) that other kinds of fruit are preferred, as our observations also show. Thus Szijj (1956–7) reported that, in Hungary, Starlings feed on privet fruit in February and March, and Schneider (1957) saw Waxwings eating partly dried-up privet fruit on 12 April. More surprisingly, in the Pillnitz area near Dresden, Creutz (1953) recorded Robin, Nuthatch and Great Tit taking privet fruit on 1 June (the last two species presumably as seed- or pulp-predators). These can only have been fruit left over from the previous year. It is at first sight hard to reconcile with other observations a report from Saxony by Heymer (1966) of 10 Yellowhammers, a male Chaffinch, 3 Blue Tits, a Great Tit, a Hawfinch, 10 Blackbirds and a Redwing perched on a laden privet bush and eagerly eating the fruit on 14 September (again, some of these species presumably as seed- or pulp-predators). One would like to know something about the circumstances, but it seems most likely that this was fruit that had ripened very early, and that little alternative food was available, which together suggest the end of a hot dry summer. This example makes one cautious of concluding that generalisations valid for one area are necessarily applicable to another area where conditions are very different. Alternatively, one wonders whether an introduced species of privet, possibly with a different fruiting season, may have been involved.

Buckthorn *Rhamnus catharticus*

In our study area buckthorn is a locally common shrub or small tree on the Chiltern escarpment; in the flat country of the Vale of Aylesbury it is much less common, but present in some undisturbed hedgerows and bushy places. It flowers in May–June and the early fruits ripen in the first half of September. The black drupes, globular and about 8.5 mm in diameter, are borne on short stalks close to the stiff twigs, and so are easily accessible to birds. They remain on the plant, if they are not eaten, until they dry up and wither in late December or January.

Our 551 records of buckthorn fruits being eaten by dispersers (Table 26) extend from 30 September, when a Starling was seen taking the first fruit from a bush near the base of the Chilterns, to 17 January, when a Redwing took some of the last remnants of fruit on the Chiltern escarpment in cold weather; most records were from the second half of October to mid December. There is much variation in the time when different bushes are stripped. Some have been cleared by the end of October while, close by, others have been virtually untouched. In the Vale

TABLE 26 *Buckthorn: monthly summary of feeding records*

	Sep	Oct	Nov	Dec	Jan	Totals
Dispersers						
Blackbird		29	56	119	24	228
Song Thrush		14	3	18	2	37
Mistle Thrush		11	3	26	1	41
Redwing		24	8	76	24	132
Fieldfare			3	8		11
Robin		1		2		3
Blackcap		2	1	6		9
Starling	1	83	6			90
All dispersers	1	164	80	255	52	551
Seed-predator						
Bullfinch		2	22	23	3	50

Note. There were also three records of Bullfinches taking seeds from dried-up fruits in February.

of Aylesbury all buckthorn fruit has been taken by the end of November, while on the Chilterns, in places where buckthorn is abundant, fruit often remains until some time in December. In addition, we had a few records of buckthorns being protected by Mistle Thrushes (p. 128) and so retaining their fruit after neighbouring unprotected plants had been stripped.

All the thrushes eat buckthorn fruit regularly, as do Starlings, and we had a small number of records for Robins and Blackcaps. We had no records for any of the Corvidae. Only one seed-predator was recorded at buckthorn, the Bullfinch; it was regular, with a total of 50 records from several sites.

Almost everywhere that Buckthorn grows in our study area, it has hawthorns growing close by, and usually more abundantly. As they have much the same fruit seasons they must compete for the attention of the fruit-eaters. Blackbirds, for which we have the most records, and Fieldfares generally prefer haw to buckthorn, the latter strongly, and Redwings probably do so (Table 51). There are fewer data for Mistle Thrush and Song Thrush and no preference is obvious. Starlings much prefer buckthorn to haw, almost certainly because they find buckthorn fruits easier to pluck and swallow. Sorensen (1983), experimenting with juice from the pulp of different fruits, found that buckthorn came at the bottom of the list of palatability of the 11 different fruits tested, including haw (see p. 197 for further discussion of these results). Nevertheless we had a few instances of thrushes taking buckthorn fruits in preference to ripe haws near by, the most striking being a group of Redwings stripping a buckthorn on 31 October and having to flutter and sally to reach the last terminal fruits, while Blackbirds, and a few other Redwings, were feeding on hawthorn close by.

BUCKTHORN ELSEWHERE

There seems to be very little information on the eating of buckthorn fruit elsewhere. Schuster (1930) lists five thrushes, Robin and Waxwing as eating the fruit in continental Europe, but without details. Goodwin (1943), dealing with buckthorn in the series 'The biological flora of the British Isles', simply says that the fruit is dispersed by birds but is 'not favoured', and is not taken until other fruit is gone, and then only in severe weather, which is incorrect. Hartley (1954), in his survey of fruit-eating by British birds, had no records. Sorensen (1981) apparently had no records from her study area near Oxford. She stated later (Sorensen 1983) that buckthorn fruits are avoided by frugivores, and suggested that the unpalatable flavour of buckthorn (and some other) fruits, which she demonstrated experimentally, has evolved so that thrushes should positively *avoid* eating them. The whole question of palatability and unpalatability is a complex one and is discussed more fully on p. 197. Here it need only be said that, at least in our study area, thrushes by no means avoid buckthorn fruits, nor do they eat them only in severe weather.

Spindle *Euonymus europaeus*

Spindle is a locally distributed shrub or small tree in our study area. It is fairly common in places on the Chiltern escarpment, and is so abundant in one small steep-sided valley that we called it 'spindle valley'. It is sparsely distributed on the chalk at the foot of the Chilterns, and is not uncommon on the north side of the Vale of Aylesbury in and on the edges of old oak woodland on clay. It flowers in May and June and the fruits ripen in November. The fruit is a four-lobed capsule which on ripening turns deep pink and splits to reveal, in each lobe, a single seed (or occasionally two) covered by a bright orange aril. The arillate seed remains partially exposed and in good condition for some weeks, but by the end of the winter, if not eaten, it dries up and is then hard to remove from the capsule. Each arillate seed, which for convenience we refer to as a 'fruit' because it is the unit taken by a feeding bird, is about 4.5 mm in diameter and 6 mm long, smaller than other native fruit in our study area. But the pulp (aril) is richer in nutriment than the pulp of any native fruit, having a low water content and being rich in fats and proteins (Appendices 2 and 4). Because the seed makes up a very high proportion of the mass of the fresh fruit (58%, highest of any native fruit), the 'relative yield' – the amount of dry nutritive matter in relation to the fresh weight of the whole fruit – is not very high. This means that the fruit is a rich source of nourishment if the seed can be quickly got rid of, a point that is more fully discussed on p. 217. Cooper & Johnson (1984) list spindle among the poisonous wild British fruits. The fruit can cause severe effects in man. What part of the fruit is poisonous is not stated, and apparently is not known; presumably it is the seed, as no seed-predators take the seed although it is easily opened.

Dispersers. Robins are the principal dispersers of spindle fruit, accounting for nearly half of our feeding records, followed by Blackbird, Blackcap and Song Thrush (Table 27). At the end of the season nearly all our records are for Robins, which have to sally to take the last remaining fruit, probably because they can exert a more effective pull in this way. Even so, they often cannot remove the more dried-up fruits from their capsules. Further details are given on p. 144. In 'spindle valley' the wintering population of Robins seems to be largely dependent on spindle fruit in hard weather.

TABLE 27 Spindle: monthly summary of feeding records

	Nov	Dec	Jan	Feb	Totals
Dispersers					
Blackbird	34	46	12	2	94
Song Thrush	4	9	6		19
Robin	23	54	33	20	130
Blackcap	6	25	5		36
All dispersers	67	134	56	22	279
Pulp-predators					
Great Tit	28	11		2	41
Marsh Tit	2	12	1	4	19
Long-tailed Tit			8		8

Notes. Omitted from the table is one Great Tit record in late October. All the November records for dispersers were in the second half of the month, and all the February records in the first half.

Pulp-predators. Great Tits and Marsh Tits are regular pulp-predators of spindle fruit, and we have a few records for Long-tailed Tits. Great Tits pick a fruit and take it to a perch near by, where they hold it under the foot and pick off part of the aril, letting the seed drop. They are rather wasteful feeders, usually removing only half to two-thirds of the aril before they drop the seed; and they generally perch with the fruit in the spindle tree itself, so that the seed falls beneath the parent plant. Hence they are probably not effective dispersers. The behaviour of Marsh Tits is similar. Long-tailed Tits peck out pieces of aril from the fruit in situ.

SPINDLE ELSEWHERE

The only published record that we have found of spindle fruit being eaten by dispersers in Britain is by Sorensen (1981). In her study of fruit-eating in oak woodland near Oxford she had records of Robin, Blackbird, Marsh/Willow Tit and Long-tailed Tit eating the fruit, the last two presumably as pulp-predators (and almost certainly Marsh rather than Willow Tit). Mountfort (1957) recorded Hawfinches as seed-predators of spindle in September and October, which means that they must take the seeds from unripe fruit. This seems to be the only record of a seed-predator attacking spindle fruit, except for Turček (1961), who also lists the Hawfinch. Sandring (1944) recorded Long-tailed Tits eating spindle fruit in East Prussia, without giving details of how they do so.

In Germany, in the years immediately after the last war, there was a minor controversy over whether or not Robins eat spindle fruit. Ringleben (1949) noted that references to Robins eating the fruit all went back to Naumann, and that several later observers had been unable to confirm it. Naumann had written that Robins are very fond of spindle fruit, and that this was so well known that the fruit had the local name *Rotkehlchenbrot* (Robin's bread). Ringleben acknowledged that some botanists gave good evidence for its being eaten by Robins: Schmeil (1911) noted that the distribution of spindle in continental Europe is exactly the same as that of the Robin, Christiansen (1914) said that the Robin is the disperser of spindle seeds, and Stresemann (1927–34) mentioned that 80% of spindle seeds in Robins' droppings germinate (this last observation is puzzling, as in our experience Robins regurgitate the seeds). Ringleben concluded that spindle fruits were for some reason no longer attractive to Robins and suggested possible reasons for the change, none of them very plausible. Sunkel (1950) followed this with a note confirming that spindle fruits were still called *Rotkehlchenbrot* in the Rotenburg/Fulda district (Hessen), and that when he was a boy he used to feed captive Robins (and Gardens Warblers) on spindle fruits. In his extensive data from the Pillnitz district, near Dresden, Creutz (1953) had no record of any bird eating the fruit, though he noted that the name *Rotkehlchenbrot* was still used; and there the matter rested. Without doubt the controversy was attributable to nothing more than lack of persistent field observation at the right time and in the right place.

84 The Fruits

Wild currant *Ribes rubrum*

Wild currant is a very local plant in our study area. All our observations were made on one group of plants growing in a wooded bog not far from a village. Although these plants are of wild type with small fruits (white, not red), they may well have descended from cultivated stock dispersed from the village. They flower in April and May and the fruit ripens in late June or early July, different fruits ripening over a period of about two weeks on each plant. Ripe fruits are variable in size, with a maximum diameter of about 8 mm, and being easy to pluck are available to all frugivores.

The ripe currants are much sought after and most are taken as soon as they ripen. By the end of July the bushes have been stripped. We have records of Blackbird, Song Thrush, Robin, Blackcap and Garden Warbler taking the fruit, the Blackcap accounting for half of all records (Table 28).

TABLE 28 *Wild currant: monthly summary of feeding records*

	2nd half June	1st half July	2nd half July	Totals
Blackbird	1	4	4	9
Song Thrush			2	2
Robin		1	2	3
Blackcap	3	8	6	17
Garden Warbler			2	2
All species	4	13	16	33

WILD CURRANT ELSEWHERE

There are many literature references to red and white currants (and fewer to black currants) being eaten by birds, but almost all of them refer to cultivated fruit. We have found no published record of wild currants being eaten in Britain. Siivonen (1939) had records of Song Thrushes eating the fruit in woodland in southern Finland in late July and late August (thus about a month later than our records). Other records, apparently referring to cultivated currants, indicate that warblers and the smaller members of the thrush family are the main eaters of the fruit. Turček (1961) lists the Willow Warbler and Wood Warbler among other warblers, one of the few records of *Phylloscopus* spp. taking fruit. There are several references to warblers and other birds feeding their young on currants: Common Whitethroat (Kemmerer 1921), Lesser Whitethroat (Wetzel 1878), Nightingale (Braes, quoted by Schuster 1930), Black Redstart (Schuster 1903) and Spotted Flycatcher (Zeddelof 1921). Diesselhorst (1972) reported that, when red and white currants were available, Blackbirds first ate the red ones, ignoring the white until the red were finished. In size, taste and softness they were similar, and the Blackbirds' choice was apparently based on colour alone (see p. 196). Gilbert White (1789) had noticed the same much earlier, but did not think that the red and the white were similar in taste: 'Birds are much influenced in their choice of food by colour; for though white currants are much sweeter fruit than red, yet they seldom touch the former till they have devoured every branch of the latter.'

Wild gooseberry *Ribes uva-crispi*

Wild gooseberry is a local plant in our study area. It is abundant very locally in the undergrowth of some woods on the Chiltern escarpment; elsewhere we have found a few plants in hedges and along woodland edges, some of which may be descendants of cultivated plants from gardens near by. They flower in March–May and the berries ripen from early July to early August. Ripe fruits are yellowish and translucent, becoming purple if they are in sunny positions; they range up to 14 mm in diameter, averaging about 12.5 mm, and are thus close to the upper limit that can be swallowed whole by Blackbirds.

We have failed to see birds eating wild gooseberries in our study area. It seems probable that Blackbirds are the main dispersers, and we have five records of Blackbirds taking wild gooseberries in the Lake District in early September. Counts of fruits over three successive years, on marked branches of a gooseberry clump growing in a hedge, showed that each year the bushes were stripped within a quite limited period, which varied from early July to early August. In 1982, when the fruit ripened early after hot weather in June, practically all the fruit went between one inspection on 4 July and the next on 6 July.

We have found no published references to wild gooseberries being eaten by birds elsewhere.

86 *The Fruits*

Woody nightshade *Solanum dulcamara*

Woody nightshade is an abundant climber in our study area. It dies down to the base in winter, grows rapidly in spring, and flowers from June onwards. The first fruits ripen in the second half of July, and fruit from later flowers continues to ripen throughout August and September. The loose cymes of oval, translucent red fruits, hanging from their supporting vegetation, are conspicuous throughout the autumn. Ripe fruit lasts in good condition for very variable periods, if not eaten: in dry exposed sites it may begin to wither and shrivel after only a month, while plants in damp sheltered places may have fruit in good condition until December.

The fruits average about 8.5 mm in diameter, and can be swallowed by all the dispersers. They are very watery – water constitutes about 90% of the pulp – and consequently their 'relative yield' is low (Appendix 2). They are also thin-skinned, and this is doubtless one of the factors accounting for their tendency to dry up. Woody nightshade berries are very poisonous to man (Cooper & Johnson 1984), but evidently not to birds.

Dispersers. Blackcaps accounted for 38% of all our records of dispersers eating woody nightshade fruits, followed in order of numbers by Blackbird, Song Thrush, Robin, Starling and Spotted Flycatcher (Table 29). We had records from the second half of July to late December, but they were by no means evenly distributed through this period. There was a main peak at the beginning, in late July and early August, then a falling-off of records (only three from the end of September to mid November), then a resurgence to a second, smaller peak in late November and early December. As the number of records in any month is affected by the time spent in watching, a more accurate idea of the seasonal pattern is given by the following figures of feeding records per hour of observation:

	hours of observation		records per hour	
July		9.6		2.1
August		25.9		2.0
September		19.7		0.8
October		10.3		0
November		12.9		0.7
December		11.7		0.9

This unusual seasonal pattern is explained by the fact that in the middle of its fruiting season woody nightshade suffers in competition with other fruits, especially elderberries, as discussed more fully elsewhere. As already mentioned, woody nightshade fruits are not very nutritious, but they are sought after early in the season, especially in dry weather, before elder fruit is available. In mid November, when the elder fruit crop is mostly finished, birds turn again to what is left of the woody nightshade crop. We suspect that at this time flocks of Starlings quickly strip the plants growing in hedges in open agricultural country, where a lot of woody nightshade fruit often remains until November and then suddenly goes, but we have not observed this directly.

Seed-predators. Bullfinches are the only important seed-predators of woody nightshade in our area. They extract the seeds in situ, leaving the shredded skin hanging, and locally may destroy a high proportion of the fruit crop. We also have single records of a Woodpigeon taking whole fruits (October), a Blue Tit eating the seeds of a fruit in situ (December), and a Chaffinch eating the seeds of dry fruit hanging on a plant in cold weather in February.

WOODY NIGHTSHADE ELSEWHERE

We have found no published references to birds eating the fruit in Britain that add significantly to our findings; in fact records of any sort seem to be extremely few. There are scattered records from continental Europe which indicate that

TABLE 29 *Woody nightshade: monthly summary of feeding records*

	Jul	Aug	Sep	Oct	Nov	Dec	Totals
Dispersers							
Blackbird	3	17			3	3	26
Song Thrush	4	15				1	20
Robin	1	1	3		5	4	14
Blackcap	10	15	12		1	3	41
Starling	3			3			6
Spotted Flycatcher		3					3
All dispersers	21	51	15	3	9	11	110
Seed-predators							
Woodpigeon				1			1
Bullfinch		2	1	1	1	8	13
Blue Tit						1	1

Note. A record of a Chaffinch taking the seeds in February is omitted.

88 *The Fruits*

August and September are the main months when the fruit is eaten. As in our area, the Blackcap seems to be the main disperser of the fruit (several records), with single records for Blackbird, Garden Warbler, Stonechat and Wheatear (Heinroth 1907, Heim de Balsac 1927, Madon 1927, Möhring 1957); and there is a single record for the Jay in January (Rey 1910).

Black nightshade *Solanum nigrum*

Black nightshade, a weed of cultivation, seems to be local in our area, but it grows on at least one vegetable plot in our village. An annual plant presumably adapted to a warmer climate than ours (it is doubtfully native in much of northern Europe), it does not flower until July and the first fruits ripen in the second or third weeks of September. Ripe fruits are black and globular, about 8 mm in diameter, a little smaller than those of woody nightshade. They are watery, like those of woody nightshade, and do not last long if not eaten, tending to burst open and then dry up. Fruit continues to ripen as long as the weather is mild, through October and November, but the plant is killed by the first frosts. According to Cooper & Johnson (1984), unripe berries are poisonous to man, but ripe ones only mildly so; ripe fruits are the least poisonous part of the plant.

We have no direct records of black nightshade berries being eaten by birds, in spite of four hours of observation; but counting and marking of fruit showed that ripe fruits were taken in the second half of September and in October. We suspect that Blackbirds and Robins ate them, as these were the only two frugivores regularly seen foraging on the ground in the immediate area.

BLACK NIGHTSHADE ELSEWHERE

Heim de Balsac (1927) found that near the coast in the lower Loire, in France, black nightshade fruits were much sought after by Blackbirds; the stomachs of several birds that he examined at the end of September were full of the berries. Blackbirds are also regular eaters of black nightshade berries in southern Spain, the season being evidently much more prolonged than in northern Europe (Herrera 1981c). The only other records that we have found from continental Europe are for the Starling, which Havlin & Folk (1965) recorded taking black nightshade fruit in July and September in Czechoslovakia (indicating much earlier ripening than in England), and for the Grey Partridge, which Vertse *et al* (1955) list as an occasional eater of black nightshade fruit in Hungary. Black nightshade is an introduced plant in New Zealand, and McCann (1953) found that its fruits were eaten by Blackbirds, which he thought were probably the main dispersers – a case of an introduced plant being dispersed by an introduced disperser.

Deadly nightshade *Atropa belladonna*

Deadly nightshade is a rather rare plant in our study area, occurring locally in bushy places on the Chiltern slopes. It flowers in June and July and the first fruits ripen in August. The ripe fruits do not last long on the plant in good condition; they soon become over-ripe and then dry up, sometimes as early as late August. Some late-ripening plants in shady places, however, retain fruit in good condition until October. The plant itself dies back to the rootstock in November.

When ripe, the berries are shiny black and contain numerous very small seeds scattered through the pulp. They are very variable in size, with diameters up to 20 mm, and of an unusual 'squat' shape, wider than they are long. Medium-sized fruits (11–16 mm) contain around 50 or 60 seeds (mean of five counted, 59); large fruits may contain 100 or more. As is well known, the fruits are very poisonous to man, but it does not seem to be known whether the poison is concentrated in the pulp or in the seeds, or is present equally in both (Cooper & Johnson 1984).

In spite of six hours of observation of deadly nightshade plants with ripe fruit, we have never seen a bird take them. In the case of one plant, whose fruits were counted and regularly checked from July to the end of October, out of 150 fruits that ripened only three were plucked, *ie* removed from the calyx, and so possibly taken by a bird. All the rest dried up, attached to the calyx, about three weeks after ripening. All the fruits of another clump of two plants, similarly monitored, went mouldy in October, and not one had been taken. Many alternative fruits were available in the areas where we watched and monitored these plants, and the negative results may merely indicate that deadly nightshade fruits do not do well in competition with other fruits for dispersers.

The great size of the larger fruits – well beyond the size range of most European fruits adapted for bird dispersal (Fig. 7) – might suggest that they are adapted for dispersal by some other agent. However, the fact that the very small seeds are distributed throughout the pulp means that a bird could be an effective disperser by picking pieces out of the fruit; there is no need to swallow it whole. The very variable size of the fruit also suggests that size has not been constrained by the need to be swallowed whole. Moreover the berries are certainly poisonous to some (all?) mammals, but there are several records of birds eating them. Hence it may be concluded that deadly nightshade is a bird-dispersed fruit, but an unusual one.

DEADLY NIGHTSHADE ELSEWHERE

We have seen only one record of deadly nightshade berries being eaten by a bird in Britain. Campbell (1985a) found the seeds in the gizzard of a pheasant. There are more records from continental Europe. Hess (1927) reported seeing Blackcaps several times eating the fruit on the plant (presumably picking pieces out, but this is not stated), and once a Song Thrush. In the latter case he mentioned that berries damaged by the bird's beak were lying on the ground in large numbers, suggesting that the thrush was plucking them and taking them to the ground, as they do with large fruits. Möhring (1957) also gave a record for Blackcap, in September, and Schuster (1930) a record for Blackbird. It is a pity that none of these reports gives details of how the birds dealt with the fruit.

White bryony *Bryonia dioica*

White bryony is a fairly common climber in our study area, growing mainly in hedges and bushy places. It dies down to the ground in winter, and in spring grows rapidly, beginning to flower in May and continuing to flower and set fruit through the summer as the climbing stems lengthen. The first fruits ripen in the second half of July and later fruits continue to ripen until early September. The whole plant begins to wither by late September, or even earlier if the weather is dry, and any remaining fruit dries up but stays attached to the plant and may persist in this state until December. The smooth, globular, red berries average about 8.5 mm in diameter (Appendix 2) and can easily be plucked and swallowed by all dispersers. The fruit contains toxins and may be fatal if eaten by poultry (Cooper & Johnson 1984), but whether the toxins are concentrated in the seeds, or present in the pulp also, seems not to be known.

Our 57 records of dispersers eating white bryony fruits are nearly all concentrated in a period of a few weeks from late July to early September (Table 30). Blackcaps account for 54% of all records, and three other *Sylvia* warblers for another 9%. When the fruits first ripen, especially if the weather is dry, they are taken rapidly. Later, as other fruits become available, especially elder, white bryony tends to be neglected. Thus the hourly rates of feeding visits to white bryony plants watched in July, August and September were, respectively, 3.6, 1.6 and 1.0. When the plants are withered and the fruits dry, from October onwards, they are rarely taken by dispersers, except in hard weather. Our latest record was of a Song Thrush taking dried-up fruit on 9 December, in very cold weather.

We had a few records of tits exploiting white bryony fruits after the main dispersal season (Table 30). Marsh Tits may act as dispersers as well as seed-predators – on the two occasions in September when a bird was seen taking fruits it took them about 20 m and stored them, on each occasion making three trips

TABLE 30 *White bryony: monthly summary of feeding records*

	Jul	Aug	Sep	Oct	Nov	Dec	Totals
Dispersers							
Blackbird	3	11		1			15
Song Thrush	4					1	5
Robin		1					1
Blackcap	17	9	4	1			31
Garden Warbler	2	1					3
Common Whitethroat	1						1
Lesser Whitethroat			1				1
All dispersers	27	22	5	1	1	1	57
Seed-predators							
Marsh Tit			2		3		5
Willow Tit				1			1

Note. Marsh Tits may also act as dispersers – see text (p. 185).

and taking one fruit at a time. The single record for a Willow Tit was of a bird taking seeds from dried-up fruits and hiding them. Many of the dried-up fruits examined in late autumn showed evidence of having been attacked by tits.

WHITE BRYONY ELSEWHERE

The only published reference that we have found to white bryony fruits being eaten by dispersers in Britain is by Campbell (1985b), who had records for Garden Warbler and Lesser Whitethroat in Oxfordshire. Sorensen (1981) saw Marsh/Willow Tits, and occasional Blue Tits, exploiting white bryony fruits in woodland near Oxford in September–November. She gave no details of how they dealt with the fruit, but presumably they were taking the seeds.

White bryony is widespread in western and central Europe, but records of the fruit being eaten by birds are few. Schuster (1930), who summarised the European literature up to the time of his writing, could cite only Snell (1866), who stated that white bryony fruits are a favourite food of Nuthatches (presumably seed-predators like the tits) in Germany, and that boys used to trap them by using the fruit as bait, and Noll (1870), who found that a caged Song Thrush ate white bryony fruits offered to it, but not willingly. We have found three later records. Blackcaps were observed eating the fruit in September (Möhring 1957) and in 'autumn' (Maas 1950) in Germany; and analysis of stomach contents showed that Starlings take white bryony fruit in July in Czechoslovakia (Havlin & Folk 1965). Lübcke & Furrer (1985) list the related *Bryonia alba* among the fruits recorded as eaten by Fieldfares.

Black bryony *Tamus communis*

Black bryony is a very common climber in the southern part of our study area, on the Chiltern slopes and the flatter country along the base of the Chilterns; elsewhere it occurs more locally, in hedges. It dies down in winter, like white bryony, and grows rapidly in spring and summer, flowering from May to July. The fruits ripen from early September and are all ripe by the end of the month. If they are not taken, some begin to dry up from mid November, but many remain in good condition until January. The bright red oval berries are rather variable in size, 9–11.5 mm in diameter and 11–14 mm long (Appendix 2). They are mostly above the size that can be swallowed by warblers and around the maximum size that Robins can swallow, but can easily be plucked and swallowed by the thrushes.

We had rather few observations of birds eating black bryony fruits, all from a two-month period from the first half of November to the first half of January (Table 31). Counts of marked plants confirmed that this is the period when nearly all the fruit is taken. Our observations showed that the rate at which the fruit crop is depleted increases in the course of the autumn and winter. Thus in 8.7 hours of observation in September and October we had no records, while in the months November (14.2 hrs of observation), December (12.1 hrs) and January (5.7 hrs) the hourly rates of feeding visits by dispersers were, respectively, 0.2, 1.5 and 2.1. Nevertheless, in spite of the increasing rate of consumption by birds we found that in some places a good deal of the fruit crop was not taken, and it dried up still attached to the withered plant. Whether or not a particular plant is stripped or retains some or much of its fruit until it is withered probably depends on the availability of alternative fruits. In competition with other fruits black bryony seems to be relatively unattractive, as much higher hourly feeding rates were recorded for other autumn and winter fruits.

Blackbirds accounted for 72% of our records, the rest consisting of 1–5 records each for Song Thrush, Fieldfare, Robin and Blackcap. A Robin swallowed one fruit, which was clearly at its upper size limit, only at the third attempt. The single record for Blackcap was of a bird pecking pieces out of a fruit.

We had no evidence that black bryony fruits are attacked by seed-predators.

BLACK BRYONY ELSEWHERE

The only published record that we have found of black bryony fruit being eaten by a disperser elsewhere in Britain is by Campbell (1985b), who mentioned that a Blackcap trapped in midwinter in Oxfordshire regurgitated black bryony seeds, which indicates that not all of the fruits are too large for Blackcaps to swallow. In her experiments on the preferences of four captive Blackbirds and one Song Thrush for the flavours of 11 different wild fruits (see p. 196), Sorensen (1983) found that black bryony came second in order of preference. Sorensen wrongly states in her discussion that black bryony fruits are avoided by thrushes in the field and suggests that this is because the plants have 'a delicate growth form which prevents thrushes from perching to remove fruits'. Those that we saw feeding on black bryony fruits had no special difficulty in taking them.

Black bryony has a southern and western distribution in continental Europe. There seem to be no published records of its fruit being eaten by birds except in Spain, where the Blackbird, Robin and three *Sylvia* warblers are reported to take it (Jordano & Herrera 1981, Herrera 1984a, Pérez-Chiscano 1983). Records for the Blackcap were all in the months October–November, with none in December–January (Jordano & Herrera 1981), indicating a distinctly earlier season than in England. It is also noteworthy that Blackcaps in southern Spain swallow black bryony fruits whole, but only after mandibulating them to reduce their size (C. M. Herrera, *in litt*).

TABLE 31 *Black bryony: monthly summary of feeding records*

	Nov	Dec	Jan	Totals
Blackbird	3	15	8	26
Song Thrush		1	4	5
Fieldfare		1		1
Robin	3			3
Blackcap		1		1
All species	6	18	12	36

Spurge laurel *Daphne laureola*

Spurge laurel is a local plant in our study area. We have found it in four places: in two places in the undergrowth of beech woods on the Chilterns, in a shrubby area on the steep slopes of a Chiltern valley and, surprisingly, in a roadside hedge in open farmland in north Buckinghamshire not far from old oak woods and perhaps persisting as a relict of former woodland. In each of these places the spurge laurels are concentrated in a limited area.

The inconspicuous greenish flowers appear very early in the year, in early February, and the fruits ripen, turning from green to purple-black, at the end of June and first ten days of July. The small oval drupes are about 7 mm in diameter, and contain a single seed. All *Daphne* fruits contain an acrid, irritant juice, which is also present in other parts of the plant; the fruits of spurge laurel are known to be poisonous to mammals (Cooper & Johnson 1984).

Robins are the only dispersers that we have seen taking spurge laurel fruits. In one Chiltern woodland area, where we did most of our watching, they took the fruit regularly; but their visits were well spaced, and often very brief, so that a rather long period of observation, totalling 18.4 hours, produced only 16 feeding records. A Robin would approach a plant, hover at it for a moment to pluck a fruit, then take it to a neighbouring perch to swallow it, repeat the process once or twice more, and then leave; or sometimes a Robin would pluck a single fruit and fly off with it, almost certainly to feed a young one. All our records were in July, nearly all in the first half of the month. For no other plant did we record such a short season of fruit dispersal. In the two years when we watched consistently, 1983 and 1984, the plants had been virtually stripped by 19 July.

Of the 65 spurge laurel plants in the roadside hedge, all but one plant lost their seeds while they were still unripe, in June. Greenfinches were almost constantly present and were seen feeding on them. The patch of spurge laurels on the bushy Chiltern slope also lost all their fruits while they were still green, between the last week of May and mid June. We suspected that Great Tits were taking the seeds but unfortunately had no direct evidence; only once were Greenfinches seen in the area when these spurge laurels had unripe fruit.

SPURGE LAUREL ELSEWHERE

We have found only two references to spurge laurel fruits being eaten elsewhere in Britain or on the Continent. In his monograph on the Hawfinch, Mountfort indicates that Hawfinches are occasional seed-predators (Mountfort 1957, Appendix 3). He had records in August and September, when all the fruit in our study area had already been eaten. Turček (1961), without giving details, lists Blackbird, Blue Rock Thrush *Monticola solitarius* and Hawfinch as eating spurge laurel fruit (the Hawfinch possibly based on Mountfort's records).

Mezereon *Daphne mezereum*

Mezereon is scarce as a wild plant in our study area, being found very locally in the Chilterns. We know of only a single plant growing in a nature reserve, and it may well have come from seed dispersed from some garden near by. It is not uncommon in gardens and all our observations were made on a plant in our garden. The attractive pink flowers come out in early spring, while the plant is still leafless, and the orange fruits (drupes) ripen in late June or July. The fruits are ovoid, about 8 mm in diameter, and contain a single seed. They are poisonous to mammals (Cooper & Johnson 1984), and Lübcke & Furrer (1985) mention that human beings have been poisoned eating thrushes that have fed on the fruit, which indicates that the poison must be present in the pulp.

Blackbirds are the only birds that we have seen taking the fruits. All our records are in late July. In the two years when our garden mezereon fruited well, 1982 and 1983, their fruits lasted only a few days once Blackbirds had begun to take them; the last fruit was taken on the afternoons of 27 July 1982 and 28 July 1983.

MEZEREON ELSEWHERE

We have found no record of *Daphne mezereum* fruits being eaten by any disperser in Britain except for the note by Aplin (1910) of a Lesser Whitethroat eating the fruit in July, in Oxfordshire. On the Continent, where the plant is evidently much commoner than in England, there are records for a number of dispersers, including Pheasant, Hazel Grouse, Robin, Blackcap, Garden Warbler and Lesser Whitethroat (Naumann 1897–1905, Schuster 1930, Möhring 1957, Heymer 1966), and – surprisingly – Fieldfares (Lübcke & Furrer 1985). Möhring gives a period of June to September for his Blackcap records, much longer than we have found, and Heymer records a Robin eating the fruit on 6 October in Saxony.

At least locally in Britain, Greenfinches are important seed-predators of mezereon, taking the seed while the fruit is unripe. As Pettersson (1956) wrote: 'A pair of birds, usually, will visit and strip the bush in May or early June, and while the fruits are still green. They can then crack the immature stones and devour the large seed.' Pettersson thought that this was a new habit that was rapidly spreading; 42% of the cases reported to him were within the two previous seasons. Much earlier, however, Butterfield (1910) had noted that Greenfinches in Yorkshire ate mezereon seeds, discarding the pulp; so perhaps the impression that the habit was new and rapidly spreading was, at least in part, the result of more attention being paid to it. In a later note, Pettersson (1961) wrote that the habit seemed to have originated between one and two centuries ago. This case has been cited as an example, similar to that of tits opening milk bottles, of a new habit spreading through a population by 'cultural diffusion' (one bird learning from another), but one cannot help wondering whether the evidence is good enough for any certain conclusion.

96 *The Fruits*

Juniper *Juniperus communis*

Juniper is a very local plant in our study area, being confined to a few places on the Chiltern escarpment where it grows scattered in semi-open busy areas. A dioecious plant, it flowers in May and June and the fruit ripens in September or October in its second or third year (Clapham, Tutin & Warburg 1973). The bushes that we monitored began to produce ripe fruit by the end of August, and ripe fruit was available from then until the end of February. Throughout this period there was also green fruit (presumably due to ripen in a later year) and uneaten ripe fruit that had dried up and become hard.

The fruits, which are formed from the flower scales, are fleshy but rather dry compared with other fleshy fruits. They are blue-black with a pale bluish bloom, variable in size, and contain up to three seeds; full-sized fruits average about 8.5 mm in diameter. Mainly because of the low water content of the pulp, the 'relative yield' of juniper fruit is very high (Appendix 2) – higher than that of any other wild fruit in our study area. The pulp is also quite rich in fats (Appendix 4), so that it should be an excellent winter food for birds. Its small size makes it available to all fruit-eating birds that are present in winter.

For some reason, over the four years for which we monitored it, a great deal of the juniper fruit in our study area remained and dried up on the plants. In spite of over 29 hours spent watching junipers with ripe fruit, spread over nearly all months of the year but concentrated in late winter and early spring, we obtained

TABLE 32 *Juniper: monthly summary of feeding records*

	Jan	Feb	Totals
Blackbird	1	1	2
Mistle Thrush	5	3	8
Robin		1	1

only 11 records of birds taking the fruit (Table 32), and all of these records were obtained in one small area where a Mistle Thrush defended one of the juniper bushes. Over 20 hours spent in another area where the junipers were much more abundant, including observations during very cold spells with snow cover, yielded not a single record. This lack of records is puzzling in view of the fact that juniper fruits are an important winter food for fruit-eating birds elsewhere.

JUNIPERS ELSEWHERE

Several species of juniper occur in Europe but only three are widespread: *J. communis* which is generally distributed throughout; *J. oxycedrus*, a red-fruited species, in southern Europe; and *J. phoenicea* from the Mediterranean region west to Portugal.

Juniperus communis. This is the only British species. According to Witherby *et al* (1938), Mistle Thrushes and Fieldfares have been recorded eating the fruit, but we have found no detailed, dated records. Gilbert (1980), who studied the ecology of juniper in Upper Teesdale, had the following to say about the dispersal of the fruit: 'In Upper Teesdale female bushes laden with "berries" are abundant in all stands. By late summer, most ripe fruits [presumably having ripened the previous autumn] have been plucked by birds which have removed the fleshy interior together with two or three seeds. Intact ripe fruits with beak marks usually contained seeds with incompletely developed embryos or embryos damaged by the insect *Megastigmus bipunctatus* Swed. Bird faeces containing up to 20 seeds each can be found on horizontal branches, perching stones, fence posts, etc. . . . Frequently seeds dispersed in bird droppings were found to have been subsequently eaten by small mammals. Trapping showed woodmice (*Apodemus sylvaticus*) were probably responsible.'

We have found a number of references to the eating of the fruit by birds on the Continent. It is noteworthy that the Fieldfare's name in German is *Wacholderdrossel* (juniper thrush), but Lübcke & Furrer's recent (1985) monograph on the Fieldfare does not suggest that juniper fruit is an important food for the bird in Germany today. In southern Norway, Meidell (1937) found that Fieldfares eat juniper berries in spring and a little in autumn. In southern Finland, Siivonen (1939) recorded Song Thrushes eating juniper berries from mid September to early October, and listed juniper among the fruits recorded as taken by Song Thrushes elsewhere in Europe.

Schüz (1933) reported that during the 1931/32 Waxwing invasion into central Europe, juniper was the most important fruit in their diet after rowan. It may be noted that two birds that very rarely take fruit have been recorded eating juniper berries, the Cuckoo (Müller & Müller 1887, Link 1889) and Dunnock (Heymer 1966), the latter in early April in the Alps, in very cold weather.

Juniperus oxycedrus. In the mountains of southern Spain, at 1,150 m, Herrera (1984a) found that ripe fruit of this juniper was present from mid November to early March and was eaten by Blackbirds, Robins and Blackcaps; but it was not much sought after and most of the fruit crop remained undispersed in all of his study seasons. Elsewhere in Spain, Mistle Thrushes have been recorded eating the fruit (Pérez-Chiscano 1983); and, in Dalmatia, Mistle Thrushes, Blackbirds and Song Thrushes (Tutman 1950, 1969).

Juniperus phoenicea. Fieldfares have been recorded eating the fruit of this juniper in Dalmatia in winter (Tutman 1962). In the Atlas Mountains of North Africa it

98 *The Fruits*

evidently provides a most important source of food for wintering thrushes. Thus in the High Atlas at about 2,000 m, over the midwinter period, Juana & Santos (1981) found old stands of the juniper covered with fruits. Parties of Ring Ouzels were everywhere, often accompanied by Song Thrushes and Redwings, and also ('bizarrement' – but why they should have been unexpected is not clear) by Mistle Thrushes. Droppings composed mainly of the remains of juniper fruits were to be found under almost every tree. *Juniperus thurifera*, growing higher up at 2,500 m, was evidently not attracting thrushes to the same extent.

Lords and ladies *Arum maculatum*

This is a common plant in our study area, occurring in woods, under thick hedges and in shady gardens. It flowers in April and May and the fruits ripen in July (mainly the second half) and early August. They remain in good condition for a few weeks, until some time in September or October, or earlier if the weather is very dry, after which the stem shrivels and the whole plant collapses. The orange-red berries, clustered at the top of the fruiting spike, are very variable in size, ranging from very small to about 11 mm in diameter. They are poisonous to mammals but it seems not to be known whether the poisons are concentrated in the seeds or present in both pulp and seeds.

With the decrease in persecution and the subsequent spread of deer in our area (Fallow *Dama dama* and the introduced Muntjac *Muntiacus reevesi*) lords and ladies

TABLE 33 *Lords and ladies: monthly summary of feeding records*

	Jul	Aug	Sep	Totals
Blackbird	1	40	1	42
Robin		8		8

fruit in woodland has greatly decreased because the deer eat the flower spikes, the leaves being poisonous (Prime 1960). A survey of a wood in the Chilterns in early May showed that 70% of 117 flower spikes had been eaten; those remaining were only just emerging from the ground and may well have been eaten later, as no ripe fruit could be found in this wood on 5 August.

We have records of Blackbirds and Robins eating the fruits, mainly in August (Table 33) when counts of fruits have shown that most are taken. Individual Blackbirds may temporarily defend a patch of fruiting spikes and prevent other birds from feeding at them. Blackbirds, the main eaters of the fruit, either reach up to take them or make short flight sallies from the ground. Of 85 fruits plucked by Blackbirds, 9% were discarded for reasons that mostly were not apparent; in one case the fruit was very squashy and so perhaps unpalatable. Robins mainly take them in flight sallies and have difficulty in swallowing the larger fruits. Partly eaten spikes tend to have the lower part stripped, leaving the remaining fruit at the apex; but when they are much in demand (in very dry weather or when alternative fruit is scarce) the fruits are taken as they ripen, from the top first.

LORDS AND LADIES ELSEWHERE

Prime (1960), in his monograph on lords and ladies, mentions that small mammals dislike the pulp of the fruit and leave the seeds alone; hence mammals probably play no part in their dispersal. He was able to cite only a single instance of a bird eating the fruit, a record of a Blackbird by W. D. Campbell. Kear (1968) quoted observations by H. Lancum of a Common Whitethroat taking the fruit, and (surprisingly) a Willow Warbler, which could hardly have swallowed any but a very small fruit whole. We have found no other published references to the fruits being eaten by birds.

Some minor fruits

In addition to those dealt with in the preceding sections, there are two bird-dispersed wild fruits in our study area for which we have no records of their being eaten, and two that are introductions from elsewhere in southern England for which we have a few records. (The two latter may occur locally in the area, but we have not found them.)

Wild strawberry *Fragaria vesca* is a local plant found on well-drained open slopes and along wood-edges, mainly in the Chilterns and, on lower ground, along railway cuttings and embankments. It flowers from April to midsummer and the fruit ripens in June and July. The small red strawberries, 9–13 mm in diameter, contain a large number of very small seeds embedded on the surface of the fruit (botanically, they are achenes embedded on the surface of the fleshy receptacle). Weighing about 0.4 mg each, they are easily the smallest of the seeds of fleshy fruits in our area.

We watched for a total of $8\frac{1}{2}$ hours at places where wild strawberries were locally abundant, but never saw any bird taking them, although on some occasions known fruit-eaters were foraging on the ground quite close by. Our lack of records is probably attributable to the fact that even where wild strawberries are locally abundant they provide, at best, a sparse fruit supply. The plants are well spaced and there is not usually more than one ripe fruit per plant at any time. When wild strawberries are ripe, the wild cherry trees provide a far more substantial fruit supply and attract many fruit-eaters. Wild strawberries, however, certainly are taken by birds. Omitting records that are not explicit and probably refer to cultivated strawberries, for Britain we have found only Holyoak's (1969) general statement, based on analysis of stomach contents, that some of the crows take them. There are several records from Germany. Blackbirds and Song Thrushes are recorded eating wild strawberries in the Erz Mountains (Schlegel & Schlegel 1965). In Saxony, Heymer (1966) recorded Song Thrushes taking wild strawberries, apparently to a nest, and Chaffinches picking at them, presumably extracting the seeds. Möhring (1957), summarising data on the Blackcap's diet, listed wild strawberries as eaten from June to August.

Jurik (1985) studied the reproductive ecology of wild strawberries *F. vesca* (also the North American *F. virginiana*) near Ithaca, New York, over three years, and found no evidence of how the fruits might be dispersed (T. W. Jurik, *in litt.*). Most of them disappeared about eight days after ripening; some dried up in situ. In view of their scent and the fact that they grow at ground level, and do not present themselves in such a way as to be conspicuous to a bird searching from above, one wonders whether the seeds of wild strawberries may be dispersed more by mammals than by birds, and possibly also by slugs, which eat the flesh of cultivated strawberries.

Dewberry *Rubus caesius* grows sparsely on road verges, banks and wood-edges in our study area. A low-growing, inconspicuous plant compared with blackberry, it flowers from June onwards and the fruits ripen from July, a good deal earlier than blackberry. They are 'polydrupes', much like blackberries, but the drupelets are fewer and larger and have a bluish bloom. The drupelets are considerably more watery than blackberry and raspberry and, mainly for this reason, their 'relative yield' is the lowest of all the wild fruits in our study area (Appendix 2).

We found no concentrations of dewberries large enough to be worth watching and had no casual records of their being eaten. We assume that much the same species of birds that eat blackberries eat dewberries, too, mainly in dry weather when watery fruits are much sought after. Heymer (1966) recorded many migrating birds eating dewberries on the island of Wangerooge off the German North Sea coast in August and September, including eight species of warblers, Whinchat and Pied Flycatcher. Evidently the fruit was abundant there and provided a valuable energy source for the migrants. This, and Heymer's observation of a Robin eating dewberries in Saxony in 'summer', are the only published records that we have found.

Stinking iris *Iris foetidissima* occurs locally as a wild plant in southern Buckinghamshire. It is not native in our area, but has been introduced into some gardens; we have plants in our garden from Dorset. It flowers from May to July. The fruits appear to ripen in October, when the pods split to reveal up to 30 seeds covered with bright orange, fleshy arils. The arils may remain in good condition through to the following spring, or may, perhaps as a result of very cold weather, go bad and blacken at some time from December onwards.

Regular counting of the fruits of several plants over three winters gave no evidence that any were eaten before the end of December. They were than taken mainly during spells of very cold weather, individual plants being stripped in January or February. We only observed the fruit being taken by birds on two occasions; once by a Blackbird and once by a Song Thrush, each of which made little upward jumps to snatch the arillate seeds from the pods. On other occasions footprints in the snow around plants from which the fruit was being taken indicated thrushes.

Stinking iris fruit is evidently relatively unpopular with fruit-eaters except when food is very short. We have found no published records. In a large garden on the outskirts of Southampton, in April following the mild winter of 1980–81, we saw plants loaded with fruit in perfect condition.

Solomon's seal *Polygonatum multiflorum*. We have not found Solomon's seal growing wild in our study area, but it apparently occurs very locally in some Chiltern woods. The plants on which we have made observations were introduced into our garden from a wood near the Berkshire-Hampshire border. It flowers in May and June and the fruits ripen in September. The blue-black berries, about 10 mm in diameter, hang on long stalks beneath curving stem and leaves, and so are accessible only to a bird approaching from below; but by October the leaves have decayed and the fruit becomes more widely conspicuous.

We have only two observations of a bird taking the fruit. On two occasions on 24 October, a first-year male Blackbird (almost certainly the same individual) snatched the fruits with upward jumps of 45–50 cm from the ground below, taking two fruits at the first visit and three at the second visit 24 minutes later. We suspected that it was significant that he was a young bird. The garden was being dominated by the resident adult male; but the Solomon's seal, growing in a shady corner, could be visited by the young male who arrived inconspicuously, took his meal, and left.

We have found no published reference to the fruits of Solomon's seal being eaten by birds elsewhere. Observation would be more difficult in the plant's proper habitat, woodland, where birds are shier and sparser than in gardens.

102 The Fruits

Introduced and cultivated plants

In gardens, parks and other public places in towns and suburbs, native fruit-bearing plants are often outnumbered by introduced species and cultivated varieties. These non-native plants, though numerous, are not, strictly speaking, relevant to a study of natural ecological interactions between birds and plants; but they are important to the birds of urban and suburban habitats, and so need to be considered in a study of how birds and plants interact in the largely man-made conditions of today. In such habitats they may well enable larger populations of thrushes to be maintained than would otherwise be the case.

A distinction must be made between plants that are grown mainly for ornamental reasons and plants grown to provide food for human consumption. Ornamental plants may be natural species introduced from elsewhere, hybrids between such species, or cultivars (cultivated varieties produced by artificial selection). In this country all fruits grown to produce food for man are cultivars and are fed on by birds, sometimes on such a scale that the birds are considered pests; but in our study area there is not much commercial fruit-growing and we are not concerned with this economic aspect. Flocks of Starlings and parties of Mistle Thrushes and Blackbirds descend on the ripe strawberries of a fruit farm within our study area, mainly in July, but the owners evidently consider the fruit superabundant and do not try to keep the birds away. Thrushes and other birds feed on apples and pears in gardens and orchards, mainly in hard weather in winter. Starlings feed much on apples, whether fallen or still on the tree, in late summer and autumn. The number of records for birds feeding on fruit of this kind is not in itself very meaningful. Fallen apples and pears in neglected domestic orchards are very important locally in tiding birds over spells of severe weather in winter (see also p. 134), but they are only marginally relevant to general aspects of fruit-eating. The more widely distributed ornamental trees and shrubs in gardens and parks that bear bird-dispersed fruits are of greater interest, and the sections that follow refer only to them.

THE SEASONAL SUCCESSION OF FRUITS

Our records for introduced/ornamental plants are summarised in Table 34. It is noteworthy that all but five of these are in the family Rosaceae, and the winter fruit supply is provided entirely by the Rosaceae. The numbers of different kinds available in each month are compared with native plants in Figure 5. In the

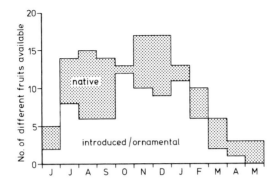

Figure 5 Numbers of different species of introduced/ornamental fruits available per month, compared to native fruits.

TABLE 34 *Introduced/ornamental plants: summary of feeding records, all dispersers*

	Jan	Feb	Mar	Apr	May	Jun	Jul	Aug	Sep	Oct	Nov	Dec	Totals
Cotoneaster 'cornubia'	149	24	6	1						27	63	231	502
C. horizontalis	12	3						14	10	14	5		58
C. distychus	11	2					1			4	3	12	33
C. simmonsii	5											2	7
C. bullatus								3	1				4
C. franchetii	1										6		7
C. microphyllus	3		1										4
Pyracantha coccinea	83	8						2	24	49	114	12	292
P. rogersiana	1									94	20	6	121
P. crenatoserrata	4	5				11	2						22
Crataegus crus-galli										2			2
C. × prunifolia										5	66		71
C. × lavallei	21	13											34
Amelanchier laevis						7	219						226
Prunus laurocerasus								8	17	2			27
P. lusitanica										1	4		5
P. padus 'watereri'						1	3						4
Rosa rubrifolia								2		7	5	1	15
Sorbus latifolia											9		9
Sorbus intermedia										16			16
Malus sieboldii	38											8	46
Mahonia aquifolium							32	21				1	54
Berberis darwinii							29						29
Morus nigra							23	34	2				59
Lycium barbarum							1						1
Euonymus latifolius										1			1

autumn and winter (October–January) the number of different kinds of fruits of ornamentals available for birds is nearly three-quarters of the number of native plants. This is entirely due to the autumn-ripening species and hybrids of *Cotoneaster*, *Pyracantha* and *Crataegus*; without them there would be few kinds of non-native fruits available in winter. In summer and early autumn the number of non-native fruits available is less than half the number of native species. But although the number of different kinds of introduced/ornamental fruits in summer is not very high, those that are available are in great demand, especially during periods of drought. Thus, during watches totalling 12 hours at snowy mespil *Amelanchier laevis* in July, we recorded an average rate of 18.3 feeding visits per hour, comparable to the highest rates of feeding at pyracantha and cotoneaster in hard weather in winter (pyracantha, January, 19.2; cotoneaster, December, 25.5).

FRUIT CHARACTERISTICS

Full details of 'design components' are given in Appendix 2. In size the fruits of ornamentals are much the same as native fruits, about three-quarters of them having diameters of 6–11 mm. This is not surprising, as they are mainly bird-dispersed plants from temperate regions where the fruit-eating birds are of much the same size as in Britain. In 'relative yield' (p. 14) such fruits all cluster around the same commoner values for native fruits of 11–20%. Few of them have been analysed for the nutritional content of their pulp, but those that have are mostly similar to native plants (Appendix 4). Thus there is no apparent reason why they should be less sought-after by birds than are native fruits.

However, horticulturists endeavour to breed strains or cultivars, particularly among the easily hybridising pyracanthas and cotoneasters, with long-lasting ornamental fruits which they claim are unattractive to birds. One avenue open to them is to alter the season of ripening to a period when there is an excess of wild fruits (it is well known that a raspberry developed to ripen in September receives far less attention from birds than those ripening in July); another is to interfere with the pluckability of the fruit. In principle it should also be possible to alter the nutritive quality of a fruit by artificial selection, so as to make it less attractive to birds; and in this connection it is perhaps significant that the little information that we have suggests that some cotoneasters may have relatively unnutritious fruits.

Unusual colours may also make a fruit relatively unattractive. Thus it seems that a colour such as white, represented in our native flora only by one not very common plant, mistletoe, may not act as an effective signal to fruit-eaters unused to the colour. This is possibly the reason why the white fruits of some exotic *Sorbus* species remain uneaten for a long time, and why the white fruits of the snowberry (*Symphoricarpus rivularis*) seem to be relatively unpopular with birds (Radford 1978, 1980). The significance of fruit colour is discussed in more detail on pp. 194–6.

EXPLOITATION OF INTRODUCED AND ORNAMENTAL FRUITS BY THRUSHES

In our area Blackbirds benefit most from the introduction of ornamental fruits into gardens and parks; 69% of the 1,649 feeding records (Table 34) are for this species, from all months of the year except May. All but two of the 107 records for the three *Crataegus* species/hybrids are of Blackbirds, and their proportion of the total for the cotoneasters was 75%.

By contrast the Song Thrush accounts for only 5% of the records shown in Table 34, a high proportion of the records being for the summer months July–August (48%; cf. 43% for the four months November–February). This appears to reflect the effectiveness of the Blackbird's defence of winter fruit territories compared with the relaxation of its territorial aggressiveness in late summer (Snow 1958). Song Thrushes were recorded eating the fruit of only 13 introduced/ornamental fruits, compared with 24 for Blackbirds.

Mistle Thrush records account for 7% of the total, most records being for *Cotoneaster 'cornubia'*, which Mistle Thrushes defend as a winter food resource. If all introduced and cultivated fruits are included (*ie* those developed for man's consumption, omitted from Table 34, such as cherries, apples, pears and strawberries), 55% of the yearly total of Mistle Thrush records are for July. Most of these are from the early-ripening *Amelanchier* and from strawberries grown in fields; but surprisingly, we also saw Mistle Thrushes taking the fruit of such low-growing shrubs as *Mahonia aquifolium* and *Berberis darwinii* in this month.

The two winter thrushes, Redwing and Fieldfare, took little introduced/ornamental fruit. All but one of our 47 Redwing records were from *Cotoneaster 'cornubia'*, and all of our 11 Fieldfare records were from the same cotoneaster; together they account for only 3.5% of the total shown in Table 34. Analysis of winter records of birds feeding on apples and pears in gardens and small orchards adjoining farms, however, gives very different proportions: of 462 such records, 50% were for Fieldfares, 42% for Blackbirds, 5% for Redwings and 2% each for Mistle Thrushes and Song Thrushes.

Notes on some of the more important ornamentals

THE COTONEASTERS

Cotoneasters are widely planted as ornamentals. Of those shown in Table 34, all but the *'cornubia'* hybrids are species, unaltered by horticulturists, which have been introduced from China or the Himalayas in the last century. No doubt as a result of dispersal by birds, several species have successfully become naturalised in Britain. *C. microphyllum* and *C. simmonsii* are treated as feral in the *Atlas of the British Flora* (1962), and we have found *C. horizontalis* and *C. bullatus* self-sown among native vegetation, and *C. distychus* self-sown in our garden. Between them, these cotoneasters produce a fruit supply from August to March or even later. The earliest ripener is *C. bullatus*; a feral example, monitored over six years, had a peak of fruit removal in September or October, and was stripped between mid October and the end of November. In the same locality, about 300 m from the *bullatus*, several *C. distychus* planted among native vegetation were also monitored. This cotoneaster has a winter ripening season with a peak of fruit removal, as shown by feeding records and by sample counts of fruit, in December and January. Its fruit is capable of remaining edible into July, when we recorded a Blackbird eating it; regular monitoring showed a steady but slow rate of fruit removal from the end of April through to July.

We collected most records from cotoneasters which we list as *C. 'cornubia'* in Table 34. This is a composite of two parental species, *C. frigida* and *C. salicifolia*, and hybrids derived from them. We have noted no differences in their fruiting seasons and no obvious differences in their fruits, and so have lumped together records from all of them. Because they grow into small to medium-sized trees and are common in gardens, they locally provide abundant sources of fruit. Mistle Thrushes may defend them in autumn and winter, and we had a case of a female Blackbird defending one for five weeks, from mid December to mid January.

Of the other cotoneasters, we had most records for *C. horizontalis*, the prostrate species with small leaves and herring-bone arrangement of twigs that is often grown on banks in gardens or trained against house walls. Most of them grow within the winter territory of a Blackbird, and this helps to conserve their fruit crop because although the fruit is ripe by September, and is taken then, many plants retain fruit until late winter. Blackbirds account for 86% of our feeding records.

PYRACANTHAS

We have records for three species of *Pyracantha* (Table 34), the majority being for *P. coccinea*, a native of southern Europe. The other two species, *P. rogersiana* and *P. crenatoserrata*, are native to western China; although described as botanically barely distinct, we found their seasons of ripening very different.

P. coccinea is the orange-fruited species commonly trained against walls. Its fruit is ripe by the end of August and may be available until February, mainly because it is often defended by a territorial Blackbird. *P. rogersiana*, a small erect tree up to about 3 m tall with orange berries, had most of its fruit taken by Starlings in October; we did not record Blackbirds defending it. The only *P. crenatoserrata* that we found, a small shrub of 2 m trained against a garden wall, was successfully defended in the winter by a female Blackbird, who fed at it in January and February. In June of the same year the fruit was still edible and was being collected by a male Blackbird for its young; and an independent young bird, probably one of the same brood, spent much time during a dry spell, at the end of June and in July, in or below the bush, eating the fruit.

For Starlings, we found a marked difference in their eating of pyracantha and cotoneaster fruits. We had many records of their taking the fruits of *P. coccinea* and *P. rogersiana*, comprising 77% of all our records of Starlings feeding on fruits of introduced/ornamental plants. In striking contrast, we had no record of a Starling successfully eating a cotoneaster fruit, although they are very similar to pyracantha fruit (and botanically the two genera are barely distinguishable). The difference seems to be attributable to the somewhat larger size of most cotoneaster fruits. Starlings have narrow gapes, and fruits about 9 mm in diameter are at the upper limit for them to swallow. We recorded a success rate of 80% for Starlings feeding on the fruit of *P. coccinea*, of which small ones are about 7.5 mm in diameter and large ones up to 9 mm or even occasionally 10 mm; most of the failures were due to birds plucking and then dropping the fruit, probably because they were too large to swallow. For the slightly smaller fruit of *P. rogersiana* the recorded success rate was 86%. For *Cotoneaster 'cornubia'*, with fruit averaging about 9 mm in diameter, we had a few records of Starlings plucking a fruit and dropping it, and not a single record of a fruit being swallowed. If this explanation is correct, we would expect Starlings on occasion to feed on small-fruited cotoneasters such as *C. horizontalis*, but we had no record of their doing so.

ORNAMENTAL CRATAEGUS SPECIES

The commonest of these, *C.* × *prunifolia* (a possible hybrid, of unknown origin – Mitchell 1974), has large red haws at about the upper size limit for Blackbirds to swallow whole. Our records of their being eaten, almost entirely by Blackbirds, are nearly all in November; before then they seem to be unavailable, possibly being too firmly attached to be plucked. Song Thrushes are attracted to them but are not very successful at eating them, being able to swallow only the smaller fruits.

The only other exotic *Crataegus* for which we have significant records, *C.* × *lavallei*, is less common in our study area. Its orange, brown-spotted fruits are even larger (diameter about 19 mm), far too large for any of the thrushes to

swallow whole. Blackbirds eat them in late winter, mainly January and February, plucking the fruits whole and flying down with them to the ground, where they eat them piecemeal as they do other fruits that are too large to swallow.

EXOTIC AND ORNAMENTAL PRUNUS SPECIES

Only two exotic species are common in our study area, the cherry laurel *P. laurocerasus* and the Portugal laurel *P. lusitanica*. The former has a well-defined season, the fruits ripening in late August and being eaten mainly in September. Ripening of the fruit of the Portugal laurel is later and more erratic, with good crops in some years and failures in others, probably depending on the warmth of the preceding summer and autumn. They are taken mainly in November.

The ornamental form of the bird cherry *P. padus 'watereri'*, with larger fruits than the wild type, tends to produce a sparse fruit crop which ripens later than the wild type. They are eaten mainly in August. (Other fruiting *Prunus* varieties in our study area are cultivars, grown for human consumption though now often not used; their fruits are eaten by birds mainly in June and July (cultivated cherries), August and September (cultivated plums), or some time in the winter (cultivated damsons or damson × bullace hybrids).

ROSES

Of the many introduced species of rose we made observations only on *Rosa rubrifolia*, a native of the mountains of central and southern Europe. Its fruit ripens by September, much earlier than our native dog rose (*R. canina*) which ripens in December; in consequence, most of the fruit is taken by a seed-predator, the Greenfinch. In $6\frac{1}{2}$ hours of watching at *R. rubrifolia* we accumulated 51 records of Greenfinches taking the seeds, compared with 15 records for dispersers, mostly Blackbirds. Thus the fate of its fruit parallels that of our native field rose (*R.*

arvensis), whose early-ripening fruits are also largely destroyed by seed-predators (p. 44).

MAHONIA AND BERBERIS

Like *Amelanchier*, the fruits of *Mahonia* and *Berberis* are much sought after by birds in summer, especially in dry weather and very largely by juvenile Blackbirds. For *Mahonia* we have records only for *M. aquifolium* (Oregon grape), all in July and August. Most bushes are stripped by the end of August, except for those growing in public places where the fruit may remain throughout the autumn and winter. For *Berberis* we have records only for the Chilean species, *B. darwinii*, all being in July, by the end of which month most plants are stripped.

Although *Mahonia aquifolium* and *Lycium barbarum* (Duke of Argyll's tea-tree) are included here under ornamentals, both are now naturalised in Britain and became so not long after their introduction in the last century. Possibly their July fruiting, when there is often a high demand for fruit, has aided their dispersal.

2: THE FRUIT-EATERS

Fourteen species of 'legitimate' fruit-eaters, *ie* seed-dispersers, regularly eat fruit in our study area: five thrushes (Blackbird, Song Thrush, Mistle Thrush, Fieldfare, Redwing), the Robin, four warblers (Blackcap, Garden Warbler, Common and Lesser Whitethroat), Starling, and three crows (Carrion Crow, Magpie and Jay). We have a small number of records for other species that occasionally eat fruit (Moorhen, Collared Dove, Green and Greater Spotted Woodpecker, Spotted Flycatcher), and for a regular fruit-eater that is a rare passage migrant in our area, the Ring Ouzel. More prolonged observation would probably add a few to the list of occasional fruit-eaters, because there are records in the literature of birds as unlikely as Cuckoo and Black-headed Gull eating fruit; but these occasional fruit-eaters cannot play a role that is of any significance in a wider ecological context. Unfortunately, as there were no Waxwing invasions during the course of our study, we have no records for that interesting and, of all European birds, most specialised frugivore.

Another common species regularly eats fleshy fruits, swallowing them whole like a disperser but playing a mixed role in its interaction with plants – the Woodpigeon. It is one of a large group of pigeons which are mainly seed-predators, with grinding gizzards, and it is well known as a seed-predator both on farmland, where it eats grain, and in woodland where it feeds on acorns and beechmast; but the seeds of several kinds of fleshy fruits pass through its gut intact, and for them it is a disperser.

Four kinds of finches are seed-predators of fleshy fruits in our area, two of them – Bullfinch and Greenfinch – being especially important and the other two – Chaffinch and Brambling – much less so. (A fifth species, the Hawfinch, is rare in our area and we had no records of it.) Three species of tits – Great, Blue and Marsh – regularly exploit fleshy fruits and play a varied role according to the kind of fruit, being either seed-predators or pulp-predators; Willow and Long-tailed Tits also do so less regularly.

In the following chapters we give species by species accounts of these fruit-eaters, based on our observations. Each account is followed by a shorter section dealing with the more interesting or relevant information available for the species from other parts of Britain and Europe. For completeness, a short chapter is included on the Waxwing, based on published information.

Blackbird *Turdus merula*

Blackbirds are versatile fruit-eaters. As they are one of the more abundant resident bird species in our area, and easily the most abundant of the three resident thrushes, we obtained more data for them than for any other bird. The monthly records are summarised in Table 35.

FRUIT-EATING ACCORDING TO SEX, AGE AND SEASON

Blackbirds of either sex and all ages (except small nestlings) eat fruit. In each of the months December–March, males made up 59–63% of our records. This excess of records for males may have been influenced to a slight extent by their greater conspicuousness (though we made a point of recording all birds seen visiting a fruit source) but is probably due mainly to a real excess of males in the winter Blackbird population. Random counts of Blackbirds in Britain in winter tend to show a considerable excess of males (Simms 1978), and this is probably attributable in part to their greater conspicuousness; but unbiased samples too show an excess. For instance, males made up 58% of Blackbirds living in Oxford in the months September–February (Snow 1958), and 59% of a large number of Blackbirds caught at a winter roost in the outskirts of London in the same months (Simms 1978). Hence our male percentages of 59–63% over the winter months suggest that the two sexes feed on fruit to about the same extent. A survey of the diet of the related American Robin, based on analysis of large numbers of stomach contents, showed no difference in the proportion of fruit in the diets of males and females in any month (Wheelwright 1986).

In April and May males made up, respectively, 77% and 78% of our records. There is a slight excess of males in the breeding population, but it is not nearly great enough to account for the great preponderance of male fruit-eating records. At this time females are spending much time on nests, whereas males are doing

TABLE 35 Blackbird: monthly summary of feeding records

	Jan	Feb	Mar	Apr	May	Jun	Jul	Aug	Sep	Oct	Nov	Dec	Totals
Evergreens													
Holly	49	64	61	16	27	76	18			59	72	170	612
Yew	3						1	95	150	135	85	11	480
Ivy	180	137	86	243	203							18	867
Mistletoe		2								1	1		4
Juniper	1	1											2
Rosaceae													
Hip *Rosa canina*	224	200	14								7	158	603
Hip *R. arvensis*											1		1
Haw	64	64	4					8	118	344	695	262	1559
Rowan							56	373	70		6		505
Whitebeam									33	57	123	44	257
Cherry					110	278	6						394
Bird cherry						87	2						89
Sloe	84								54	3	38	89	268
Blackberry							6	121	10	1			138
Raspberry						3	2						5
Crab-apple	21	100	7									3	131
Caprifoliaceae													
Elder							35	107	22	5			169
Guelder rose	3										11		14
Wayfaring tree						3	33	4					40
Perfoliate honeysuckle						36	42						78
Common honeysuckle							19	1					20
Others													
Dogwood							1	2	34	25	10		72
Spindle	12	2								34	46		94
Privet	56	28								13	13		110
Buckthorn	24								29	56	119		228
Currant					1	8							9
Woody nightshade							3	17			3	3	26
White bryony							3	11		1			15
Black bryony	8										3	15	26
Lords and ladies							1	40	1				42
Daphne mezereum							8						8
Stinking iris		1											1
Solomon's seal										2			2
No. *different fruits*	13	10	5	2	2	3	13	15	11	12	16	15	
Total monthly records	729	599	172	259	230	187	505	690	661	697	1167	973	6869

most of the feeding of fledged young, and this seems an adequate explanation of the figures. In June, July and August the male percentages fell to 61%, 54% and 44% respectively (August being the only month with more female than male records). The fall in the male percentage may be attributed to the fact that in the latter part of the breeding season, when females are not re-nesting, they take a larger share than males in feeding fledged young (Snow 1958, Edwards 1985), and continue to do so in August, after breeding has finished.

Newly independent young Blackbirds feed a great deal on fruit, especially in dry weather when other kinds of food are less easily available. In July and August, 50% and 56% respectively of all our Blackbird fruit-eating records refer to juveniles, the percentage falling to 26% in September, by which time many young birds are in first-winter plumage.

METHOD OF TAKING FRUIT, AND RANGE OF FRUIT TAKEN

Blackbirds can swallow whole fruits up to about 13 mm in diameter, which include the great majority of wild fleshy fruits in southern England and most ornamental fruits (Fig. 7). Their gape width, defined as the maximum diameter

TABLE 36 *Blackbird: incidence of aerial sallying to pluck fruit*

	No. feeds	% with sallies
Yew	472	13(± 3)
Ivy, Dec–Mar	421	20(± 4)
Apr–May	446	30(± 4)
Hip	602	6(± 2)
Haw	1557	6(± 1)
Rowan	503	1(± 1)
Whitebeam	257	7(± 3)
Cherry	393	6(± 2)
Sloe	267	5(± 3)
Elder	169	1(± 2)
Dogwood	72	3(± 4)
Spindle	91	3(± 4)

Note. Figures in parentheses are 95% confidence limits. The incidence of sallying to take ivy fruits is significantly higher in April–May than in December–March, probably because the more easily accessible fruits have been taken.

TABLE 37 *Blackbird: success rate according to method of taking fruit*

	Sally		Perched	
	No. attempts	% success	No. attempts	% success
Holly	32	88(± 11)	547	76(± 4)
Yew	52	90(± 8)	245	89(± 4)
Ivy	240	97(± 2)	905	83(± 2)
Hip	25	80(± 16)	227	54(± 6)
Haw	49	98(± 4)	524	86(± 3)
Rowan	7	[100]	517	82(± 3)
Cherry	10	[70]	434	41(± 5)
Sloe	8	[100]	175	54(± 7)
Elder	9	[100]	274	93(± 3)
All native plants	509	94	4412	82

Note. Figures in parentheses are 95% confidence limits. Differences in success rates between sally and perched feeding are significant when there is no overlap between them, taking into account confidence limits, *ie* ivy, hip and haw. Success rates are square-bracketed and confidence limits are not calculated if the number of observations is 10 or less. Confidence limits are not calculated for all plants combined (bottom line) as the samples are very heterogeneous and far from random.

of a round or ovoid object that can pass between the rami of the lower jaw (see p. 189), is a little over 11 mm, which indicates that the rami are bowed outwards to some extent when very large fruits are swallowed. A fruit that is at the upper size limit may be held in the bill for some time while the bird makes repeated attempts to swallow it; soft fruits may thus be squeezed to a more manageable size. In the case of even larger fruits, if the Blackbird can pluck them it usually flies down to the ground with them and there pecks out pieces of pulp (*eg* some ornamental *Crataegus* spp. with very large fruit); while the largest fruits of all – of wild fruits, only crab-apples – can only be dealt with when they have fallen to the ground.

Fruits are usually taken from a perched position, less often with short flight sallies (Table 36). Blackbirds are adept at plucking a terminal fruit by seizing it in passing flight, and at making upward sallies, often from the ground, to take otherwise inaccessible fruit. Probably because more force can be exerted on the fruit in flight, the success rate is generally higher than when fruit is taken from a perch (Table 37), but flight sallies must be energetically much more demanding than perched feeding. They are used only when fruit is difficult to reach from a perch, except that very often the last fruit of a feed is seized on the wing as the bird flies off.

We have records of Blackbirds taking 33 different kinds of wild fruits (Table 35) and 23 introduced/ornamental fruits, a greater number than for any other bird. The numbers of records for the different fruits are influenced by several factors: their abundance in our study area; the amount of time we spent watching each kind of fruit; the birds' choice between different fruits, when more than one kind was available; and various factors affecting the birds' access to different kinds of fruit. We can quantify some of these factors but not others. In any case the study area is very diverse, and fruits that are important in some parts of it are absent in other parts. One may say that the numbers of records for different fruits roughly reflect their relative importance for Blackbirds in our study area, but they should not be taken as unbiassed quantitative measurements.

DOMINANCE RELATIONSHIPS

There is one factor of particular importance affecting the fruits taken by Blackbirds as well as by other thrush species: Blackbirds are in the middle of the size range of the five thrushes that regularly occur in our area (Table 76). Dominance is related to size, and so Blackbirds are subordinate to Mistle Thrushes and Fieldfares but dominant to Song Thrushes and Redwings. Table 38 summarises the results of aggressive interactions at fruit when Blackbirds were involved with Mistle Thrushes, Fieldfares and conspecifics.

All the large mistletoe clumps in our area, and many hollies, are protected by Mistle Thrushes through the winter (see pp. 127–131). This is the reason for the very small number of records of Blackbirds taking mistletoe berries, and it undoubtedly reduces the number of records for holly far below what it would be if Mistle Thrushes were absent. Usually, as shown in Table 38, a defending Mistle Thrush prevents an intruding Blackbird from feeding; the cases where the Blackbird managed to take some fruit before being chased off were often attributable to special circumstances. The defence of fruit sources by Fieldfares (p. 135) is not common enough to have any general effect on Blackbird feeding records; but it may have a local effect, for example in Tring Park during very cold weather in January 1982, where a Fieldfare successfully defended one of the few remaining clumps of hips.

Some adult territory-holding Blackbirds, both males and females, defend fruit

TABLE 38 *Records of Blackbirds being chased from fruit*

Attackers	No. records	Result (%)	
		Blackbird prevented from feeding	Blackbird took some fruit before being chased
Mistle Thrush	191	69	31
Fieldfare	37	85	15
Blackbird	268	46	54

Note. The majority (72%) of the cases where the Blackbird took some fruit before being chased by a defending Mistle Thrush were when the Mistle Thrush was in the process of being overwhelmed by a flock of intruding fruit-eaters, or when the Mistle Thrush was defending yew or ivy, fruits which they have difficulty in defending successfully (p. 128).

Most of the cases (73%) of Blackbirds chased by Fieldfares were from easily defendable clumps of hips (p. 135).

sources within their territories during the winter. We have recorded this mainly in gardens, where Blackbird territories often contain a compact, defendable fruiting plant; but in one case a female Blackbird, in an area of farmland with hedges, temporarily defended a holly after it had for some reason been abandoned by the defending Mistle Thrush. When a *Pyracantha coccinea* growing against our house fruited prolifically in 1980, the resident male Blackbird began to defend it soon after mid November and continued to do so until February 1981. His mate was first allowed to feed at it on 13 February. In another case, a female Blackbird defended a *Pyracantha crenatoserrata* shrub in a Tring garden through January 1981 and until mid February; it was not seen to be fed at again until June when, thanks to the Blackbird's earlier defence, it still had fruit which her mate took to their young.

Apart from preventing conspecifics from feeding on the fruit, the defence of a fruit source by Blackbirds has an important effect on Song Thrushes, which may be almost completely prevented from feeding on the best fruit sources in their immediate area, as discussed later.

MEAL SIZES, INTERVALS BETWEEN MEALS, AND PROCESSING OF FRUIT

A Blackbird that is taking fruit as a main part of its diet makes regular visits to a fruit source, has a 'meal' of fruit, and then leaves for a period before returning to feed again. We define 'meal size' as the number of fruits taken at a single visit. During the interval between feeds the fruits are broken up in the gizzard, the pulp is passed down the digestive tract, and the seeds are regurgitated or passed down the gut with the pulp. If seeds are regurgitated, the last act before the next feed is frequently the regurgitation of any remaining seed from the previous feed. Often, after returning to the fruit source, a Blackbird waits a few seconds, regurgitates a seed (thereby failing to disperse it), and then immediately begins to feed.

Meal sizes vary considerably, as would be expected (Table 39). Presumably, a bird sometimes has food still in the stomach; sometimes a bird will leave a food source, for reasons that may not be apparent, before it has had a full feed (if a bird was driven or frightened away while feeding, we have not included it in

TABLE 39 *Blackbird: meal sizes for different fruits*

	Mean wt. of fruit (g)	No.	Meal sizes Range	Meal sizes Mean	Wt. of largest meal (g)
Holly, winter	0.44	17	5–15	9.1	6.6
spring/summer	0.44	32	1–12	4.5	
Yew	0.70	19	5–15	8.4	10.5
Ivy	0.33	80	2–19	8.3	6.3
Hip, to mid-Jan	1.8	40	1–4	1.6	7.2
after mid-Jan	1.8	23	1–4	2.6	
Haw	0.63	67	3–13	7.1	8.2
Rowan	0.49	32	3–16	8.7	7.8
Cherry	1.3	22	1–5	3.0	6.5
Sloe	1.9	30	1–7	2.7	13.3
Buckthorn	0.36	21	9–23	14.5	8.3

Notes. Meal sizes of 6–10 g represent about 6–10% of Blackbird's body weight. As sloes are rather variable in size, the figure in the last column may be unduly high; other fruits with variable sizes (*eg* privet, lords and ladies) have been omitted. The differences between meal sizes for holly and hip at different times of year are statistically significant (holly, median significantly greater in winter than in spring/summer; $U=76$, $P<0.05$, Mann-Whitney U-test): hip sample before mid-January highly skewed, proportions of meals of a single fruit significantly higher before than after mid-January (0.63 and 0.09 respectively).

our 'meal size' records). The calculations given in the last column of Table 39 indicate that the largest meal a Blackbird can take is 6–10 g, representing some 6–10% of the bird's weight. For holly and hip we had evidence that meal size may show seasonal variation. Holly meals averaged twice as large in winter as in spring and early summer, presumably because winter birds were more dependent on fruit for their energy needs. Hip meals averaged larger after mid January than before mid January, evidently because the Blackbirds were more dependent on hips in late winter when few other fruits were available, especially in cold spells when most observations of hip-eating were made (see p. 42).

We have records of 75 intervals between successive meals (omitting cases where a Blackbird was dominant over a fruit source and stayed at or near it between feeds). Intervals were 9–28 minutes, with a mean of 17.9 minutes; 81% of the total were 12–23 minutes.

Whether a seed is regurgitated or defaecated depends mainly on its size: large seeds are more often regurgitated, smaller ones mostly defaecated. It seems also to depend on how easily the pulp is detached from the seed. We have records for Blackbirds regurgitating the seeds of ten native fruits; the most commonly regurgitated (number of observations in relation to total of feeding records) were: lords and ladies (10%), wild cherry (1.8%), yew (1.5%), and ivy (1.3%). All these are large or moderately large seeds and are easily separated from the pulp. For haw and holly, with pulp more firmly attached to the seed, we observed regurgitation relatively less often (0.3% each). The advantage of regurgitation over defaecation as a method of getting rid of seeds is discussed on page 217.

The mean values for meal sizes and intervals between meals, when combined with the nutritive contents of the fruits involved, allow rough calculations to be made of the energy intake of Blackbirds, and other species, derived from fruit-eating. This is discussed on pages 236–7.

FEEDING OF FRUIT TO YOUNG

Blackbirds regularly feed their young on fruit, especially ivy berries in April and May and cherries in June and July. From April to August we often saw them collecting and carrying away fruit for their young, and recorded 216 visits to 14 different plants (Table 40). In many cases fledged young were near by; in a few cases the parent birds were seen to deliver the fruit to a nest. There were many instances of Blackbirds coming from some distance and we could not tell whether the fruit was destined for nestlings or fledged young. A male Blackbird flying about 160 m for holly fruit, and another male going 300 m for one cherry, are examples of the distances that they may travel for this purpose.

Parent Blackbirds normally carry fruit to their young in the beak, as do other thrushes, so that the number of fruits taken per visit to the fruit source is small. They can carry up to four ivy berries (mean of 40 records, 2.0), 3 rowan fruits (10 records, mean 1.8), 3 holly berries (13 records, mean 1.9), 3 Amelanchier fruits (11 records, mean 1.8), and 2 wild cherries (29 records, mean 1.3). It is noteworthy that the wayfaring tree was the only wild species for which we recorded more than four fruits being carried, and up to 8 may be carried (3 records, mean 6.3). Wayfaring tree fruits are oval and flattened, conformably with the single flat seed, and this obviously makes them easier to pack between the mandibles. We twice saw a male Blackbird carrying ivy berries concealed in its mouth, in addition to those visible in the beak. That this method is not used more frequently is surprising, because when holding one fruit in the beak Blackbirds (and Song Thrushes) have difficulty in plucking further fruits, often making many attempts before succeeding at a cost sometimes of dropping the one fruit that they had.

TABLE 40   *Records of Blackbirds feeding fruit to young*

	Apr	May	Jun	Jul	Aug	Totals
Holly	1	5	23	4		33
Ivy	24	37				61
Yew					2	2
Haw					2	2
Rowan				9	21	30
Cherry			12	29		41
Wayfaring					3	3
Perfoliate honeysuckle				3	2	5
Common honeysuckle					1	1
Introduced/cultivated						
Amelanchier			6	21		27
Strawberry				1		1
Pyracantha crenatoserrata			7			7
Cotoneaster bullatus					2	2
Daphne mezereum				1		1
All species	25	42	48	68	33	216

Male and female Blackbirds each take fruit to their young, as would be expected, but not equally throughout the breeding season. For young born in the first part of the season (fed in April–June) 67% of our records are of the male parent providing fruit; for young born in the second part of the season (fed in July–August) only 45% are records of males. This difference, which is statistically significant ($\chi^2 = 5.40$, 1 d.f.), is to be expected since, as mentioned earlier, males take a greater part in caring for the fledged young of early broods but the last broods of the season are tended more by the female than by the male.

BLACKBIRDS ELSEWHERE

There are more published records of Blackbirds eating fruit than there are for any other British bird. Many records are more or less casual; some refer to unusual fruits and are cited in the sections dealing with different fruits. Most of them simply show that Blackbirds eat a wide variety of wild and cultivated fruits confirming the generalisation made earlier that fruit-eaters eat any kind of fruit that they are physically able to obtain and eat. We have not found any information from other parts of Britain based on a thorough, year-round study of fruits in the Blackbird's diet.

More complete data are available from the extreme southwest of Europe. Herrera (1981c) studied the fruit part of the diet of Blackbirds in two areas of natural sclerophyll scrub in southern Spain, one montane (1,150 m altitude) and the other lowland. In the lowland area, where ripe fruits were available all year round, Blackbirds were frugivorous at all times except perhaps in the breeding season. In the highland area, ripe fruits were available only from September to January, so that the Blackbird's fruit-eating was confined to these months. Altogether 14 different kinds of fruit were eaten, all except two (blackberry and black nightshade) being primarily southern species not found in our study area. In Herrera's study areas the Blackbird is the only thrush that occurs in large numbers.

In winter in some Mediterranean regions Blackbirds are considered to be pests because they feed on ripe, cultivated olives. Thus, in the Dubrovnik area of Yugoslavia, Tutman (1969) reported that Blackbirds occur in the macchia (where presumably they feed on the many kinds of wild fruits), but from late January or early February they enter the olive groves in large numbers and feed on fallen olives, continuing to do so until the end of March. (It may be assumed that in addition to eating fallen olives they also take them from the trees.) Olive-eating is also reported in winter from Italy and Spain.

The Fruit-eaters

Song Thrush *Turdus philomelos*

The Song Thrush is a common resident in our study area. It is known also to be a partial migrant in southern England and part of our breeding population presumably leaves in the autumn; and there are apparently, at least at times, influxes of birds from elsewhere in winter, probably from farther north in Britain or, less likely, from the Low Countries (Goodacre 1960). Thus the number of birds present must fluctuate in the course of the year in addition to the changes brought about by recruitment to, and losses from, the breeding population. The number of our fruit-feeding records for Song Thrushes is about 30% of that for Blackbirds, and this probably gives a fair measure of the overall relative abundance of the two species. In the following account we concentrate on a comparison of the fruit-eating of the Song Thrush with that of the Blackbird as described in the previous section.

Song Thrushes are considerably smaller than Blackbirds and have smaller bills, and their gape width is a little less (10.5 mm, *cf* 11.1 mm for Blackbird). Hence they would be expected to be less well able to grasp large fruits or to pluck a fruit that is firmly attached; nor are they able to swallow fruits as large as Blackbirds can. But they appear to compensate to some extent for their smaller size and strength by the vigorous twisting movement with which they try to pluck firmly attached fruits, a movement that looks similar to that used when they are dealing with a snail. These physical differences affecting the ability to handle fruit seem to account for some of the differences in fruit-eating between the two species.

Another consequence of their smaller size in that Song Thrushes are subordinate to Blackbirds and also to Mistle Thrushes and Fieldfares (but not to Redwings, which are a little smaller than Song Thrushes). Mistle Thrushes protect fruit sources throughout the winter, and territorial Blackbirds are often aggressive towards other birds trying to exploit fruit sources in their territory. Song Thrushes, too, are highly territorial in the breeding season and some individuals, probably old males (Davies & Snow 1965), remain so throughout the winter. Although not specifically defending fruit sources these territorial Song Thrushes drive off intruding conspecifics that come to feed on fruit within their territory, often very vigorously and persistently. Thus Song Thrushes are frequently chased when they attempt to feed on fruit. We have 189 records of Song Thrushes being chased at such times (84 by Blackbirds, 84 by Mistle Thrushes, 21 by other Song Thrushes), representing nearly 10% of all feeding records for Song Thrushes. In 41% of the instances the Song Thrush failed to feed before being chased off.

As a result of their liability to be attacked, Song Thrushes often approach a fruit source inconspicuously. When feeding on small trees and shrubs they more often take fruit from the lower, more concealed parts of the plant than do Blackbirds. They tend to feed solitarily more often than Blackbirds, perhaps simply as a consequence of their lower population density. When several are

TABLE 41 *Song Thrush: monthly summary of feeding records*

	Jan	Feb	Mar	Apr	May	Jun	Jul	Aug	Sep	Oct	Nov	Dec	Totals
Evergreens													
Holly	9		1	2	1	9	2			5	5	23	57
Yew	43						6	55	120	109	124	53	510
Ivy	46	76	36	138	78	1						4	379
Mistletoe	1		1										2
Rosaceae													
Hip	4	2										1	7
Haw	6	4							4	36	48	15	113
Rowan								20	2				22
Whitebeam									16	16	4		36
Cherry						8	40						48
Bird cherry							21						21
Sloe	76									2	130	102	310
Blackberry									1	4			5
Crab apple	2	11											13
Caprifoliaceae													
Elder								31	123	28	5		187
Guelder rose	76										19		95
Wayfaring tree									6				6
Perfoliate honeysuckle							5	6					11
Common honeysuckle								2					2
Others													
Dogwood			1						4	9	29	16	59
Spindle	6										4	9	19
Privet	10										1	1	12
Buckthorn	2									14	3	18	37
Currant							2						2
Woody nightshade							4	15				1	20
White bryony							4					1	5
Black bryony	4											1	5
Stinking iris		1											1
No. different fruits	13	6	3	2	2	3	8	7	7	9	10	14	
Total monthly records	285	95	38	140	79	18	84	135	270	223	353	264	1984

exploiting a much favoured fruit source, however, it is not uncommon to see up to six birds synchronising their visits and feeding in company.

In spite of these limitations to their fruit-eating, Song Thrushes feed on a great variety of wild fruits; we have records for 27 species (Table 41). In comparison

TABLE 42 *Song Thrush: incidence of aerial sallying to pluck fruit*

	No. feeds	% feeds with sallies
Holly	57	7($\pm$7)
Yew	510	27($\pm$4)
Ivy	379	31($\pm$5)
Haw	113	4($\pm$4)
Sloe	310	1($\pm$1)
Elder	187	2($\pm$2)
Dogwood	59	5($\pm$6)

Note. Figures in parentheses are 95% confidence limits.

TABLE 43 *Song Thrush: success rates according to method of taking fruit*

	Sally		Perched	
	No.	% success	No.	% success
Holly	2	[100]	30	80($\pm$15)
Yew	131	98 ($\pm$2)	245	88($\pm$4)
Ivy	128	92 ($\pm$5)	379	77($\pm$4)
Haw	5	[100]	92	73($\pm$9)
Sloe	2	[100]	134	37($\pm$8)
Elder	4	[100]	284	97($\pm$2)
Guelder rose	–	–	66	89($\pm$8)
All native fruits	280	95	1436	77

Note. Figures in parentheses are 95% confidence limits. See footnote to Table 37 for significance of differences.

TABLE 44 *Song Thrush: meal sizes for different fruits*

	Mean wt. of fruit (g)	Meal sizes			Wt. of largest meal (g)
		No.	Range	Mean	
Yew	.70	14	4–14	8.6	9.8
Ivy	.33	28	6–14	8.8	4.6
Sloe	1.9	16	1–4	2.5	7.6
Elder	.14	6	12–62	25.5	8.7
Guelder rose	.38	6	6–13	8.8	4.9

Note. Meal sizes of 7.6–9.8 g represent about 9–12% of the Song Thrush's body weight. Mean meal sizes are 4–7% of body weight.

with Blackbirds, which take a similar range of fruits, they show clearer tendencies to prefer some and avoid others (see also p. 221). The most evident preferences are for yew, sloe, elder and guelder rose, which make up respectively 26%, 16%, 9% and 5% of our Song Thrush records, compared with 7%, 4%, 2% and 0.2% of Blackbird records. Some preference for ivy is indicated by the figures of 19% for Song Thrush compared with 13% for Blackbird. The clearest cases of relative avoidance of fruits by Song Thrushes are hip, rowan and haw, which make up 0.4%, 1% and 6% of Song Thrush records, compared with 9%, 8% and 23% of Blackbird records. All these differences are highly significant. In the case of hips, the firmness of attachment and size of the fruit make it difficult for Song Thrushes to pluck and swallow them; the observed success rate was 19%, compared with 54% for Blackbird. But rowan berries and nearly all haws are within the size range suitable for Song Thrushes. The fruits that they prefer either have pulp that is separated easily from the seed(s) (yew, ivy), or a soft pulp (sloe), or both (elder). It may be that haws, with rather dense pulp firmly attached to the seed, are harder for Song Thrushes to process internally; but this does not account for their relative avoidance of rowan fruit, which remains puzzling. It will be seen from Tables 37 and 43 that, although Song Thrushes show a preference for sloes, they are less efficient at plucking them than are Blackbirds (success rates of 37% and 54% respectively, for sloes taken from a perched position; a significant difference).

Song Thrushes take most of their fruit from a perched position; but will regularly sally to pluck ivy and yew berries (Table 42), especially when the crop has been depleted and most of the remaining fruits are terminal or, in the case of yew, hanging beneath the branches. Thus when stripping the last fruits from a yew they may do so almost entirely by making short upward sallies from a lower branch to take the sparse fruit hanging from the branch above.

Meal sizes for some of the more important fruits are given in Table 44. The maximum meal sizes recorded represent about 9–12% of the Song Thrush's body weight, compared with 6–10% for Blackbirds. We have rather few records for intervals between successive feeds; they suggest on average shorter intervals than for Blackbirds, ie 9–19 minutes (mean 15.3) for 10 instances of birds feeding on sloe, 13–15 minutes (mean 14.5) for four instances of birds feeding on yew. The records are too few to show statistical significance.

Song Thrushes regularly take fruit to their young, either in the nest or, more often, after they have fledged. We have 23 records in April and May, all for ivy berries (the main fruit available in those months), and 16 records in June–August for a variety of fruits: holly, rowan, woody nightshade, wild cherry, bird cherry and the introduced snowy mespil *Amelanchier laevis*, whose early-ripening fruit is used by all three resident thrushes for feeding to their young. Two was the maximum number of any fruit seen carried to their young by parent Song Thrushes. This includes ivy, rowan, woody nightshade and bird cherry; and there was a nearly equal number of instances of their taking single fruits of these species, as well as single fruits of the much larger wild cherry. Blackbirds can carry considerably larger numbers of fruits to their young (p. 117), thanks probably to their much greater bill-length.

SONG THRUSHES ELSEWHERE

From a survey of wild fruits in the diets of British thrushes, Hartley (1954) gave records of Song Thrushes taking only seven kinds of fruits, of which yew figures most prominently with 76% of all records, followed by elder, haw, ivy and holly. His records were nearly all from Oxfordshire, Hampshire and South Wales, and included one for whortleberry from South Wales, a fruit not present in our study area. In her study of fruit-eating in woodland near Oxford, Sorensen (1981) had records of Song Thrushes eating only four different kinds of fruit: elder, haw, ivy and blackberry. Clearly these studies failed to give a complete picture of the range of fruits taken by Song Thrushes. For the reasons mentioned above, Song Thrushes tend to be inconspicuous when feeding on fruit, so that it is easy to fail to get records of fruits that are taken in small quantities. We have found no data from elsewhere in Britain on the seasonal pattern of fruit-eating that add significantly to our findings. Radford (1967) made an interesting observation on a juvenile Song Thrush's method of eating cherries. The bird held the cherry by its stalk, beat it on the ground and then ate bits of the pulp, thus treating the fruit in the way Song Thrushes treat snails.

For continental Europe, all the data available up to 1939 were summarised by Siivonen (1939), and nothing very significant seems to have been added since then, except for the data from southern France mentioned below. Siivonen's own records, which are of most interest, were from woodland in southern Finland, where Song Thrushes are summer migrants. There, in spring and early summer, they eat cowberries from the previous year's crop that have been preserved through the winter under the snow (see also p. 137). Later in summer they take a variety of fruits as they ripen: bilberry, crowberry, red currant, mountain currant and bird cherry. In autumn, before leaving, they eat rowan, juniper and cowberry (the fruit of the new crop).

Siivonen lists a large number of fruits recorded as eaten by Song Thrushes in central Europe, including a number of species not found in our study area. For southern Europe he lists six fruits, all except ivy absent from our area. One that he does not mention is madder; we saw a Song Thrush eating this fruit in southeastern France in mid October, in dry scrubby woodland. At this time we also saw parties of migrant Song Thrushes, noticeably larger and rather greyer-looking than British birds and so certainly of the northern European breeding population (the nominate race), feeding on the fruit of dogwood and haw in the limestone mountains of the Vercors. Other fruits recorded as eaten in southern Europe include *Olea europea*, *Pistacia lentiscus*, *Phytolacca decandra*, *Myrtus communis* and *Arbutus unedo* (Siivonen 1939, Tutman 1969, Pérez-Chiscano 1983). This list has recently been much amplified by a thorough study by Debussche & Isenmann (1985b) of the autumn and winter diet of Song Thrushes in 'garrigue' (scrubland) around Montpellier in southern France. They found that the diet of these wintering Song Thrushes consisted predominantly of fruit, especially grapes in autumn and juniper (three species) in winter. Fleshy fruits of 16 other plants were eaten, and Debussche & Isenmann conclude that Song Thrushes play a major role in the dissemination of fleshy-fruited plants in their area.

Mistle Thrush *Turdus viscivorus*

Mistle Thrushes are common residents in our study area, but not ubiquitous like Blackbirds and Song Thrushes. They were, it seems, originally birds of open, park-like woodland, and in southern England today gardens and parks provide the nearest approximation to such habitat. A high proportion of our records of Mistle Thrushes eating fruit are from village and suburban gardens; others are from parkland and farmland with hedges, and only a few from woodland.

We have records of Mistle Thrushes taking 18 different kinds of native fruits (Table 45). Four evergreens – holly, yew, ivy and mistletoe – account for 69% of the total, and eight species of Rosaceae make up 27%. Six species make up the remaining 4%, and of these only buckthorn can be considered at all important. We also have records for a few introduced and ornamental plants (*Amelanchier, Cotoneaster, Mahonia* and *Berberis*).

The Mistle Thrush is the largest of our thrushes, and it is presumably a consequence of its size and strength that it is more successful at plucking fruits than the other thrushes. Thus we recorded a success rate of 68% for Mistle Thrushes plucking sloes from a perched position (Table 46), compared with 54% for Blackbird and 37% for Song Thrush; and 90% when plucking holly berries from a perched position, compared with 59–80% for Blackbird, Song Thrush and Redwing. In spite of its size, when necessary it readily takes fruit by aerial sallying, a method that it uses increasingly as a fruit crop is depleted and the remaining fruits are those on the more inaccessible terminal twigs. We have seen Mistle Thrushes fly up 90–120 cm (3–4 feet) vertically from the ground to take otherwise inaccessible fruit from the lower branches of holly and yew. The seasonal increase in the percentage of ivy and mistletoe fruits taken with flight sallies is shown in Table 47.

126 The Fruit-eaters

TABLE 45   Mistle Thrush: monthly summary of feeding records

	Jan	Feb	Mar	Apr	May	Jun	Jul	Aug	Sep	Oct	Nov	Dec	Totals
Evergreens													
Holly	135	63	39	23	21	5	2		1	8	14	59	370
Yew	10	1						2	26	68	44	54	28	233
Ivy	21	26	25	26	21								119
Mistletoe	57	70	18	16	6	6				2	30	46	251
Juniper	5	3											8
Rosaceae													
Hip	6	5										1	12
Haw	28	47								15	33	24	147
Rowan							5	21	4	2			32
Whitebeam									41	9	14	5	69
Cherry					8	32							40
Bird Cherry						28							28
Sloe	6									13	14	20	53
Crab-apple	1	12											13
Others													
Elder							1	3					4
Guelder rose	1	2										1	4
Dogwood										1			1
Buckthorn	1									11	3	26	41
Privet	3	2											5
No. different fruits	12	10	3	3	3	3	5	3	5	9	7	9	18
Total monthly records	286	242	82	65	49	19	69	48	117	105	161	207	1450

TABLE 46   Mistle Thrush: success rates according to method of taking fruit

	Sallies		Perched	
	No.	% success	No.	% success
Holly	8	[88]	443	90(±3)
Yew	6	[100]	125	94(±4)
Ivy (not juv)	73	100	168	91(±4)
Mistletoe	56	91(±8)	262	93(±3)
Whitebeam	–	–	37	70(±15)
Wild cherry (not juv)	–	–	37	35(±16)
Haw	9	[100]	178	92(±4)
Sloe	–	–	34	68(±16)
All native plants				
adults	154	96(±4)	1578	90(±1)
juveniles	8	[50]	66	32(±11)

Notes. The figures for ivy and wild cherry are for adults only (*ie* juveniles excluded); for the other fruits in the table there were no records for juveniles. Figures in parentheses are 95% confident limits. The only significant difference in success between sally and perched feeding is for ivy; adults are significantly more successful at perched feeding than juveniles (see footnote to Table 47).

TABLE 47 *Mistle Thrush: incidence of aerial sallying to pluck fruit*

	No. feeds	% feeds with sallies
Mistletoe		
Oct–Dec	74	1(±2)
Jan–Mar	168	14(±5)
Apr–Jun	29	31(±17)
Ivy		
Jan–Mar	72	18(±9)
Apr–May	47	43(±14)
Holly	350	3(±2)
Yew	226	18(±5)
Haw	140	13(±6)

Notes. Figures in parentheses are 95% confidence limits. Sallying increases towards end of season, for mistletoe and ivy, as the fruit crop is depleted and terminal fruits are left (see text). Total feeds for holly and haw are slightly lower than shown in Table 45 due to exclusion of a small number of records of birds picking fallen fruit off ground.

Another consequence of its large size is that the Mistle Thrush is dominant over the other thrushes, and almost certainly this has been the main factor determining its most important behaviour pattern related to fruit-eating: long-term defence of a winter fruit supply. Not only is this behaviour one of the keys to understanding the Mistle Thrush's ecology, it also affects the ecology of several other species. We became aware of the behaviour in the course of the 1980–81 winter, and for the next three winters made a point of watching Mistle Thrushes at their defended fruit trees (Snow & Snow 1985). If we had been more familiar with early ornithological literature, we might have known that this characteristic had been noted 200 years ago, and had given the Mistle Thrush one of its local names (p. 132).

For a period of about three months after breeding has finished Mistle Thrushes become highly nomadic. They move about in parties from half a dozen up to 30 or so strong, presumably consisting of one or several family parties, foraging both on open ground and on fruit trees. It is at this period that we have records of them taking the fruits of yew, rowan, elder, whitebeam and bird cherry, and also of a concentration of up to 20 birds feeding on the strawberries of a local fruit farm. In October the nomadic parties break up and a change takes place to a settled existence of pairs or individual birds centred on a source of fruit which, if all goes well, will be defended through the coming winter. Since, with some exceptions, the same fruit trees are defended each winter, it is probable that there is continuity of defence from year to year by established adults, for as long as they survive, but as our birds were unringed we are not sure of this. Ringing recoveries indicate that well over half of young Mistle Thrushes fledged in southern England migrate in their first autumn, most going to France (Snow 1969). The period when the system of defended fruit sources is set up, in October and early November, is marked by persistent chasing between Mistle Thrushes in the neighbourhood of fruit trees that are suitable for defence, until the winter pattern of defended trees is established. Occasional song is heard at this time.

Hollies are by far the most important of the plants defended by Mistle Thrushes in our area. Hawthorns are next in importance. Mistletoes are regularly defended, but they are very local in our study area. Yews are sometimes defended, but in most of the cases that we observed defence was rather short-lived, in early winter. Ivy is regularly defended, but again only for a short period: we had three cases where Mistle Thrushes gave up defence of a holly in early spring and switched to an ivy near by when its fruits ripened. We also had a few cases of buckthorn being defended, and one of juniper – a single bush, on a scrub-grown slope of the Chilterns, which was guarded by a single Mistle Thrush in two successive years. The numbers of defended plants at which we made observations were: holly 30, hawthorn 7, hawthorn with buckthorn 4, mistletoe 5, yew 3, ivy 3, juniper 1. These figures may be biased somewhat against holly, since we paid more attention to the less usual kinds of fruit once it had been found that they, too, might be defended.

The number of Mistle Thrush-defended plants we observed increased during the course of the study. This was mainly due to a striking fact that became apparent in the course of the 1980–81 winter, that the presence of fruit on a holly or hawthorn after a certain date was an almost infallible sign that it was being defended by Mistle Thrushes. Once this was appreciated, we had only to notice a patch of fruit remaining in an otherwise stripped area to find another Mistle Thrush food territory. Casual observation at the trees themselves does not necessarily reveal the situation since – as we shall see – a Mistle Thrush may feed at its defended tree only briefly and at long intervals, while other birds living in the immediate area know that it is protected and tend to avoid it. Thus one may wrongly suppose that for some reason the fruit is unattractive to birds.

To be suitable for defence a holly tree should be of moderate size, standing free of other trees but preferably with other perches (trees, wires, etc) near by from which a look-out can be kept. Sparse foliage is more suitable than thick, bushy growth, since if the tree is too large and thick a Mistle Thrush may be unable to see intruders coming from all directions, or see them feeding if they manage to fly in unperceived, or deal with several intruders coming in at the same time from different sides. Many of the protected hollies in man-made habitats were adjacent to areas of grassland, suitable for foraging, from which the defended tree was in view. (See p. 25 for further discussion of the characteristics of protected hollies). Hawthorns with similar characteristics of size and location are also suitable for defence, but one of the defended hawthorns, and the three growing with buckthorn, were parts of overgrown hedges with adjacent trees and bushes. The three defended yews all had slender terminal spikes well laden with berries; one had spreading lower foliage which was not defended. This was the only case of a Mistle Thrush not attempting to defend the whole of a tree. The defended mistletoes consisted of large clumps of several closely grouped plants, and were the only such clumps that we found in the study area. Single plants or small clumps are apparently not defended.

When fruit territories are first established, in autumn, each defended tree may be guarded by a single Mistle Thrush or, more often, by a pair. We had three cases where two Mistle Thrushes began by defending separate trees quite near one another (60–180 m) and later, when one of the trees was stripped as a result of onslaughts by Fieldfares and Redwings in very severe weather, the two (presumably a pair) jointly defended the tree that still had fruit. We had a few other cases of single Mistle Thrushes defending a fruit tree for limited periods. Nearly all trees successfully defended throughout the winter were jointly defended by a pair of Mistle Thrushes.

Mistle Thrushes in possession of a fruit supply spend most of the day in its immediate vicinity throughout the middle part of the winter. Where we had an unobstructed view of the surroundings, most of our watches in November–February showed that the birds were near the tree for the whole of the period of observation. They spend a considerable time perched in the fruit-tree, often preening, and longer periods in trees or at other vantage points near by from which they keep a look-out for intruders. Any thrush approaching the tree is immediately attacked and driven away. Even if the Mistle Thrush is at the periphery of its territory, perhaps 100 m away from the fruit tree, it is usually aware of birds approaching the fruit and will fly rapidly back and drive them off if necessary.

Behaviour towards other species depends mainly on whether they are potential food competitors. Other thrushes are invariably attacked, and Robins are usually driven off. Blackcaps, however, though sometimes attacked, can avoid a Mistle Thrush by slipping among the foliage, and will often feed effectively in a defended holly. Starlings are normally ignored by Mistle Thrushes guarding hollies and hawthorns because they do not (or in the case of hawthorn, seldom) eat the fruits, but Starlings are attacked by Mistle Thrushes defending ivy, one of the main fruits eaten by Starlings. Woodpigeons are attacked vigorously if they try to feed at a defended holly, but not always successfully. Bullfinches, which eat the seeds of buckthorn fruit, are recognised as competitors by Mistle Thrushes defending buckthorn and driven off. Other species are usually tolerated. Greenfinches and Great Tits are chased from defended yews, but not from holly. The main competitors, thrushes of other species, are often chased and chivvied when feeding on the ground near the guarded tree even if they are not trying to get at the fruit. Thus a zone round the guarded tree is kept, for most of the time, clear of other thrushes.

Established Mistle Thrushes usually drive off intruders silently. When one Mistle Thrush is in conflict with another the attacking bird churrs persistently, and churring often accompanies attacks on other species if the attacks are for any reason unsuccessful or frustrated. When perched between attacks, or simply guarding fruit at times of potential attack, a Mistle Thrush will frequently make quick downward flicks of the wings and tail.

It is evident that fruit is defended as a continuing food resource. The general behaviour of Mistle Thrushes guarding fruit is unlike that of birds keeping other birds away from food on which they are actually feeding. Much of the defence takes place at times when Mistle Thrushes are mainly taking other food. In mild weather they spend much time feeding on grassland within 100 m of their fruit trees. If there is another fruit source available near by, as is often the case in autumn, they tend to feed there more than at the guarded fruit. Thus in 15.5 hours of observation of Mistle Thrushes guarding hollies in October and November, we recorded them feeding on the holly berries only nine times, and on other undefended fruit near by 18 times. When they feed on the defended fruit they often take only a small number, far below their full capacity (Table 48), and will quite often visit the defended tree without feeding at all.

As long as the winter weather is not exceptionally severe, Mistle Thrushes are normally able to safeguard their defended fruit supplies until some time in March, when they begin to spend most of their time on nesting activities. We have not seen a Mistle Thrush chase another bird from its defended fruit supply after 30 March. If the fruit tree has been prolific, much of the crop may still remain and, as described later, will become an important supplement to the Mistle Thrushes' (and other birds') food supplies in late spring or summer, especially in times of

TABLE 48 *Mistle Thrush: meal sizes for different fruits*

	Mean wt. of fruit (g)	No.	Meal sizes Range	Mean	Wt. of largest meal (g)
Holly	0.44	48	2–16	7.8	7.0
Yew	0.70	10	5–15	9.9	10.5
Ivy	0.32	16	5–15	9.3	4.8
Mistletoe	0.40	53	2–19	7.6	7.6
Haw (def.)	0.63	42	2–14	5.9	8.8
Haw (undef.)	0.63	3	9–16	12.0	10.1
Whitebeam	0.68	3	4–6	5.3	4.1
Guelder rose	0.38	2	12, 15	13.5	5.7
Buckthorn	0.36	9	7–36	17.7	13.0

Notes. Meal sizes of 10–13 g represent about 7–9% of the Mistle Thrush's body weight. Records for haw are divided into defended and undefended bushes; the larger meal sizes at undefended bushes (though few) are typical – see text.

drought. But when there is a very severe cold spell in winter, especially with snow cover making ground foraging impossible, the defended fruit may be an object of intense pressure from other thrushes and, if the cold weather lasts, defending Mistle Thrushes may be unable to keep invaders away and the fruit store may be overwhelmed and quickly consumed.

We first saw this in the hard winter of 1981–82. From 8–28 December there was severe cold weather with snow, interrupted by a brief thaw on the 14–15th. After the thaw, mild weather continued until 7 January, when another extremely severe spell began and continued until the 15th, during which period there was deep snow cover. Large autumn flocks of Fieldfares, Redwings and Blackbirds had stripped the hawthorns by late November and many had apparently moved away. The smaller flocks of Fieldfares and Redwings that remained in December turned to other fruit when ground-foraging became impossible, and the Mistle Thrushes were unable to withstand the sustained pressure of these flocks. In the period 20–26 December most of the defended hollies were overwhelmed and stripped. Of nine defended hollies on which we were making regular observations, seven were stripped, as were nearly all other hollies which we were watching less regularly. Five defended supplies of hawthorn or hawthorn with buckthorn all survived, apparently because they were rather small and inconspicuous fruit-clumps in areas that had otherwise been stripped and abandoned by the migratory flocks. At the end of the second, shorter spell of severe weather in January, one of the two surviving hollies was invaded by a group of at least nine Mistle Thrushes, which behaved like the invading Fieldfares and Redwings in the previous cold spell. They overwhelmed the two defending Mistle Thrushes and the tree was almost stripped within 24 hours; but just before it was completely stripped the thaw came and the invading Mistle Thrushes – presumably birds that had lost their own hollies in the previous cold spell – moved away. Whilst these hollies were being overwhelmed the owning Mistle Thrushes attacked the intruders almost continuously until they were exhausted and gave up, whereupon they fed with the invading birds until the fruit was finished.

The overwhelming of Mistle Thrush defence in the 1981–82 winter was one of the more spectacular happenings in the five years of field work. The 1982–83 winter, like 1980–81, was comparatively mild and none of the Mistle Thrushes lost their fruit supplies (four that had been defended the previous year were not defended, perhaps because Mistle Thrush numbers had been reduced by the very hard 1981–82 winter). In the 1983–84 winter, only one Mistle Thrush pair (whose holly was probably too large for effective defence) lost their fruit supply to a flock of Redwings in late January, when the weather was only moderately cold. In the 1984–85 winter there were two very severe spells, in the first half of January and from 9–20 February. One holly was overwhelmed by Redwings in the first cold spell, and two by mixed parties of Fieldfares and Redwings in the second.

The most obvious effect of the Mistle Thrush's behaviour on the other, smaller thrushes is that (except on the few occasions when the Mistle Thrushes' defences are overwhelmed) it deprives them of fruit which they would otherwise exploit. The locally resident Blackbirds and Song Thrushes soon learn that they risk being driven off if they approach the fruit, and most of the time they do not attempt to do so. The most aggressive Mistle Thrushes, which chase other thrushes even when they are feeding on the ground some distance from the fruit, deprive the other birds of some potential feeding ground as well as of the fruit. It is not uncommon, however, for Blackbirds, Song Thrushes and Robins to sneak in for a quick feed when the guarding Mistle Thrush is temporarily absent (sometimes as a result of human disturbance). Sometimes this happens so soon after the guard has left that they must have been watching and waiting for the opportunity. These clandestine visits are too few to make any significant inroads into the fruit supply.

If a holly has been defended successfully and still has fruit in the following spring, the owning Mistle Thrushes regularly use it to feed their young (we have records from 12 April to 5 June); and since by this time they are no longer guarding the fruit, other birds make similar use of it. In the most striking case an unusually large and prolific holly that had been successfully defended through the winter had a very large number of berries in the following spring. In dry weather that spring it was fed on by Blackbirds, who took the fruit to their young, and it continued to be fed on until mid July by Song Thrushes, adult Blackbirds and recently independent young Blackbirds.

All but 13 of our records of Mistle Thrushes collecting fruit for their young were from holly (29 records) and mistletoe (11 records) that they had defended through the winter. The exceptions were ivy (3), bird cherry (1) and *Amelanchier* (9 records). As Mistle Thrushes usually nest near their defended fruit supply, only short journeys are necessary. Although they are large birds, Mistle Thrushes have shorter beaks than Blackbirds and Song Thrushes, and this probably accounts for the fact that we did not record a Mistle Thrush carrying more than two fruits at a time in the beak.

MISTLE THRUSHES ELSEWHERE

When we found in the course of 1980–81 winter that Mistle Thrushes defend a fruit supply throughout the winter, we thought we had made a new discovery. We had seen references to Mistle Thrushes guarding clumps of mistletoe in France, but it was not clear that the fruit was a long-term resource to see them through the winter; and the standard British books (*eg* Witherby *et al* 1938, Simms 1978) made no mention of it. But we have since found that over 200 years ago William Borlase was well aware of the behaviour. In *The natural history of Cornwall* (1758), quoted by Penhallurick (1978), Borlase wrote of 'the missel-bird or shrite, which we in Cornwall call the holm-thrush; the Cornish call the Holly-tree, Holm, and this the holm or holm-thrush, because, as I imagine, in winter it feeds upon holly-berries, each bird taking possession of his tree, keeping constant to it, as long as there is fruit, and driving away all other birds'. More recently Williamson (1968) briefly described the behaviour in his account of the great yew wood at Kingley Vale: 'Resident Mistle Thrushes, some using spring and early summer breeding territories, maintain winter territories throughout, guarding for themselves favourite yew bushes or lone holly trees, fighting off foreign birds if they attempt to land and eat the berries. They will not eat the berries until later winter when there is little food, and the farming or guarding work then really pays off.' Williamson is mistaken in saying that they do not eat the guarded fruit until later winter; but as we have seen, their feeding visits may be few and far between, and easily missed.

Doubtless the Mistle Thrush's winter guarding of fruit is the same all over southern Britain as in our study area. In the 1981–82 winter, Derek Goodwin in Kent saw Mistle Thrushes trying to defend hollies and being overwhelmed by flocks of other thrushes in very cold, snowy weather at the same time as our birds, 60 miles away. In Sussex, Leverton (1983) saw the thing happening; on 13 December, during the same severe spell: 'Twenty Fieldfares are feeding on a couple of hawthorns still with berries, which a Mistle Thrush seems to regard as its own personal property. Every time it sees a Fieldfare take a berry it attacks furiously, but the birds are tired and hungry and will not leave the tree, nor can the Mistle Thrush take on all twenty at once, though it flies frantically hither and thither, chittering its anger.'

It would be interesting to discover how far north in Britain Mistle Thrushes defend winter fruit supplies. Not many ringing recoveries are available for birds from Scotland and the Borders, but those that there are indicate that these northern birds are highly migratory, not only in their first winter but in later winters too; so it may well be that there is a northern limit to the fruit-guarding behaviour somewhere in northern England.

On the continent of Europe the Mistle Thrush is closely associated with mistletoe, which is generally more abundant than in England. The most thorough observations are those made by Heim de Balsac (1928b) and Heim de Balsac & Mayaud (1930) in the course of a detailed study of the dispersal of mistletoe seeds in central France. Their main findings are summarised on p. 39. Evidently many Mistle Thrushes remain centred on clumps of mistletoe plants, which they defend against other thrushes, from autumn through to March. This is confirmed by Labitte (1952), who on the basis of many years' observation, also in central France, recorded that Mistle Thrushes may remain even in the coldest spells if there is a supply of mistletoe berries, after all the other thrushes have departed. In Germany, too, Mistle Thrushes defend mistletoe (Naumann 1897–1903). Creutz's (1952) account of Mistle Thrushes attempting unsuccessfully to defend their mistletoe clumps against a large flock of Waxwings in early March, referred to on p. 40, indicates that in this particular case the Mistle Thrushes had not been present through the winter but had recently returned, presumably from winter quarters further south. Wijngaarden (1986) has given an account of a ringed Mistle Thrush, of unknown origin, which defended a juniper in a garden in Holland in five successive winters, arriving in October.

In southern Europe there seems to be no record of Mistle Thrushes defending winter fruit supplies. In the Cantabrian Mountains of northern Spain, Guitian (1984) found that in winter Mistle Thrushes eat holly and rowan fruits at low altitudes but, surprisingly, are absent from the extensive holly woods higher up, around 1,400 m, whose fruit provides the staple diet of large flocks of Redwings.

In Extremadura, in west-central Spain, the Mistle Thrush was the only species that Pérez-Chiscano (1983) found eating the fruit of *Juniperus oxycedrus*. In Dalmatia, Tutman (1969) recorded that in winter Mistle Thrushes feed mainly on the fruit of two kinds of juniper, *J. phoenicea* and *J. oxycedrus*. It would be interesting to know if they guard juniper bushes in southern Europe, as we found one bird doing in the Chilterns. As juniper berries are long-lasting and nutritious, they should be very suitable for long-term defence.

Fieldfare *Turdus pilaris*

Fieldfares usually arrive in our study area in numbers in October, and depart in April. Within the six-month period when they are present they move about a good deal and the numbers fluctuate in response to weather and feeding conditions. We have records of fruit-eating for seven months of the year; in one year we had records for some late-departing birds in early May. They usually feed gregariously, often in company with other thrushes; but, as noted below, individuals may temporarily defend suitable clumps of fruit against other Fieldfares and birds of other species. This behaviour is much less well developed in Fieldfares than in Mistle Thrushes.

Of all the thrushes in our area, Fieldfares feed on the smallest number of different fruits; we have records for 12 native species, of which only four are

TABLE 49 Fieldfare: monthly summary of feeding records

	Oct	Nov	Dec	Jan	Feb	Mar	Apr	May	Total
Evergreens									
Holly			56	3	25				84
Ivy					1	1	101	7	110
Rosaceae									
Hip			96	73	262				431
Haw	40	504	201	48	197				990
Sloe		8	3	22					33
Whitebeam		3							3
Crab-apple					7				7
Others									
Elder	1								1
Spindle			2						2
Privet		1		1	1				3
Buckthorn		3	8						11
Black bryony			1						1
No. different fruits	3	5	7	5	6	1	1	1	
Total monthly records	41	519	367	147	493	1	101	7	1676

really important (Table 49). The early flocks feed predominantly on haw, with a peak in November. The greater part of the haw crop is finished by late November or December, and hip is then the main fruit taken, especially in February. (The large number of haw records in February is mostly from one site, where during a severe cold spell a flock of Fieldfares overwhelmed the defences of a pair of Mistle Thrushes which up to then had successfully defended the fruit.) Finally, before the Fieldfares' departure in spring, ivy berries may be an important food; but we had records for ivy in only two years, when Fieldfare flocks were present in an area with abundant ivy-covered trees. Holly berries may be an important food in very severe weather in mid and late winter, when the haw crop is finished. Nearly all our holly records were from the two very hard winters of 1981/82 and 1984/85.

These four fruits – haw, hip, ivy and holly – account for 96% of all our feeding records. Of the other eight fruits, sloe is the most important with 2% of the total, all the others being less than 1%. Except in very severe weather Fieldfares keep to open country and feed gregariously. This probably accounts for the small number of records for fruits that grow in areas with a more or less thick tree cover, and in particular for the complete lack of records for yew. Yew berries are fed on mainly in autumn, with a peak in October, when the Fieldfare flocks are feeding on haw in field hedgerows.

Fallen apples and pears in neglected orchards are a locally very important source of food for Fieldfares in hard weather. Big flocks collect at such places, and may stay around for days on end. We made only occasional observations on

Fieldfares and other birds at such places; the number of records depends entirely on how often one visits them, and no realistic comparison can be made with those of birds feeding on wild fruits. The only other fruits, other than wild species, that we saw Fieldfares eating were those of the cultivated varieties of cotoneaster for which we use the name 'cornubia' (p. 105). Our few records for these extend from the mid December to late February.

Fieldfares seem less agile than the other thrushes when feeding on fruit. They rarely sally to pluck a fruit in flight; only 1.2% of the total records included sallying, compared with 7–14% for the other thrushes. Nearly all fruit is taken from a perched position, and compared with the other thrushes (11.2% of records, compared with 2.0–3.7% for the other thrushes) Fieldfares more often use the perched-with-flutter method, i.e. they perch but beat their wings to maintain position. Like the other thrushes, they sally most frequently when feeding on ivy berries, which often can only be reached in this way; but even at ivy the percentage of sallies by Fieldfares (14%) is considerably lower than for the other thrushes (25–49%). Because they rarely sally, our data are inadequate for comparing sallying success rates with those when perched. For perched feeding on haw and ivy, success rates were 83 ± 7 and $80 \pm 8\%$ respectively (percentages followed by 95% confidence limits); birds feeding on hips were significantly less successful ($57 \pm 17\%$).

Fieldfare meal sizes for the three fruits for which we had records are shown in Table 50. The largest meals recorded (21 ivy berries, 10 haws, 4 hips) amount to approximately 6% of the Fieldfare's body weight. As is the case for Mistle Thrushes too, Fieldfares feeding at defended fruit sources tend to take smaller meals but probably feed more often. Six recorded meal sizes of a Fieldfare defending a hawthorn averaged 4.0 (range 1–7), compared with 6.9 (range 4–10) for Fieldfares feeding on undefended haws. We have few recorded intervals between successive feeds; at defended fruits seven intervals averaged 14 minutes (range 7–20), and at undefended fruit three intervals averaged 21 minutes (range 17–26).

We recorded 11 cases of Fieldfares, always single birds, temporarily defending a fruit supply. Apart from one in late November in quite mild weather, all were during the coldest part of the winter, from mid December to late February, and at times when the weather was severe. The fruits concerned were haws (4 cases), hips (4), a mixed clump of haw, hip and black bryony (1), and the common garden cotoneaster *Cotoneaster 'cornubia'* (2). In no case did the defence last very long. The longest period recorded was 14 days, and since we made observations at intervals of a few days it might have been a little longer. This is in sharp

TABLE 50 *Fieldfare: meal sizes for different fruits*

	Mean wt. of fruit (g)	No.	Meal sizes Range	Mean	Wt. of largest meal (g)
Ivy	0.33	3	15–21	17.3	6.9
Haw	0.63	22	4–10	6.9	6.3
Hip	1.8	14	1–4	2.6	7.2

Notes. Meal sizes of 6.3–7.2 g amount to about 6% of the Fieldfare's body weight. Records for haw exclude those for a Fieldfare feeding at a defended tree (see text).

contrast to the Mistle Thrush, which defends its fruit supplies for months on end (p. 127). In ten of the eleven cases the Fieldfares appeared, from their plumage, to be adult males.

A Fieldfare defending a fruit supply behaves very like a Mistle Thrush in similar circumstances. It mounts guard over the fruit, either on the plant itself or on a strategically placed perch near by. When engaged in active defence, it holds its wings slightly lowered so that the white under-wing coverts are visible as a white spot at the bend of the wing, and may flick the wings and tail downwards, as a Mistle Thrush often does when defending its fruit. All other thrushes will be chased off if they approach the fruit, including the larger Mistle Thrush; and it is striking that the Fieldfare's call, when thus engaged, changes from the usual *chuck chuck* flight call to a more continuous churring very like the Mistle Thrush's churr.

In addition to the shortness of the periods for which Fieldfares defend fruit, another indication that the behaviour is less highly developed than in the Mistle Thrush is the fact that they do not, as Mistle Thrushes do, chase seed-predators from their defended fruit, although seed-predators rob them of their fruit just as effectively as other thrushes, and tend to destroy more fruits per feeding visit than a thrush does. This was very strikingly seen during a half-hour watch on a small patch of hips defended by a Fieldfare in Tring Park during very cold weather one January, with the ground all around snow-covered. During the watch the Fieldfare fed twice and chased other birds away 16 times (Mistle Thrush 1, Fieldfare 3, Blackbird 12). Throughout the watch a Greenfinch feeding on the seeds was perched close to the Fieldfare on the defended hip clump, and was ignored.

FIELDFARES ELSEWHERE

Little quantitative information is available on fruits eaten by Fieldfares elsewhere in Britain. In his survey of wild fruits in the diet of British thrushes Hartley (1954) had records for Fieldfares taking only haw, holly and crab-apple. Sorensen (1981), studying fruit-eating in woodland near Oxford, recorded only haw (83% of records) and sloe (17%). Williamson (1978) mentioned that large numbers of Fieldfares (with other thrushes) exploit the haws and yew berries in the ancient yew wood at Kingsley Vale in Sussex, which shows that our lack of records for yew is not because it is in any way unsuitable for Fieldfares.

Tye (1982) studied the feeding ecology of Fieldfares in open country in Cambridgeshire, where haws were the main wild fruits available. From his observation of feeding rates and analysis of the food value of the fruits, he calculated that haws, if easily available, can provide ample energy for a Fieldfare's daily energy needs (see also p. 236), as also can fallen apples, the other fruit source available to Fieldfares in his study area. By experimentally laying out apples on pasture fields where Fieldfares fed, Tye (1986) was able to make individual birds change from gregarious feeding to defence of a fruit supply, and showed how increasing intrusions by other Fieldfares and Starlings eventually made such defence uneconomical.

Swann (1980) investigated the weights of Fieldfares and Blackbirds in Inverness-shire during the 1978–79 winter. He found that during a prolonged severe spell Fieldfares lost weight and many died, whereas Blackbirds put on weight. He attributed the difference to their feeding habits. The Blackbirds showed their usual versatility, feeding on bread, fat, cheese and other food put out by man; but the Fieldfares fed almost exclusively on cotoneaster berries (species not stated), which declined sharply in numbers. In the autumn both species fed on sloes. It seems from Swann's account that it was the depletion of the food supply that affected the Fieldfares adversely, rather than that the cotoneaster berries were an inadequate diet. Hope Jones (1962) gave an account of a flock of about 150 Fieldfares feeding on sea buckthorn fruit in Anglesey during a very cold spell in January. Many became weak and died, which perhaps indicates that the sea buckthorn fruit was an inadequate diet in such conditions. Lübcke (1980) reported that, during very severe weather on the German North Sea coast in February 1978, Fieldfares began to die about eight days after the beginning of the cold spell, although they had an ample supply of apples; and Jedicke (1980) reported that exclusive feeding on rowan berries by Fieldfares results in loss of condition and death. On the other hand, Pénzes (1972) found from analysis of stomach contents that Fieldfares in Hungary were apparently living exclusively on nine different fruits from the beginning of December to the beginning of March; and Dieberger (1982) found that in Austria, after a heavy snowfall on 18 December 1981, from then until 20 February 1982 Fieldfares had their digestive tracts filled almost exclusively with mistletoe berries, which are known to be among the most nutritious of wild fruits. Hence it seems probable that Fieldfares can survive lengthy periods of severe weather on a diet of fruit, but only if it is of high enough nutritive quality (see pp. 231–4, where this question is discussed further).

There are many casual records of fruits taken by Fieldfares wintering in Germany and other parts of central Europe. Two that are worth mentioning (Bettmann 1953, Zedler 1954) refer to flocks of Fieldfares stripping the fruit of guelder rose in hedges during the autumn migration season. This is an interesting instance, the only one known to us, of a fruit crop being eaten with a marked difference in the season between our English study area and continental Europe at a similar latitude.

Meidell (1937) studied the Fieldfare's food in a subalpine birchwood and mountain area in southern Norway. In spring the fruits of juniper and cowberry (which had survived the winter under the snow) were important components of the diet, and in autumn bilberry and crowberry. The most important fruit for Fieldfares in Fenno-scandia is, however, the rowan. Tyrväinen (1975) studied the winter movements of Fieldfares in Finland in relation to the crop of rowan berries. He showed that when the rowan crop is plentiful, part of the Fieldfare population remains in Finland after the normal time of migration. The factor that stimulates migration is apparently the decrease of the fruit crop to a few berries per inflorescence. This seems to be the only study showing quantitatively a relationship between fruit abundance and Fieldfare movements, although it has generally been assumed that the two are intimately linked, and that changing fruit supplies are the main factor in the Fieldfare's irregular and almost nomadic migratory behaviour.

138 The Fruit-eaters

Redwing *Turdus iliacus*

Redwings arrive in our study area in late September or October. The main flocks, usually in association with Fieldfares, at first feed on haws along field edges, and locally on buckthorn, dogwood and yew (Table 51). When the bulk of the haw crop is finished many of the large flocks leave our area, presumably moving on to the south or west. Those that remain turn to other fruit, especially holly which is the main fruit eaten in December and January. Whitebeam, which fruits irregularly, is locally an important fruit in late autumn on the Chiltern slopes in years when there is a good crop. In very severe weather sloe is eaten, and may locally be important in farmland areas where there is little other fruit once the haws are finished. When they ripen early in the new year, ivy berries are eaten by those Redwings that are still present, and later they are taken in greater quantity by Redwing flocks which arrive in our area in March or April, before departure to their northern breeding grounds.

The Redwing is the smallest of our thrushes and is unable to feed efficiently on some of the larger fruits that the bigger thrushes can take. Thus they hardly feed at all on hips; we have only four records, of birds taking them in very hard weather in January, either picking pieces from attached fruit or taking a hip to the ground to eat. Sloes are also at or above the upper size limit for Redwings; they feed on them mainly by picking bits of pulp from the fruit in situ, or will pluck a fruit and take it to the ground to deal with it. Even some haws are too large for them to swallow: 48% of the failed attempts recorded were cases where the fruit was plucked and then dropped apparently because it was too large. By contrast, almost all (96%) of the 81 recorded failures to eat holly berries were

The Fruit-eaters 139

failures to pluck them, the rate of failure being particularly high in November, when the berries are more firmly attached than later in the winter. The Redwing's overall success rate (67%) for plucking and swallowing fruits from a perched position (Table 52) is lower than the rates for the other thrushes. Like the other thrushes, when they sally for fruit they are more successful, and do so frequently

TABLE 51 Redwing: monthly summary of feeding records

	Oct	Nov	Dec	Jan	Feb	Mar	Apr	Total
Evergreens								
Holly	1	11	326	204	32	8		582
Yew	34	19	1	1				55
Ivy				8	7	123	30	168
Mistletoe		1						1
Rosaceae								
Hip				4				4
Haw	63	310	118	45				536
Whitebeam	1	147	22					170
Sloe	1	15	14	30				60
Crab-apple					1			1
Others								
Elder	1							1
Guelder rose				1				1
Dogwood	31	51	19					101
Buckthorn	24	8	76	24				132
Privet			3	4				7
No. different fruits	8	8	8	9	3	2	1	14
Total monthly records	156	562	579	321	39	131	30	1819

TABLE 52 Redwing: success rates according to method of taking fruit

	Sallies		Perched	
	No.	% success	No.	% success
Holly	4	[50]	189	57($\pm$7)
Yew	15	100	15	80
Ivy	17	100	44	96($\pm$6)
Haw	3	[100]	124	61($\pm$9)
Whitebeam	1	[100]	23	70($\pm$19)
Dogwood	–	–	22	95($\pm$9)
Buckthorn	–	–	11	100
All native plants	40	95($\pm$7)	462	67($\pm$4)

Notes. Figures in parentheses (for samples of 20 or more) are 95% confidence limits. Totals for all native plants (perched feeding) include some records for plants for which sallies were not recorded and some plants not included separately in the table; the comparison between sally and perched is thus based on heterogeneous samples, but the difference is so great that it is undoubtedly significant.

140 The Fruit-eaters

TABLE 53 Redwing: records of sallying to pluck fruit

	No. feeds	% feeds with sallies
Holly	581	2
Yew	55	29
Ivy	168	49
Haw	535	2
Sloe	59	2
Dogwood	101	1
Buckthorn	132	2

TABLE 54 Redwing: meal sizes for different fruits

	Mean wt. of fruit (g)	No.	Meal sizes Range	Mean	Wt. of largest meal (g)
Holly	0.44	10	5–10	7.0	4.4
Yew	0.70	4	2–8	4.8	5.6
Ivy	0.33	3	9–14	11.7	4.6
Haw	0.63	7	5–9	6.9	5.7
Whitebeam	0.68	7	1–3	2.0	2.0

Note. Meals of 4.4–5.7 g represent about 6–8% of the Redwing's body weight (mean about 68 g).

when feeding on ivy and quite often when feeding on yew, but rarely for other fruits (Table 53). Meal sizes are summarised in Table 54.

Although most Redwings are highly gregarious in their feeding, especially in the few weeks after they arrive in autumn, in the course of the winter individual birds or small groups separate themselves from the main flocks and frequent tree-grown gardens where they feed inconspicuously on holly berries, and remain behind after the big flocks have left. These Redwings may be little in evidence as long as the weather is mild, but in severe weather, once all unguarded hollies have been stripped, they become more conspicuous, gathering in larger groups at the Mistle Thrush hollies, often in company with Fieldfares. Though chased by the Mistle Thrushes when they approach the trees, by sheer persistence and force of numbers many eventually succeed in feeding, and, as described on p. 130, they may overwhelm the Mistle Thrushes' defences and demolish the fruit supply.

REDWINGS ELSEWHERE

We have not found any comprehensive records from other parts of southern England with which our data can be compared. In the survey of fruit-eating by British thrushes analysed by Hartley (1954), 86% of Redwing records were for haw, and holly was the only other fruit for which there were more than 10 records. For a woodland area near Oxford, Sorensen (1981) recorded Redwings taking only three kinds of fruits: haw (93% of records), sloe (6%) and elder (1%). Our records show haw making up over 50% of the records in October and November, falling to 20% in December and 14% in January; which suggests that

Hartley's and Sorensen's data are biased towards an excess of autumn records, which are of course the easiest to obtain as it is then that large Redwing flocks are present.

Observations by Swann (1983) in the Scottish highlands show a picture very different from southern England. In Glenurquhart, Inverness-shire, Redwings arrive in large numbers in autumn, feed mainly on rowan fruit and then pass on, individual birds apparently staying less than 24 hours. The passage lasts from late September to mid November and is at its peak between 10 and 24 October. Few birds remain later. Essentially, they make a re-fuelling stop, with rowan fruit the chief energy source.

For most of continental Europe there seem to be only casual records of fruit-eating, involving a good number of different fruits but no quantitative data. However, near the southwestern extreme of the wintering range there is good information from the Cantabrian Mountains of northern Spain (Castroviejo 1970, Guitian 1984). At the western end of the range (Sierra de Ancares) holly is very abundant and is the only evergreen tree, replacing conifers and providing conditions which enable many birds and mammals to over-winter. Redwings winter in huge numbers in these holly woods, feeding mainly on the fruit. Being the only fruit-eating birds present in numbers, they apparently have the bulk of the holly berry crop to themselves, at least from January to March.

142 The Fruit-eaters

Robin *Erithacus rubecula*

Robins take a very wide range of native fruits; we have records for 29 species (Table 55). The wild cherry is the only fruit for which we have records for Blackbird and Song Thrush but not for Robin, and this is clearly because wild cherries are too big for Robins to swallow. Robins also take fruits from many introduced plants; we recorded them feeding on 15 species and more intensive observation would certainly have extended the list.

In view of the Robin's abundance, which is greater than that of the Song Thrush and far greater than that of the Mistle Thrush in our study area, the total number of feeding records at native fruits is comparatively low compared with the records for the resident thrushes (628, compared with 1,431–6,868). This is largely because Robins will tolerate only their mates taking fruit in their territories, whereas the thrushes usually feed gregariously at fruit. Hence a watch in one place is likely to yield records only for one or two Robins. Also their feeds are very rapid: as mentioned in more detail below, they very often take only a single fruit, with a movement that is so quick that it is likely to be missed unless one is carefully watching the Robin beforehand.

We had some evidence, which unfortunately we have not been able to quantify, that Robins occupying adjacent territories tend to synchronise their bouts of fruit-eating, and that they do so by uttering the ticking call just before they move to the fruit source. It would be of interest to try to confirm this by more detailed study, which would need the coordinated observations of at least two observers. As Robins tend to expose themselves more when visiting fruits than when foraging for invertebrates within cover, it might be advantageous for neighbouring birds to do so at the same time, and thus keep a more effective, joint watch for approaching predators. As discussed on p. 228, efficient anti-predator behaviour is very important for fruit-eaters.

In our study area there are native fruits available for Robins in every month of the year except June, and Robins with fruit in their territories feed on it as

TABLE 55 *Robin: monthly summary of feeding records*

	Jan	Feb	Mar	Apr	May	Jun	Jul	Aug	Sep	Oct	Nov	Dec	Totals
Evergreens													
Holly	2	5								4	2	2	15
Yew	5							2	9	5	4	1	26
Ivy	6	11	17	26	12							4	76
Mistletoe	1		1	1									3
Juniper		1											1
Spurge laurel							16						16
Rosaceae													
Hip												1	1
Haw											12	7	19
Rowan							7						7
Whitebeam									1		1		2
Bird cherry							11						11
Sloe	8										1	12	21
Blackberry								3	22	1			26
Raspberry							3						3
Crab-apple			13										13
Caprifoliaceae													
Elder								22	61	12	5		100
Guelder rose	1											2	3
Wayfaring tree								21	7				28
Common honeysuckle								16	2				18
Perfoliate honeysuckle								6					6
Others													
Dogwood	2							1	22	8	13	7	53
Spindle	33	20									23	54	130
Privet	7	9										3	19
Buckthorn										1	2		3
Currant							3						3
Woody nightshade							1	1	3		5	4	14
White bryony								1					1
Black bryony											3		3
Lords and ladies								8					8
No. of different fruits	9	6	2	2	1	0	5	11	8	6	10	12	
Total monthly records	65	59	18	27	12	0	34	88	127	31	69	99	629

TABLE 56 Robin: meal sizes for different fruits

	Mean wt of fruit (g)	No.	Meal sizes Range	Meal sizes Mean	Wt of largest meal (g)
Spindle	0.08	23	1–5	3.3	0.40
Elder	0.14	35	1–9	3.1	1.26
Dogwood	0.21	10	2–5	2.7	1.05
Spurge laurel	0.25	6	1–3	1.8	0.75
Pyracantha coccinea	0.26	10	1–3	1.9	0.78
Honeysuckle	0.28	8	1–5	2.3	1.40
Wayfaring tree	0.28	4	1–2	1.8	0.56
Privet	0.30	3	1–3	2.0	0.90
Ivy	0.33	16	2–4	2.3	1.32
Guelder rose	0.38	2	1–3	2.0	1.14
Holly	0.44	3	1–2	1.7	0.88
Lords and ladies	0.49	3	1–2	1.3	0.98
Woody nightshade	0.50	5	1–3	1.4	1.50
Haw	0.63	6	1–2	1.5	1.26
Yew	0.70	3	1	1.0	0.70

Note. Fruit weights and mean meal sizes are inversely highly correlated ($r_s = -0.823$, $P<0.01$; Spearman Rank correlation, 1-tailed). Meal sizes of 1.3–1.5 g amount to 7–8% of the Robin's body weight.

regularly in late summer as in winter. Thus we had 215 records in August and September (0.58 feeds per hour of observation), when temperatures are high and invertebrates abundant, compared with 168 records in December and January (0.47 feeds per hour). We had 56 records of the interval between successive feeds by the same Robin at the same fruit source. The mean interval was 11.9 minutes, 84% being in the range 6–18 minutes (overall range 5–25). We recorded Robins taking three kinds of fruit to their young: ivy, spurge laurel and rowan; and on two occasions saw a male Robin courtship-feeding his mate with an ivy berry.

Robins take most of their fruit in flight, with a rapid sally from a neighbouring perch or (in the case of very low-growing fruit) from the ground. Overall this was the method used for 68% of the 839 fruits that we observed being plucked, but for some fruits the percentage was considerably higher: 91% for privet, 86% for ivy, 77% for dogwood and honeysuckle. In the case of spindle the percentage increased through the season from 69% taken with sallies in November to 84% in February, and this was obviously because as the crop was depleted the remaining fruits were those in terminal positions which could only be reached in flight. (The difference is on the border-line of statistical significance, 95% confidence limits for the two percentages being 7.9 and 8.4 respectively.) Three abundant fruits, haw, holly and woody nightshade, were mainly taken from a perched position (sallies 40%, 36% and 29% respectively); none of these is a preferred fruit of the Robin, each representing only 2–3% of our total records. Although plucking a fruit in a flight sally is energy expensive, it is the more successful method: thus 93% of 372 sallies resulted in a fruit being plucked and eaten, as against 76% of 138 attempts from a perched position. The difference is statistically highly significant.

As Robins are a lot smaller than thrushes there are several native fruits which thrushes can swallow and Robins cannot. As might be expected, Robins feed largely on the smaller fruits that are available; thus 75% of all records are from fruits with a mean diameter of 8 mm or less. They often have difficulty in swallowing haws, which are very variable in size, large ones being above their upper limit. Sloes are too big for them to swallow; Robins feed on them by picking out bits of pulp from the fruit in situ, and to do this they must perch. Wild cherries are also too large to swallow, and it is presumably because their long stalks make them inaccessible to a perching Robin that we have no record of Robins taking the pulp in situ. The other native *Prunus*, bird cherry, has fruits small enough for Robins to swallow, and all our records are of Robins taking them with flight sallies. Blackberries are also too large to swallow whole; Robins either sally and pluck one or two drupelets, or pluck the whole fruit and take it to the ground and deal with it there.

Table 56 summarises meal sizes of Robins when feeding on different kinds of fruit, and Figure 6 illustrates the strong inverse correlation between meal size and the weight of individual fruits. As already mentioned, often only a single fruit is taken; the mean number for nearly all fruits is between about $1\frac{1}{2}$ and 3. The small fruits of elder are the only ones of which a much greater number can be taken at a single feed (two records of 7, one of 9), but even for them 3 or 4 fruits per feed are most usual. It will be seen from the table that the largest feeds that we recorded amounted to 7–8% of the Robin's body weight.

We have records of Robins regurgitating the seeds of ivy and spindle, and they must also regurgitate other large seeds such as those of haw and bird cherry. The small seeds of elder and blackberry are passed through the gut; but pellets of small seeds may also be regurgitated – see below. Although we have no hard figures to support the suggestion, we think that Robins must be very effective seed dispersers. In contrast to thrushes and Starlings, they tend to feed solitarily, moving round their territories, taking fruits in small numbers and presumably depositing seeds thinly but more or less uniformly through their territories, which

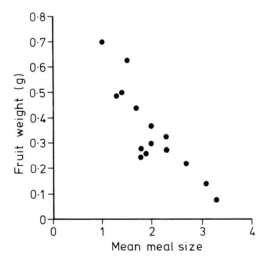

Figure 6   Mean meal sizes of Robins for 15 different kinds of fruit (see Table 56).

generally are suitable habitats for the development of the seeds. The more gregarious feeders, which also take more fruit at a time, must tend to produce much more clumped patterns of seed dispersal, especially if they also roost or rest gregariously.

ROBINS ELSEWHERE

Published records of fruit-eating by Robins from other parts of Britain do not add very significantly to what we found. Lack (1948) gave a list of fruits recorded as eaten, one of which, the whortleberry (which does not occur in our study area), is additional to those given in Table 56. He recorded that a Robin regurgitated a pellet containing 20 raspberry seeds and four currant seeds. From northern and central Europe there are more or less casual records of Robins eating many kinds of fruit, some of which do not occur in Britain. The controversy in the German ornithological literature about Robins eating spindle fruit has been mentioned under that fruit.

Two thorough studies have been made of fruit-eating by Robins in the Mediterranean region. Herrera (1981a) examined the diet of Robins wintering in two areas in southern Spain. In both areas, one lowland and the other montane and each with natural vegetation, fruit was the main food during the midwinter period, when insect food was scarcest. In the montane area, where winter is longer and more severe, Robins were more frugivorous than in the lowland area, and fed on a wider variety of fruits. In both areas the Robins showed an increase in body weight from early autumn to late winter. The final level of weight reached was higher in the montane area, apparently because the most important fruit there, *Viburnum tinus*, has a higher fat content than *Pistacia lentiscus*, the principal fruit eaten in the lowland area. With the exception of the blackberry, all the fruits eaten in Herrera's Spanish areas were different from those in our study area.

In an earlier study of Robins wintering in southern Spanish evergreen oak *Quercus ilex* woodland, a habitat that is not rich in fleshy fruits, Herrera (1977) made the surprising discovery that they feed mainly on the endosperm of acorns. Being unable to break into the acorns themselves, they are dependent on the remains of acorns left by other species, especially Nuthatches and Great Tits.

Debussche & Isenmann (1985a) studied the fruit diet of transient and wintering Robins in scrubland and evergreen oak woodland round Montpellier, southern France. They found that 23 different fleshy fruits were eaten, mostly in the size range 6–8 mm, the most important being *Rubia peregrina* (madder) and *Viburnum tinus*. Like Herrera, they found that significant quantities of acorns were eaten, especially in the midwinter period, apparently ingested as fragments that had been either crushed (by domestic animals, vehicles, etc) or damaged by previous consumers, principally field mice.

Blackcap *Sylvia atricapilla*

Blackcaps are common breeding birds of woodland, thickets and wooded gardens in our study area. They also winter in small numbers. As Langslow (1979) has shown, two populations are involved. Most of the British breeding population has left the country by the end of September, to winter mainly in the Iberian peninsula and northwest Africa. The wintering birds, which probably arrive mainly in October, are of Continental origin. The numbers wintering have been increasing in recent years (Leach 1981).

Blackcaps take a wide range of native fruits; we have records for 23 species (Table 57). Our records cover all months of the year but we had very few in March and June. In the years of our study most of the wintering Blackcaps seemed to have left us by March, and incoming migrants did not arrive until April, when

TABLE 57 Blackcap: monthly summary of feeding records

	Jan	Feb	Mar	Apr	May	Jun	Jul	Aug	Sep	Oct	Nov	Dec	Totals
Evergreens													
Holly	30	13	1								1		45
Yew							1	3					4
Ivy	5	6		82	32	1							126
Mistletoe			1										1
Rosaceae													
Hip	2												2
Rowan							1	7	4				12
Whitebeam									3				3
Cherry						2	1						3
Bird cherry							9						9
Blackberry									5				5
Raspberry							2	4					6
Caprifoliaceae													
Elder								39	51	18			108
Guelder rose											1		1
Wayfaring tree								23	2				25
Perfoliate honeysuckle							14	24					38
Others													
Dogwood									6	2	1		9
Spindle	5										6	25	36
Privet	3											5	8
Buckthorn										2	1	6	9
Currant						3	14						17
Woody nightshade							10	15	12		1	3	41
White bryony							17	9	4		1		31
Black bryony											1		1
No. different fruits	5	2	2	1	1	3	9	8	8	3	6	6	
Total monthly records	45	19	2	82	32	6	69	124	87	22	11	41	540

they began to feed a great deal on ivy berries. The number of records of Blackcaps eating ivy berries in April was easily the greatest of the monthly totals for any fruit, and in fact exceeded the annual total for all other fruits except elder. The small number of records for June is due to the lack of suitable fruit in that month, the ivy crop having mostly finished and wild cherry being too large for Blackcaps to tackle effectively. As soon as suitable summer fruits become available in quantity, in July, Blackcaps make much use of them; our records for the three months July–September make up 52% of the overall total.

Of the two major plant families that produce fleshy fruits, Rosaceae and Caprifoliaceae, the latter is much the more important for Blackcaps in our area, accounting for 32% of all records. Disregarding the single record for guelder rose, the fruits that Blackcaps take in this family are small, with mean diameters less than 7 mm, and so are easily plucked and swallowed whole. The success rate recorded for Blackcaps feeding on fruits of Caprifoliaceae was 89% (total attempts: 194). The family Rosaceae, with mainly larger fruits, accounts for only 7% of all Blackcap records, and of those taken, only the fruits of bird cherry could be swallowed whole (observed success rate 68%, of 28 attempted; a significant difference from the success rate for Caprifoliaceae, $\chi^2 = 7.21$, 1 d.f.). Our feeding records at rowan were all of Blackcaps taking pieces of pulp left on the tree by Bullfinches after removal of the seeds; we watched Blackcaps try but fail to pluck rowan fruits 16 times. Our few records at whitebeam were also of birds taking the pulp after seed-predators had removed the seeds, and eight attempts to pluck a whole fruit were seen to fail. When feeding on blackberry and raspberry Blackcaps take beakfuls of the soft drupelets, and so function as dispersers. Though our records are few, they are consistent in showing that they feed in the same way on wild cherry and hip, two fruits that are much too large for them to swallow whole. In the case of wild cherry, Blackcaps act as pulp-predators, damaging the fruit without dispersing the seed; but they may occasionally disperse the seeds of hips, by ingesting some of them with the pulp.

Holly and ivy together account for a third of our records, and are taken from January until late spring. Our holly records are all from trees defended by Mistle Thrushes, except for the one November record. One of the effects of Mistle Thrush defence is that fully ripe and therefore easily pluckable fruits are available late in the winter. Most of our Blackcap records at holly in January and February were at a tree which, in a year when it was not defended by a Mistle Thrush, had its fruit taken in October, so was obviously an early ripening strain probably with fruits that were unusually easily plucked. It is possible that the paucity of Blackcap records at holly in November and December, when most undefended holly fruit is eaten by thrushes, is because Blackcaps cannot compete then for the more easily pluckable berries. The Blackcap's success rate at holly was 64% (of a total of 67 observed attempts), close to its average rate of 66%.

Our records show that ivy is the only fruit taken by Blackcaps in April and May, the months when we obtained over half of our thrush records and most of our Starling records for ivy. Hence competition for the ripest fruits must be high at this time, and this may account for the relative difficulty experienced by Blackcaps in plucking ivy berries (observed success rate 46%, of 184 attempted), in spite of their pulling hard, with a head-twisting movement. On several occasions we saw Blackcaps employ a special technique, leaning down, grasping an ivy berry in the beak, and then "falling" off the perch. When this method failed the bird was suspended for a moment by its beak!

Of the 763 fruits of 23 species which we observed being plucked by Blackcaps, 95% were taken from a perched position, 2% while clinging with wings fluttering, and 3% with a flight sally. More than half the records of sallying were at three fruits, ivy, spindle and bird cherry, at which Robins also showed high percentages of sally-feeding. We did not see Blackcaps feeding on fruit from the ground, but had records of them coming down inside a fruiting bush to within 30 cm or less of the ground.

Like all the other fruit-eaters that we watched, Blackcaps go to a fruit source, take a complete feed of a size consistent with their weight (Table 58) and then leave, returning for further feeds at fairly regular intervals. A complete feed usually takes less than half a minute. The average interval, based on 45 occasions when we were confident that the same bird was returning, was 12 minutes (range 6–21 minutes). Though not statistically significant, the data suggest that the interval averages shorter in winter (10 records November–February, average 10.9 minutes) than during the rest of the year (35 records March–October, average 12.3 minutes). Blackcaps that we watched feeding on holly berries in midwinter did not leave the tree but went inside it after feeding, moving at intervals to the periphery where the fruit was available. The seeds of fruits eaten by Blackcaps seem usually to pass through the gut. We had only four observations of seeds

TABLE 58 *Blackcap: meal sizes for different fruits*

	Mean wt of fruit (g)	No.	Meal sizes Range	Meal sizes Mean	Wt of meals (g) Mean	Wt of meals (g) largest
Ivy	0.33	22	1–5	2.5	0.8	1.7
Holly	0.44	9	2–4	2.4	1.1	1.8
Bird cherry	0.29	6	2–6	3.2	0.9	1.7
Blackberry	1.7	5	$\frac{1}{2}$–1	0.6	1.0	1.7
Elder	0.14	31	3–11	6.3	0.9	1.5
Perfoliate honeysuckle	0.19	5	1–9	4.4	0.8	1.7
Woody nightshade	0.50	7	1–3	2.0	1.0	1.5

Note. All fruits are included for which at least five meal sizes were recorded. The bird cherry records are from Czechoslovakia, from trees whose fruits were a little larger than our English samples (means of 0.29 g, cf 0.20 g). Meal sizes of 0.8–1.0 and 1.7–1.8 represent, respectively, about 5–6% and 10% of the Blackcap's body weight.

being regurgitated, twice bird cherry seeds and twice ivy seeds, each time during or just before a meal on these fruits.

We had little evidence of aggressiveness at fruit sources affecting the availability of fruits for Blackcaps. Mistle Thrushes, who vigorously chase all other fruit-eaters from their defended hollies, soon give up chasing Blackcaps, who do not fly but hop away among the thick foliage. Male Blackcaps sometimes chase other males from holly in winter, and we had two records of their chasing female-plumaged Blackcaps from fruit in summer. We saw Blackcaps chased from ivy fruit on three occasions by Robins and twice by Song Thrushes. Gregarious feeding, a strategy against predation (p. 227), is a far more important element in the Blackcap's fruit-eating behaviour, at least in our study area. (As mentioned later, Blackcaps wintering in France and feeding on mistletoe fruit may behave very differently.)

We have defined gregarious feeding as taking fruit at the same time and from the same bush as another bird, either a conspecific or another species. Of our total of 540 records, 38% came within this definition of gregarious feeding. In April and May gregarious feeding on ivy (29% of total records) usually involved a pair feeding together, occasionally two males. In the post-breeding period we have seen up to six Blackcaps feeding together. In the months July–October gregarious feeding is usually preceded by *chack* calls, as the Blackcaps move through the cover towards the fruit; the call appears to function as a mustering signal both to conspecifics and to other species. We had no records of *chack* calls preceding fruit-feeding in winter, and only one record in spring. Hence there is probably a stronger tendency to gregarious feeding in late summer and autumn than at other times, though this is not demonstrated by our figures (41% gregarious feeding in these months, not significantly different from the overall mean).

We had ten examples of gregarious feeding with Blackbirds, ten with Robins, four with Garden Warblers, four with Lesser Whitethroats, and one with a Song Thrush. Particular observations are sometimes more impressive than statistics; for instance, during a 40-minute watch at a large fruiting patch of blackberry measuring about 30 × 10 m, the only birds that fed were a female Blackcap and a Robin which took fruit within a metre of one another at the same time. What might almost be described as mixed-flock fruit-eating is a feature of Blackcaps and other birds feeding at white bryony and wayfaring tree in July and early August before the ubiquitous elder ripens. For instance, a watch of 80 minutes on 1 August 1982 on two bushes of wayfaring tree with ripe fruit produced 16 records of fruit-eating by dispersers and one by a seed-predator, a Bullfinch. Besides Blackcaps, the dispersers were Robins, Blackbirds and a Lesser Whitethroat, and all except three were gregarious feeding records.

It is interesting to compare the fruit diet of the Blackcap with that of the Robin, the only other year-round fruit-eater of similar size. The percentages of records for plants of different growth forms are as follows:

	Blackcap	Robin
climbers	44	19
shrubs	40	61
trees	16	16
herbs/subshrubs	—	4

The climbers whose fruits figure prominently in our Blackcap records – woody nightshade, perfoliate honeysuckle and white bryony – were not taken much by Robins. For woody nightshade, Blackcap records made up 37% of the total of 110 records, more than the number for any other bird. Some of the fruits recorded as taken by Robins and not by Blackcaps, such as lords and ladies and spurge laurel, are taken from the ground or near it, below the level at which Blackcaps normally feed. Because Robins pluck most of their fruit (68%) with flight sallies, whereas Blackcaps pluck nearly all theirs perched (98%), even when the two species are feeding at the same fruit source they are likely to select fruits from different sites.

Whereas we had 51 records of Robins taking introduced or cultivated fruits of 15 different kinds, including five species of *Cotoneaster* and two of *Pyracantha*, we had only two such records for Blackcaps, one at the Duke of Argyll's tea tree (*Lycium barbarum*) and one at a cultivated cherry. This is in marked contrast to the data collected by the BTO enquiry into Blackcaps wintering in Britain in 1978/79 (Leach 1981): 77% of the 445 records of berry-eating were at introduced plants. Nearly all (95%) of the records submitted to this enquiry were from gardens, and to judge from the very large number of records of Blackcaps eating artificial foods, it seems that most if not all the gardens contained well-stocked bird tables. We spent 1,136 hours watching at introduced and cultivated fruits, but not in gardens where artificial food was also available.

The bias towards garden records in the BTO enquiry probably also accounts for another difference from our results. Defining the winter as from the beginning of November to the end of March (as mentioned, we had no evidence of migrants arriving before April in the years of our study), we had records of 50 different Blackcaps eating fruit in winter, of which 24 were male and 26 female; which suggests equality of the sexes. Records from the 1978/79 BTO enquiry, as well as from earlier enquiries, showed a preponderance of males (1.43:1). This is presumably because males, being dominant, are more often able to monopolise the rich food sources artificially provided at garden feeders.

BLACKCAPS ELSEWHERE

The ornithological literature contains many references to Blackcaps eating fruit. For the most part they are more or less casual records, many of them referring to the same fruits as we recorded Blackcaps eating. We have found references to only two wild fruits eaten in Britain other than those that we recorded, haw and sea-buckthorn (Hardy 1978, Leach 1981).

Möhring (1957) gave a full account of fruits eaten by Blackcaps in Germany, listing 38 different species, mainly eaten in summer and early autumn (June–September). He also made observations on the time needed for the digestion of fruit pulp and regurgitation of seeds (within 10 minutes of feeding, in the case of the larger seeds), and carried out an experiment on the germination of seeds from fruit eaten by Blackcaps. Hess's (1927) record of a Blackcap eating deadly nightshade berries is one of the few records of that fruit being eaten. Naumann (1897–1905) reported a case of a Blackcap found dead with a cherry stone stuck in its throat, in the process of being regurgitated.

In Anjou, in west-central France, mistletoe berries form the staple winter diet of many wintering Blackcaps. In their study of the dissemination of mistletoe, H. Heim de Balsac and N. Mayaud made careful observations on the feeding behaviour of Blackcaps and examined the gut contents of specimens that they collected. They concluded that in that area Blackcaps are the main disseminators of mistletoe seeds. The bird's usual routine is to pluck a fruit, place it on a neighbouring branch (not the mistletoe), and then with rapid bill movements separate the seed, with its enclosing mucilage, from the rest of the pulp and skin, leaving the seed sticking to the branch. With several side-to-side movements of the head the bird then breaks the threads connecting the seed on the branch from the skin and pulp held in the bill, and swallows the latter.

These Blackcaps in Anjou behave very differently from the wintering birds in our study area. Each individual takes possession of a mistletoe clump which it defends fiercely against other Blackcaps. Heim de Balsac and Mayaud found that, when they collected a bird, its place was at once taken by another, so that the local population remained the same. As mentioned earlier, the Blackcaps wintering in our study area feed to a large extent gregariously, without mutual aggression. Hardy (1978), however, noted that three Blackcaps that fed at mistletoe in a garden in North Wales were aggressive to one another. It seems likely that concentration of the fruit within a small, sharply circumscribed clump makes mistletoe especially easy to defend, by Blackcaps as well as by Mistle Thrushes.

In Majorca, in the third week of March, we found at least 20 Blackcaps feeding extensively on the fruits of two species of introduced palms (*Phoenix dactylifera* and *P. canariensis*) planted along the sea front at Puerto Pollença. The fruits of both these palms are too large for Blackcaps to swallow whole, and they were pecking pulp from the fruit on the tree. A number of male Blackcaps were defending individual trees against all other Blackcaps, of which there were many trying to intrude. While actively defending their fruit supplies the males were continually making the *chack* call, at times accelerating it to a continuous chattering. Females, who were largely unsuccessful at reaching fruit on the trees, were on four occasions seen feeding on the ground on fallen fruit. Though our observations were too brief to give more than an indication of what was happening, and the situation was often confusing as no birds were marked, it seemed that these Blackcaps were behaving in essentially the same way as those defending mistletoe clumps in France.

In southern Europe Blackcaps are predominantly frugivorous in winter. Jordano & Herrera (1981) found that in southern Spain their diet consists almost wholly of fruit from October to March; 21 different wild fruits and 8 cultivated or weedy fruits were recorded from four study areas, two of which had natural vegetation unaltered by man. Records by Pérez-Chiscano (1983) from Extremadura, in west-central Spain, included three additional wild fruits eaten by

Blackcaps. As discussed later, many of the Spanish winter fruits are very nutritious, and a mixed diet of such fruits evidently enables Blackcaps to maintain a steady weight throughout the winter. Further east, in southern Dalmatia, Tutman (1969) gave details of the diet of wintering Blackcaps. Arriving in the area in September, they eat figs almost exclusively in September and October. From November or December onwards the very nutritious fruits of *Laurus nobilis* are an important part of their diet; and later in winter olives are eaten. Blackcaps can swallow whole fruits of the wild olive *Olea oleaster* and the smaller cultivated varieties, but peck pieces of pulp from the larger varieties in situ. Finally, ivy berries are eaten from January to the end of March.

The Blackcap populations which breed in the east of the range, in eastern Europe and western Siberia, winter much further south, in tropical East Africa (Moreau 1972). Here too they are almost or entirely frugivorous in winter. In montane forest in Malawi, at the extreme south of the wintering range, F. Dowsett-Lemaire (pers. comm.) found them to be largely frugivorous and had records of their eating five kinds of fruit, including *Polyscias fulva*, a relative of the ivy.

There are several records of Blackcaps feeding their young on fruit. Wahn (1950) watched a nest, and found that the female brought a fruit (a berry of the honeysuckle *Lonicera xylosteum*) to the young for the first time when they were nine days old. She tried to feed it to a nestling, who could not swallow it, tried another with the same result, and finally ate it herself. At 11 days old the young could swallow fruit, and from then onwards they were fed mainly on fruit. Berthold (1984) studied Blackcap nestlings over many years. Although the adult Blackcaps feed on ivy berries in late spring in his area, as they do in ours, not until 1984 did he find evidence of ivy berries being fed to nestlings. In that year the spring was unusually cold and dry, and from 13 May to 17 June he found that young Blackcaps were being fed on ivy berries. Many young birds died. He concluded that, as food for young, ivy berries are a 'Notnahrung' (a food eaten when other, better food is scarce (p. 232)). At Potes (Santander) in northern Spain, on 24 May 1987, when the weather was cool but there was no evidence of a food shortage, some of the Blackcaps visiting an ivy clump with abundant ripe fruit were taking it to their young. At least three birds (a pair and another male) flew off with ivy berries in their bills a minimum of nine times during a 75-minute watch, the pair flying off to a nest (or perhaps fledged young) about 250 m away. Blackcaps were extraordinarily abundant in the area, seeming ecologically very versatile, and possibly they feed their young on ivy berries more regularly than they do in Germany. Other records of Blackcap young being fed on fruit refer to succulent summer fruits such as currants, strawberries and honeysuckle, which are less nutritious than ivy and presumably are fed to nestlings partly as a source of water. Fruit as food for nestlings is discussed further on page 234.

154 The Fruit-eaters

Other warblers and flycatchers

In addition to the Blackcap, three other warblers regularly eat fruit in our study area. Spotted Flycatchers are also occasional frugivores. As discussed on pp. 214–6, fruit is an important part of the diet of many of the *Sylvia* warblers, the genus to which all four of our fruit-eating warblers belong; European warblers of other genera are at best irregular eaters of fruit. We have no first-hand evidence, for instance, of the two common *Phylloscopus* warblers, Chiffchaff and Willow Warbler, showing any interest in fruit in our study area; but there are a few records of their eating fruit elsewhere.

In this chapter we discuss our observations on warblers and flycatchers, and summarise the published records of fruit-eating by these and other related birds from elsewhere.

Garden Warbler *Sylvia borin*

Garden Warblers take fruit regularly from when it first becomes available in July until mid September when they depart on migration. We had 32 records for eight different wild fruits: rowan (1), wild cherry (1), bird cherry (5), blackberry (2), elder (15), wayfaring tree (3), wild currant (2), white bryony (3). All of these records were of birds feeding from a perched position, but we recorded a Garden Warbler plucking a mulberry fruit with a sally. Like the other *Sylvia* warblers, Garden Warblers like to feed gregariously; 48% of our records were of birds feeding in company with conspecifics and also with Blackcap, Blackbird, Song Thrush, Robin and Spotted Flycatcher. Besides Blackcaps, Garden Warblers were the only warblers that we saw taking bird cherry fruit. With a mean diameter of nearly 7 mm, bird cherry fruits are within the size range that Blackcaps and Garden Warblers can swallow, but are probably too large for the two whitethroats, whose gapes are considerably narrower.

We had too few observations of Garden Warblers taking fruit in our study area to make meaningful comparisons of their feeding methods and success rates with those of the Blackcap. But during 12 hours spent watching a group of bird cherry trees at Pec Pod in the Giant Mountains of Czechoslovakia, in August 1980, we saw many individuals of both species feeding on the same fruit under the same conditions. Our data, summarised in Table 59, suggest that the Blackcap is the more efficient frugivore, at least on this fruit. We had many instances of Garden Warblers tugging unsuccessfully and sometimes overbalancing while trying to pluck a fruit. Sometimes a Garden Warbler would lean down, grasp a fruit in its bill, and then drop off the perch in an attempt to pluck the fruit; at times even this failed and the bird was momentarily left suspended.

TABLE 59 *Success and meal sizes of Garden Warblers and Blackcaps feeding on bird cherry*

	No. attempts	% success	No. Range	Meal-sizes Range	Mean
Garden Warbler	120	38($\pm$9)	7	2–3	2.3
Blackcap	55	76($\pm$12)	6	2–6	3.2

Notes. Data from the Giant Mountains, Czechoslovakia. The Garden Warbler's success was significantly lower than the Blackcap's (95% confidence limits given in parentheses). Meal-sizes of the two species did not differ significantly.

GARDEN WARBLERS ELSEWHERE

We have found no published references to Garden Warblers eating fruit in Britain that add significantly to what we recorded. For the continent of Europe there are many references to a variety of wild fruits being eaten; and the Garden Warbler was one of the species used in the experimental study of fruit-eating by Berthold (1976) discussed on pp. 232–3. The field evidence indicates that Garden Warblers are as strongly frugivorous in late summer, on autumn migration and in winter quarters as are Blackcaps; but Berthold found that his captive Garden Warblers did not, as his Blackcaps did, show a positive preference for fruit over insect food at any time.

Banzhaf (1932) and Kroll (1972) each studied the food of Garden Warblers passing through Heligoland on autumn migration. There they feed extensively on the fruits of elder *Sambucus nigra*, blackberry, raspberry, the hips of the rose *Rosa rugosa* (from which they must pick pieces of pulp, as these hips are very large), and the introduced snowberry. Elder fruit is preferred to hips when both are available (the hips ripen first), and the fruit of *Sambucus nigra* is strongly preferred to the red-fruited *S. racemosa*, which Banzhaf did not see being eaten at all although it was fairly abundant (see p. 68).

Madon (1927) stated that Garden Warblers on autumn passage in central and southern Europe concentrate in areas of fig production, where they gorge on the fruit and become very fat. This is confirmed by Thomas (1979) who found that in southern Portugal, in September, Garden Warblers eating figs were heavier than those taking other food. They also feed a great deal on blackberries during autumn passage through the Iberian peninsula (Jordano 1982, Pérez-Chiscano 1983, Thomas 1979).

In winter quarters in south-central Africa Garden Warblers seem to be as strongly frugivorous as Blackcaps. In montane forest in Malawi, F. Dowsett-Lemaire (pers. comm.) recorded them taking eight different species of fruit.

Common Whitethroat *Sylvia communis*

Due no doubt to their present scarcity in our study area, since the population crash of 1969, we had very few records of Common Whitethroats taking fruit: elder (3 records), blackberry (2) and white bryony (1). All fruits were taken from a perched position, and half of the records were of gregarious feeding, once with a Lesser Whitethroat. The Common Whitethroat's gape is a good deal narrower than that of the Blackcap and Garden Warbler (5.7 mm, *cf* 7.4 mm for the other two species), and bird cherry fruits are almost certainly beyond its capacity. When we were watching at bird cherries in the Giant Mountains of Czechoslovakia and recording many Blackcaps and Garden Warblers taking the fruit, we twice saw Common Whitethroats spending a considerable time in the bird cherry trees, but they ignored the fruit and fed on red elders near by.

COMMON WHITETHROATS ELSEWHERE
Apart from Kear's (1968) record of a Common Whitethroat eating the berries of lords and ladies, we have found no reference to wild fruits eaten in Britain that adds to what we observed. There are records of a number of different fruits being eaten on the continent of Europe; Emmrich (1973–74), from a literature survey, found references to 23 species. The commonest were *Rubus* spp., *Ribes* spp. and the elder *Sambucus nigra*. Emmrich made a detailed study of the food of Common Whitethroats on the island of Hiddensee off the German Baltic coast. In May no fruit was eaten; from 8 June–11 July a little *Ribes alpinum* fruit was eaten; and from 18 July–11 September fruits of four kinds were regularly eaten: *Rubus* sp., *Ribes alpinum*, sea buckthorn and elder. Emmrich found that nestlings were almost never fed on fruit, in contrast to nestling Blackcaps and Garden Warblers; he had one record, for one brood. (Kammerer (1921) recorded young being fed on red currants in June.)

Lesser Whitethroat *Sylvia curruca*

Lesser Whitethroats in our study area regularly take fruit in August and September. We had no records in July, which suggests that they may not take much fruit until the breeding season is over. We had 39 records for six different fruits as follows: elder (13 records), perfoliate honeysuckle (12), wayfaring tree (7), rowan (3), blackberry (3), white bryony (1). Lesser Whitethroats regularly came into a rural garden to feed on perfoliate honeysuckle berries, but we had no records at common honeysuckle in the same garden. Common honeysuckle fruits mainly ripen in the latter half of August, when elderberries, a preferred fruit, are ripe; the perfoliate honeysuckle, ripening a little earlier, avoids competition with elder.

All of the 39 feeding records were of birds taking fruit from a perched position, except for three that included a sally. The success rate was low (45% of the 20 attempts where we had good views of the bird), the failures being due to inability to pluck the fruit. These records were collected from birds feeding on perfoliate honeysuckle, elder and wayfaring tree. Like Blackcaps, Lesser Whitethroats took only scraps of rowan fruit pulp left on the tree by Bullfinches after they had taken the seeds. They pecked pieces out of blackberries and one was also seen pecking pieces out of a damson. The lack of records for bird cherry is probably due to the Lesser Whitethroat's small gape (5.0 mm, compared with the Blackcap and Garden Warbler's 7.4 mm). Like Blackcaps, Lesser Whitethroats frequently feed on fruit gregariously and give *chack* calls as they move towards a fruit source, probably as a signal enabling them to synchronise their feeding. 39% of our records were of gregarious feeding, mostly with conspecifics but also with Blackcaps, Robins and Blackbirds.

LESSER WHITETHROATS ELSEWHERE

One of the few British records of a disperser eating berries of *Daphne mezereum* is Aplin's (1910) note of a Lesser Whitethroat taking the fruit in Oxfordshire in July (p. 95). Otherwise most mentions of fruit-eating by Lesser Whitethroats in Britain are general or refer to cultivated soft fruit. Witherby *et al* (1938) mention blackberry and elder. From continental Europe a variety of wild fruits is recorded. Like Blackcaps, Lesser Whitethroats eat the flesh of cherries in situ, leaving the stones hanging.

Spotted Flycatcher *Musciapa striata*

We had a few records of Spotted Flycatchers eating two kinds of fruit in August and September: elderberries (4 records) and woody nightshade (3). Except for two records of a fruit plucked from a perch, all records were of birds taking fruit with a flight sally or while hovering. They had no difficulty in swallowing the small elder fruits, but the one bird, a juvenile, seen taking woody nightshade berries had some difficulty with them. Three that it plucked in flight were taken to a branch near by. One it manipulated in its beak for half a minute and then dropped; one it hit against the branch and manipulated in its beak, and then swallowed it with difficulty; and one it swallowed after manipulation in the beak. Two of the seven records were of gregarious feeding, once with a Blackcap and once with a Garden Warbler.

SPOTTED FLYCATCHERS ELSEWHERE

Spotted Flycatchers are known to be occasional eaters of fruit; Turček (1961) lists 14 different kinds. One might suppose that they would take fruit only when flying insects are scarce, as a result of bad weather. Glutz (1973) reported an observation from Switzerland of a pair of Spotted Flycatchers in late July, in cold windy weather, feeding berries of the dogwood *Cornus stolonifera* to their young. Unfavourable weather may well have been the reason for two other reports of young being fed on fruit, red currants (Zeddelof 1921) and raspberries (Epprecht 1964). But the juvenile Spotted Flycatcher that we saw feeding on woody nightshade berries, on 3 August, was doing so in very warm weather, and it seems more likely that, being inexperienced at flycatching, it was supplementing its diet with a more easily obtained food.

Starling *Sturnus vulgaris*

Starlings are abundant residents in our study area, and their numbers are increased in winter by an influx of Continental birds. In their fruit-eating they differ strikingly from the thrushes in several ways. They are even more highly social in their feeding behaviour; they are unable to deal with, or inefficient at dealing with, several kinds of fruit that the thrushes take quite easily, and so feed on a smaller range of fruits than the thrushes; those fruits that they can take easily they devour in huge quantities, so that they can rapidly strip even a very abundant fruit crop; and their fruit-eating (at least as far as wild fruits are concerned) is more seasonally restricted than the thrushes', with a major peak centred on October.

In midsummer they feed to some extent on the wild cherry (Table 60), the first important fruit to ripen, but probably concentrate much more on the sweeter cultivated varieties in gardens. When the bulk of the elder fruit crop ripens, in August, they turn to it and locally feed almost exclusively on it from late August, to early October, by which time most of the bushes are stripped. At the same time they feed a great deal on yew fruit where it is available. As the elderberry crop is depleted they turn to the late blackberries in field hedges, and to dogwood on the Chiltern slopes. One of the most spectacular sights at this time has been the swirling flocks of Starlings, circling over the dogwood clumps, swooping down again and again but nervous of landing until finally a few birds go down into the bushes and others follow, to feed hard for perhaps half a minute before they all take off and whirl away.

Equally characteristic of the behaviour of Starling flocks in autumn is the way they alternate ground-foraging with blackberry-eating. At this time, especially where there are fields with sheep, they feed in flocks on pastures. Every few minutes a group of birds will fly up together and go to hedge near by, where they feed hard on the blackberries for a minute or two before returning to their

The Fruit-eaters

ground-foraging. At this time they also feed on buckthorn fruit and to a small extent on haw. We suspect that the flocks that feed on the hedgerow blackberries also strip the woody nightshade fruits that remain in the hedges; but we have not seen this and have only a few records of Starlings taking woody nightshade fruits, in July and October. This autumn peak of fruit-eating by Starlings is very marked: 88% of all our Starling records are from the months September–November, 52% of them being in October. After the dogwood and blackberry fruit crops are finished, they feed very little on wild fruits until the ivy crop ripens. There is then a minor peak of fruit-eating in March and April.

The picture is significantly modified if cultivated fruits are added to the seasonal pattern. Great flocks of Starlings feed on strawberry fields in July and on cherries in June and July. They feed a great deal on apples and pears in summer and, later, in winter when the weather is severe.

The Starling's inefficiency at picking and eating some kinds of fruit is almost certainly a consequence of structural characters that are primarily adapted for other functions. Compared with the thrushes the Starling has a high wing-loading; the familiar triangular wing shape, tapering from a broad base at the body to a pointed tip, is very different from a thrush's more rounded wing. As discussed on p. 213, this kind of wing is inefficient for the short-range aerial manoeuvres that are involved in flight sallies to pluck fruit, and Starlings almost never feed in this way. In hundreds of observations of Starlings taking fruit we have only a single record of one sallying for an ivy berry. We have a few records of Starlings clinging and fluttering to maintain their position when plucking, or attempting to pluck, haws and blackberries, and four records of Starlings taking rather inaccessible fruits by fluttering as they took them, flying away as soon as they were taken. Otherwise all our records are of fruits being taken from a perched position. Often, especially when feeding on hanging bunches of elder-berries, Starlings cling with the feet and reach down to take fruit well below them, a manoeuvre at which they are more efficient than the thrushes.

The Starling's bill is narrow and has no hook at the tip, thus contrasting with

TABLE 60 Starling: monthly summary of feeding records

	Jan	Feb	Mar	Apr	May	Jun	Jul	Aug	Sep	Oct	Nov	Dec	Totals
Yew								32	230	445	127	3	837
Ivy	55	8	73	138	5								279
Haw									1	45	146	6	198
Rowan								30	13	20			63
Whitebeam											11		11
Cherry						46	21						67
Sloe	1										17		18
Blackberry									1	227	107		335
Crab-apple	3	1											4
Elder								66	478	172	6		722
Perfoliate honeysuckle							9	5					14
Common honeysuckle								1					1
Dogwood	6								2	1408	540	20	1976
Buckthorn									1	83	6		90
Woody nightshade							3			3			6
No. of different fruits	4	2	1	1	1	1	3	5	7	8	8	3	
Total monthly records	65	9	73	138	5	46	33	134	726	2403	960	29	4621

TABLE 61 *Starling: success rates for taking different fruits, compared with rates for thrushes*

	Starling		Thrushes
	No. attempts	% success	% success
Yew	249	61(± 6)	80–94
Ivy	72	64(± 11)	77–95
Haw	45	40(± 14)	73–92
Cherry	77	10(± 7)	35–41
Elder	388	93(± 3)	93–97
Dogwood	62	97(± 4)	95

Note. Figures refer to fruit taken from perched position (the method almost exclusively used by Starlings – see text). Most of the failures to take yew fruit (71%) were due to fruit being plucked and then dropped. Figures in parentheses are 95% confidence limits.

the broader, slightly hooked bills of thrushes and most other frugivorous birds (p. 215). It is narrow all along its length, widening at the gape less than the bills of thrushes, so that a fruit of 9 mm diameter is at about the upper limit for swallowing, compared with about 13 mm for the Blackbird. Its bill shape and associated specialisations of the skull are adaptations for the Starling's specialised foraging method of probing the ground, gaping to open up a hole, and looking into the hole so made (Feare 1984). It is almost certainly a consequence of its bill shape that Starlings are relatively unsuccessful at taking fruits and swallowing them. Table 61 compares the success rates of Starlings feeding on several different fruits with the rates for thrushes. Only for the small fruits of elder and dogwood (diameters 6.2 mm and 7.3 mm) do they have high success rates; and these are, as already mentioned, two fruits of which they are especially fond. Their failures to deal successfully with other fruits involve failure to pluck them and a tendency to drop them if they do succeed. Some fruits, such as haw, are largely avoided, apparently because Starlings find them difficult to pluck. Our records of Starlings eating haws are very patchy. We often saw Starlings trying rather perfunctorily to pluck haws, failing, and soon giving up, and our records of them plucking and eating haws were largely from certain bushes whose fruits seemed for some reason to be especially easy to detach. Yew fruits, which Starlings pluck quite easily, are often dropped (Table 61), apparently because they are often too large to swallow (mean diameter 10.3 mm). In the case of large fruits with soft pulp, Starlings compensate for their inability to pluck and swallow them whole by picking pieces from the pulp in situ; for this their pointed bills are efficient. When feeding on wild cherry (mean diameter 11.8 mm), they usually pick pieces of pulp in situ, neatly leaving the bare stone still hanging on its stalk, as they also do when feeding on larger cultivated cherries; but we had two records of Starlings swallowing wild cherries whole. If they are taking cherries for their young they pluck the fruit and carry it to the nest, where the stones accumulate either inside or outside the nest hole. We have not seen how they deal with cherries when feeding them to their young; presumably they put them down to pick pieces out of them, then either throw out the stone or leave it at the bottom of the nest. Cherries, wild and cultivated, are the only fruits that we have recorded Starlings taking to their young, being the main (and in most places the only) fruit available at the time the young are in the nest.

As already mentioned, Starlings are usually highly gregarious when feeding on fruit, and in open country often spend some time circling round before they land and begin to feed. They then feed rapidly and generally for a short time, compared with the thrushes, before flying off. A consequence of this behaviour is that Starlings take very variable but often small numbers of fruit at a feed (Table 62). The largest recorded was about 8% of the bird's body weight, much the same as for the thrushes; but most feeds were around 2–3% of body weight.

TABLE 62 *Starling: meal sizes for different fruits*

	Mean wt of fruit (g)	No.	Meal sizes Range	Mean	Wt of largest meal (g)
Yew	0.70	13	3–7	4.8	4.9
Haw	0.63	10	1–4	1.9	2.5
Dogwood	0.21	7	5–15	8.6	3.8
Elder	0.14	16	6–47	20.7	6.6
Pyracantha	0.26	21	3–10	5.9	2.6

Note. A meal size of 6.6 g represents about 8% of the Starling's body weight. The mean meal sizes range from about 1.5 to 4% of the body weight. Data for *Pyracantha coccinea* (introduced) are included to supplement the rather small numbers of data for native fruits.

STARLINGS ELSEWHERE

We have found no published information on the year-round pattern of fruit-eating by Starlings in Britain with which our data can be compared. There are obviously local differences. Thus in northern England, rowan is apparently a much more important fruit than in our study area. We have casually seen Starling flocks feeding on rowan in Cumbria in August, and Butterfield (1910) mentions that Starlings feed a great deal on rowan fruit in Yorkshire; but the records for our study area are not numerous, in spite of over 48 hours of observation.

In continental Europe ornithologists have been especially concerned with the Starling's depredations on cultivated fruit, and it is locally a pest (*eg* Bruns & Haberkorn 1960). Huge numbers of cherries may be brought to young in the nest. Löhrl (1957) recorded 783 cherry stones in a box that had been used for two Starling broods, and Volkmann (1957) counted 357, 321 and 246 stones in three boxes after use by Starlings. In some Mediterranean areas Starlings are a pest in the olive orchards in winter (Tutman 1969). There are also many records of Starlings eating wild fruits, but we have found only two thorough studies that give an idea of the seasonal succession. Havlin & Folk (1965), in a survey of the economic importance of the Starling in Czechoslovakia, had no records of fruit being eaten from March to May, and the following records for June to November: *Lycium* spp., June; elder, July–October; white bryony, July; black nighshade, July–September; rowan, October; sloe, November. Havlin & Folk give no data for the months December–February, presumably because Starlings are absent in those months.

Further south, in Hungary, Szijj (1957) studied the food of the Starling in relation to the latter's agricultural importance, and listed a number of wild fruits, several of which are southern species not found in Britain. In late summer and autumn, elder and blackberry were the main wild fruits eaten. In winter (November–March), especially when the weather was severe, the main food of overwintering Starlings consisted of the fruits of *Celtis, Elaeagnus, Ampelopsis* and *Ligustrum* (privet), only the last of which occurs in our area. This suggests that in south-central Europe fruit must be a more important part of the diet in winter than it is in England, where our data show a marked peak in fruit-eating in October and rather little from December to March; and a similar pattern to that in Hungary may well apply to southern Europe and the Mediterranean region, where many fruits are available in winter (p. 203).

The Corvidae

Our total of 154 records of corvids taking native fruits is a minimum figure, as many of our watches at fruiting plants were at too close a range for visits by these wary species. However, we often saw corvids feeding at more distant fruit, too distant for smaller birds to have been seen, and this may partly compensate for under-recording at close quarters.

We treat the corvids as dispersers, but it is possible that they destroy some soft seeds in the gut, a point discussed on pp. 30 and 32.

Magpie *Pica pica*

We recorded Magpies taking fruit during the period July–December, with a peak in September–November (Table 63). Like the other corvids, Magpies are particularly attracted to whitebeam fruits, coming to feed on them in small groups of 2–8 birds; two other preferred fruits are elder and dogwood. All fruits were taken from a perched position, occasionally with a wing-flutter to maintain balance if the fruit was below the perch. When feeding on whitebeam fruit a Magpie may pull up a whole corymb, then hold it under a foot while removing the fruit. Occasionally it takes dogwood fruit by stretching up from the ground to pluck low-growing fruit.

In the areas where we spent many hours watching rowan fruit, Magpies were plentiful, yet we saw them feeding on rowan fruit only three times. On one occasion, when details of a bird's feeding behaviour could be seen, its success rate was low, only five fruits being eaten out of 12 attempts to take them. It either failed to pluck them, or plucked a whole spray of fruit and then dropped it. Rowan fruits are smaller than whitebeam and the terminal twigs thinner and less rigid, which probably makes them difficult for Magpies to deal with.

164 The Fruit-eaters

TABLE 63 Magpie: monthly summary of feeding records

	Jul	Aug	Sep	Oct	Nov	Dec	Totals
Rowan		2	1				3
Whitebeam				15	10	2	27
Cherry	3	1					4
Bird cherry	1	1					2
Sloe					1		1
Raspberry	1						1
Elder		5	9	3	2		19
Dogwood			3	7	3		13
Privet					1		1
							71

MAGPIES ELSEWHERE

A good many different kinds of fruits are recorded for the Magpie, mainly from northern Europe: mistletoe and spindle (Schuster 1930), dog rose and haw *C. oxyacantha* (Turček 1948, who found the seeds in pellets beneath a winter roost), holly and cloudberry (Turček 1961), and sea buckthorn (Maas 1950, Heymer 1966). Holyoak (1968) recorded blackberry from stomach contents collected in England. For Spain there are records of figs and grapes being eaten (Pérez-Chiscano 1983).

Jay *Garrulus glandarius*

The Jay is the shyest of the corvids in our area. This, and the fact that in woodland fruit-eating can only be watched at close range, gave us few recoreds for this species (Table 64), although it is renowned as a fruit-eater. Like the other corvids, it often holds fruits under its foot while feeding on them.

JAYS ELSEWHERE

There are records of Jays taking many kinds of fruits. The most complete data are given by Bossema (1979) for Holland, where the fruits of *Vaccinium* spp., *Amelanchier ovalis* and cultivated cherries are eaten in June–July, *Rubus* spp. August–January, bird cherry September–October, and elder September–December. For Germany, Rey (1910) listed the following fruits, based on analysis of stomach contents: rowan, October; red elder, October–November; woody nightshade, January. An observation of a Jay feeding on yew berries, quoted by

TABLE 64 Jay: monthly summary of feeding records

	Jul	Aug	Sep	Oct	Nov	Totals
Whitebeam			1	5	1	7
Rowan		1				1
Wild cherry	7					7
Elder			2	1		3
						18

Schuster (1930), seems to be the only record of a corvid taking yew fruits, a point discussed on p. 30. For southern Europe there are records of olives being eaten in Dalmatia (Tutman 1969), holly berries in northern Spain (Guitian 1984), and *Arbutus unedo* and *Pyrus bourgeana* in southern Spain (Pérez-Chiscano 1983).

Carrion Crow *Corvus corone*

We regularly saw Carrion Crows taking native fruits, and obtained altogether 66 records for seven different fruits (Table 65). Most records were for whitebeam, followed by sloe and crab-apple. Except for crab-apples, all fruits were plucked directly from the tree or bush, at heights ranging from about 1 m (dogwood and sloe) to about 10 m (wild cherry). When feeding on whitebeam fruits, Carrion Crows frequently flutter their wings to maintain balance, and once we saw one pluck a whole panicle of fruit and fly away with it. For wild cherry, our records were of an adult accompanied by two begging juveniles, who were fed the fruit seven times during a 40-minute watch.

TABLE 65 *Carrion Crow: monthly summary of feeding records*

	Jan	Feb	Jul	Oct	Nov	Dec	Totals
Haw				1		1	2
Whitebeam				12	5	7	24
Wild cherry			8				8
Sloe				1	12	4	17
Crab-apple	6	6					12
Elder				1			1
Dogwood					2		2
							66

We recorded Carrion Crows feeding on fallen crab-apples below the tree in January and February. They chose the brown, slightly rotten apples, holding them under the foot and pecking them open to eat the pulp. On occasion the bird may act as a disperser, as we twice saw a pair of birds each pick up an apple in the bill and fly off with it for about 150 m, either to eat it or perhaps to hide it.

We had records of Carrion Crows feeding on cultivated strawberries, apples and pears. When feeding on apples and pears they would pluck a fruit from the tree and carry it some distance away; once, having done so, a bird was seen to hide an apple under a tussock of grass. (Rooks and Jackdaws were also seen feeding on fallen apples and pears in very hard weather in winter; but as they were never seen taking any native fruits we have not included them among the fruit-eaters.)

CARRION CROWS ELSEWHERE

Compared with the Magpie and Jay, there are comparatively few records of Carrion Crows eating wild fruits, but altogether a good number of different kinds have been recorded; Turček (1961) lists 22 species, but without details. Mistletoe berries (Schuster 1930) are among the more surprising. In an analysis of the diet of Carrion Crows in the Limoges area, southwestern France, based on stomach contents, Jollet (1984) reported that ivy berries were taken in April and May, and cherries (*Prunus cerasus*) in June and July; no fruits were recorded in other months. Ivy berries were apparently quite important in the diet in spring, making up respectively 7.1% and 4.3% of the items recorded in April and May. This seems to indicate a significant difference between Carrion Crow diets in southwestern France and southern England, as we had no evidence of Carrion Crows taking any interest in ivy fruits in over 200 hours of observation; indeed there does not seem to be any evidence of ivy berries being taken by any of the crow family in Britain (D. Goodwin, pers. comm.). We suggest elsewhere (p. 32) that crows may generally avoid ivy fruits because the soft seeds, which are toxic, may be damaged and so may release toxins in the gut.

Occasional or scarce fruit-eaters

A good many species of birds that do not regularly eat fruit are known to do so occasionally. We had records for three: Moorhen, Green Woodpecker and Greater Spotted Woodpecker. Many others are recorded in the literature.

On five occasions in September we watched Moorhens eating blackberries. They took them either by walking about on the plants, plucking and swallowing the fruits whole, or by reaching for them from the ground or from the water. Once, we saw a juvenile Moorhen in late August taking elderberries by climbing the bush and clambering among the outer branches to pluck the fruit. We had four records of Moorhens taking yew fruit in October, plucking the fruit as they clambered about the tree. For the Green Woodpecker we had three records of birds taking dogwood fruit, in October and November, perching on the top of a bush for some time and systematically picking and swallowing the fruits that were within reach. Our single record for the Greater Spotted Woodpecker was of a bird taking fruit from a wild cherry in July.

Ring Ouzel *Turdus torquatus*

Ring Ouzels, which like the other thrushes are regular fruit-eaters, are scarce birds of passage through our study area. We saw only three birds in the course of our field work, one of which was not seen to feed. The other two each spent several days in one place, feeding on fruit; both were very well-marked birds, apparently adult males. One was present for six days (15–20 April) in an area with an abundant supply of ripe ivy fruit, on which it fed in company with other thrushes. The other was present for five days (17–21 September) in a garden with many fruiting yew trees, on which it was probably feeding the whole time, though only three records were obtained in the time available for watching it.

LITERATURE RECORDS

For the sake of completeness, from a search of the literature we can add a few records that are marginally relevant to fruit-eating in our study area. First, there are a number of regular frugivores for which we had no records as they are either scarce or only occur in small numbers on passage in our area. This category includes several small turdids: Nightingale, Common and Black Redstart, Whinchat and Wheatear. Not only do they regularly take fruit, but most of them are also known to feed their young on fruit on occasion. The Pied Flycatcher is also an occasional fruit-eater.

Warblers of genera other than *Sylvia* rarely eat fruit, but there are some records. Reed Warblers have been recorded eating elderberries and dogwood fruit, and Sedge Warblers and Grasshopper Warblers elderberries (Turček 1961). Of the *Phylloscopus* warblers perhaps the Chiffchaff takes fruit most often, but even so records are few. Elder, red elder and *Vaccinium* sp. are recorded for northern Europe, and blackberry for Spain (Pérez-Chiscano 1983). Gooch (1984) recorded a striking case of Chiffchaffs eating persimmon *Diospyros kaki* fruit in northern Portugal in autumn. In 1979 'for many weeks, passing chiffchaffs gorged themselves on the pulp and skin of ripe persimmons, eating from early morning until evening, with only occasional gaps (not exceeding ten seconds) when none was in the tree; this continued until no fruit remained ... The white-washed lower branches of the tree and the whitened ground below suggested a crowded roost rather than a feeding place.' In 1980, fewer Chiffchaffs were present and they spent much more time feeding normally. One can only speculate about what food scarcity or abnormal concentration of migrants caused the unusual feeding behaviour in 1979. For Willow Warbler and Wood Warbler, there seem to be only a few records of fruit-eating, listed by Turček (1961): red currant, raspberry, common and red elder for Willow Warbler; and red currant and common and dwarf elder for Wood Warbler. These are all small and soft fruits, within the capacity of a small-billed warbler to pluck and swallow.

There are two old records of Cuckoos eating fruit. Müller & Müller (1887) reported their eating juniper berries, and Link (1889) juniper and alder buckthorn *Rhamnus frangula*. According to Link, alder buckthorn fruit is regularly eaten.

More interesting, perhaps, in view of their adaptability and recent spread inland, are the records of gulls eating fruit. Black-headed Gulls have been recorded eating a variety of wild fruits (Cramp & Simmons 1983). Gewalt (1986) has described what is apparently a recently developed habit of Black-headed Gulls in Austria and Germany, little reported in the earlier literature, where they now feed regularly on cultivated cherries; they come in flocks, especially in the evening, and pluck the cherries in flight. Also from continental Europe, there are records of Herring Gulls eating fruits of *Vaccinium* spp. and crowberry (Turček 1961), and olives in the south (Tutman 1969), and of Common Gulls eating wild cherry and sea buckthorn fruit (Turček 1961) and feeding their young on cherries (Wachs 1922). It will be interesting to see whether fruit-eating by gulls becomes commoner as they become increasingly adapted to inland habitats.

168 The Fruit-eaters

Waxwing *Bombycilla garrulus*

Of all European birds the Waxwing is certainly the most specialised frugivore, but for only part of the year; in summer it is almost exclusively insectivorous. It is well known to be an irregular winter visitor to Britain, with substantial invasions in some years and few or none coming in other years. Unluckily in the years of our study there was no Waxwing invasion. Almost certainly it is because it is such a specialised frugivore that it is not a regular migrant to Britain. Waxwings are very hardy; as long as they have adequate fruit supplies they can winter in very cold areas. In most winters they can subsist on the large fruit supplies available on the Continent and have no need, unlike the thrushes, to take advantage of the milder climate of Britain.

The Waxwing's specialisations for fruit-eating are structural, physiological and behavioural. For a bird of its size it has a notably wide gape, enabling it to swallow relatively large fruits – an adaptation seen in more extreme form in some tropical frugivores (p. 215). The following measurements illustrate this point:

Waxwing mean weight	53 g	gape	11.0 mm
Blackbird	95 g		11.1 mm
Fieldfare	112 g		10.7 mm

(weights from Hickling (1983); gape widths from museum specimens)

Its gut is relatively long compared with other birds of its size (Cvitanic 1970), a feature associated with adaptation to a vegetable diet, but a diet of seeds and coarse vegetation and not a specialised diet of fruit-pulp (p. 216). Its liver is relatively large and it has no gall bladder, but the significance of this is not understood (Pulliainen *et al* 1983). Hence it is uncertain to what extent the alimentary canal and digestive system show adaptations to a fruit diet. One would like to know whether there is any seasonal change in gut size or structure, as has been found, for example, in the Starling; such a change would be expected, in view of the marked seasonal change in diet, but seems not to have been investigated.

Physiologically Waxwings appears to be better adapted to a pure fruit diet than unspecialised frugivores such as thrushes and warblers. Whereas Blackbirds, Robins, Blackcaps and Garden Warblers were unable to maintain body weight and rapidly lost condition when kept in captivity and fed exclusively on fruit (Berthold 1976a; see p. 232), Waxwings maintained and in some cases increased their weight, or reduced it only moderately (Berthold 1976b). Their daily fruit intake in the wild is very high. Gibb & Gibb (1951) reported that a bird that was watched for almost a complete day in early December in England, was estimated to have eaten 600–1,000 fruits of *Cotoneaster horizontalis*, and another to have eaten about 500. If, say, 600 fruits were eaten daily, the weight of fruit would be about 90 g, approaching twice the bird's weight. This is in line with the weight of fruit eaten by tropical frugivores in captivity (Moermond & Denslow 1985).

Waxwings usually pluck fruit from a perched position, occasionally fluttering to take a terminal fruit (Glutz 1985). Most fruit is swallowed whole but they pick pieces of pulp from large fruits such as apples. Like some other fruit-eaters, they drink frequently. They habitually feed unaggressively in groups or larger flocks, and persistently return to a good food source, even if driven away. Just as Mistle Thrushes are unable to protect their defended fruit supply against persistent flocks of Redwings and Fieldfares in very cold weather, so they may be overwhelmed by Waxwing flocks. Creutz (1952) describe how Mistle thrushes that were defending mistletoe clumps in the Dresden area of East Germany were unable to keep Waxwing flocks away from the fruit. A flock of about 150 Waxwings came to the mistletoe, going from one plant to another, often 20 birds or more to a single clump. The Mistle Thrushes persistently chased them, with continuous churring. Their attacks broke up the Waxwing parties, but the Waxwings, apparently very hungry, did not move far away and soon returned. It seems possible that where Waxwings are regular winter visitors, long-term defence of a fruit supply by Mistle Thrushes may be impracticable. This may account for the fact that such long-term defence has not been reported from Germany, although many ornithologists there have been interested in fruit-eating (see also p. 40). Creutz's observations were made in early March when the Mistle thrushes had just returned to the area, presumably from winter quarters further south.

Siivonen (1941) investigated the causal factors involved in the winter movements of Waxwings. As long as fruit is available – especially the most important fruit, rowan – Waxwing flocks may remain all winter in northern Europe. Hungary is the southern limit of their regular wintering area. Siivonen concluded that moderate-sized movements are related to fruit abundance in different areas, but that the mass invasions of areas south and west of the usual wintering area are not related to fruit abundance, and apparently have a 10-year periodicity. Invasions that have occurred in the last hundred years have been in the following winters: 1892–93, 1903–04, 1913–14, 1921–22, 1931–32, 1932–33, 1936–37, 1949–50, 1956–57, 1965–66, 1970–71. Intervals of about 10 years predominated in the first few decades, but since then there has been no regular periodicity.

Waxwings have been recorded taking a great variety of fruits; a complete list would comprise all the main fruits, wild and cultivated, available in northern and central Europe in winter. During the 1913–14 invasion of Britain, haws and hips were the main fruits eaten; holly, ivy, bullace and *Cotoneaster affinis* were also recorded, and, surprisingly, elderberries in December–January (various sources). In the 1949–50 invasion, haws, *Cotoneaster* and *Pyracantha* spp. were eaten in late January, and hips and privet fruit in mid March (Gibb & Gibb 1951). During the 1965–66 invasion, in Kent, haw, hip, ivy and *Cotoneaster horizontalis* were again recorded, the haw crop being finished by mid November. In January there was a marked change to apples, including ornamental crab-apples; in February and March crab-apples were eagerly sought and some *Cotoneaster frigida* fruit was still available (Harrison 1966). Ellis (1985) reported Waxwings feeding on guelder rose in Norfolk in January 1985 (not an invasion year), and the Waxwing's fondness for guelder rose has been noted by continental observers (Berthold 1976b, Glutz 1985).

Woodpigeon *Columba palumbus*

When we began our observations we assumed that Woodpigeons are seed-predators, not dispersers of fruits. Typical pigeons of the genus *Columba* have muscular gizzards adapted for grinding food, and long narrow guts, unlike the specialised tropical fruit-eating pigeons (*Ptilinopus, Ducula,* etc) which have less muscular stomachs and short wide guts (Goodwin 1985). Gradually, however, we concluded from examinations of droppings and knowledge of the hardness of different seeds that Woodpigeons are dispersers of several important wild fruits. So much attention has been given to Woodpigeons in relation to farm crops (Murton 1965) that their original ecological role in woodland, their primary habitat, has rather been lost sight of. At the outset, therefore, we should set out our reasons for treating Woodpigeons as dispersers, or as seed-predators, of the fruits listed in Table 66, which summarises our feeding records.

Ivy	Seeds soft. Seeds of unripe fruit (very soft) certainly ground up; a proportion, at least, of seeds of ripe fruit defaecated intact and viable (Macleod 1983).
Holly	21 seeds from Woodpigeon faeces examined; 20 intact, one (the least ripe, palest in colour) broken in half.
Hip	268 seeds from Woodpigeon faeces examined; all intact, except that surface hairs were rubbed off.
Haw	76 seeds from Woodpigeon faeces examined; 74 intact with some surface wear, two broken. (Many others

	examined but not counted; noted as mainly intact with some surface wear.)
Wild cherry	Seeds presumed defaecated intact, as similar stones of cultivated cherries found intact in droppings (D. Goodwin, pers. comm.).
Whitebeam	No direct evidence, but seeds very soft and presumed digested.
Elder	A proportion of seeds of ripe fruits probably dispersed, as two intact seeds found in faeces.
Privet	Seeds presumed digested, because droppings of birds feeding on privet (with characteristic purple stain) contained no intact seeds or identifiable seed fragments.
Woody nightshade / White bryony	No evidence

Woodpigeons are surprisingly agile when feeding on the fruit of trees or climbers, clambering about among the twigs and foliage, clinging, sometimes bracing themselves with half-open wings, and even hanging upside down. We have mostly recorded them feeding singly or in pairs, less ofen in small groups. If undisturbed, they may spend many minutes feeding, presumably until they have filled the crop. We have records of birds taking 453 unripe ivy berries in 21 minutes, 150 ripe ivy berries in 7 minutes, and 59 holly berries in 5 minutes. Their success rate at plucking and swallowing fruit is high: for haw we recorded a 100% success rate (of 97 attempts), for unripe ivy berries 92% (of 1,831 attempts), the failures being mainly due to the fruit being too firmly attached, and for ripe ivy berries 89% (of 117 attempts), the failures being due to the difficulty of reaching and pulling the less accessible fruit. The largest feeds that we recorded were: 453 for unripe ivy berries, 76 and 77 for haws (all these being complete feeds), 59 for holly and 12 for elderberries (these two were incomplete feeds, the birds being disturbed). Four complete feeds of unripe ivy averaged 244 berries. These can be supplemented from the literature: Glutz & Bauer (1980) record 363 elderberries in a Woodpigeon crop, and Cramp (1985) 136 haws plus many cabbage leaf fragments, and a mass of haws weighing 121 g, equivalent to about 190 fruits.

Woodpigeons have very different feeding routines from thrushes. They fill the crop with large meals, followed by long periods of resting and digesting, staying motionless on a perch (which allows their presence to be checked). On five days in January and early February 1984, when up to six Woodpigeons were feeding on unripe ivy near a convenient window, prolonged watching gave an indication of these digesting periods. The seven watches, totalling 14 hours 35 minutes, took place between 11.00 and 15.35 hours. Ten resting and digesting periods ranged from 25 minutes to 2 hours, and averaged 1 hour 16 minutes. The 12 feeding periods ranged from 5 to 26 minutes, and averaged 12.5 minutes.

Of the records summarised in Table 66, 36% are records of Woodpigeons feeding on unripe ivy berries. This is certainly an under-estimate, as it was not until the second winter of our study that we differentiated between ripe and unripe fruit; nearly all the 80 undifferentiated records probably belong in the unripe category. The numbers for other fruits are also subject to more observational bias than is the case for most of the other fruit-eaters. Like the crows, Woodpigeons are wary when feeding and can seldom be closely watched; to compensate for this, they can be seen from a long way off, but from a distance one is more likely to see them when feeding on some kinds of fruits rather than

TABLE 66 Woodpigeon: monthly summary of feeding records

	Jan	Feb	Mar	Apr	May	Jun	Jul	Aug	Sep	Oct	Nov	Dec	Totals
Ivy (ripe)	22	67	38	10	3								140
(unripe)	179	43	15	2								17	256
(not differentiated)	53	5											80
Holly	12	2	9	3							4	2	32
Hip	6	1										9	16
Haw	1	6							2	24	60	4	97
Whitebeam										18			18
Wild cherry					1	8							9
Elder (ripe)								5	47	3			55
(unripe)						1							1
Privet											2	2	4
Woody nightshade									1				1
White bryony	2												2
													711

Note. Woodpigeons are dispersers of the following fruits: some ripe ivy, most holly, hip, most haws, wild cherry, some elder. For other fruits they are probably seed-predators. For further details, see the text.

others. We suspect that holly is under-represented in Table 66, as we had no records for some hollies that were stripped in circumstances suggesting that Woodpigeons were responsible. All of our 15 records of Woodpigeons eating hips were obtained in hard weather in late winter when the ground was snow-covered, most of them by observation from a considerable distance. It is probable that hips are a more important food at such times than the records suggest, to judge from the number of Woodpigeon feathers caught on dog rose bushes.

Our records of Woodpigeons eating fleshy fruits raise more questions than they answer. One concerns the relationship of fruit-eating to the woodland nut crops. In the period of our study, 1980, 1982 and 1984 were very good years for beechmast, and large Woodpigeon flocks fed on it in the Chiltern beechwoods in the autumn. In 1981 the beechmast crop failed entirely, and likewise in 1983, or almost so. Nearly all (88%) of our records of Woodpigeons eating fleshy fruits (mainly haws) in October and November were obtained in 1981 and 1983, which suggests that such fruits are an alternative autumn food when the preferred beechmast fails. The shortage apparently continued to have its effect into mid and late winter, as 77% of our records of Woodpigeons eating unripe ivy berries were obtained in the 1981/82 and 1983/84 winters.

Another question raised by our observations concerns the treatment of fleshy fruits in the gizzard. Woodpigeons feeding on grain grind it in the gizzard by muscular action, helped by the presence of a certain amount of grit (whether grit is always present seems not to be recorded). In examining a large number of droppings of Woodpigeons that had fed on fleshy fruits we have found only two small pieces of grit (small, sharp pieces of hard stone). Some seeds are known to be toxic – of those ingested by Woodpigeons, ivy, at least when ripe, and probably holly – and so it may be best for a Woodpigeon that has fed on such fruits to treat them less harshly than nuts and grain. Possibly, therefore, Woodpigeons adjust their grit intake (and also the muscular action of their gizzards?) to the food that they are eating. Although the food of Woodpigeons has been intensively studied for years, admittedly mainly from an economic viewpoint, it is surprising that these questions which are fundamental to the natural ecology of the species

174 The Fruit-eaters

cannot be answered. Little attention has been paid to the food once it has passed beyond the crop.

WOODPIGEONS ELSEWHERE

In East Anglia, Murton (1965) studied the food of Woodpigeons in farmland over many years and recorded 10 kinds of fleshy fruits in their diet, including some that we did not record (sloe, guelder rose, buckthorn, spindle). Colquhoun (1951), studying Woodpigeon food in the Oxford area during the war, recorded only three kinds of fleshy fruits in crop contents: haw from October to January, elder in September, and ivy (ripe and unripe) from November to April. Colquhoun's records show almost the same seasonal distribution as ours for the same fruits, which are also the three for which we have most records (Table 66). We have found no other published records that add significantly to the range or seasonal pattern of fleshy fruits eaten by Woodpigeons in Britain; but mention should be made of observations by Macleod (1983), who studied the eating of ivy berries by birds in Gloucestershire and found intact ivy seeds in Woodpigeon droppings, many of which seeds germinated when sown. Macleod's observations helped to alert us to the fact that Woodpigeons are dispersers of several kinds of fruits and not, as we had thought, seed-predators.

Essentially the same picture emerges from a survey of the literature from continental Europe: many fleshy fruits are recorded in the Woodpigeon's diet, but the emphasis is generally on ground-feeding, especially on agricultural crops. The most thorough recent survey (Glutz & Bauer 1980) lists many fleshy fruits, but with no indication of whether or not the seeds are digested. Heim de Balsac (1928a) mentioned Woodpigeons eating unripe ivy berries in eastern France, where it seems that they may destroy even more of the crop than in England, as he noted that few berries remained to ripen in the spring.

In the Canary Islands and Madeira three endemic pigeons occur, two of which are probably recent derivatives of Woodpigeon stock (Boll's Pigeon *C. bollii* and the Long-toed Pigeon *C. trocaz*) while the third (the Laurel Pigeon *C. junoniae*) is probably an earlier derivative of the same stock (Goodwin 1983). Though all of them have been reported to feed on cultivated fields, their main food, and certainly the food to which they are adapted, is the fruit of forest trees (now unfortunately much depleted), among which the laurels were most important. Probably these pigeons were dispersers of the seeds of the several endemic laurel species in the islands; there are no other likely dispersers. The laurels (family Lauraceae) tend to have soft seeds and some at least have toxic seeds, presumably as a deterrent to seed-predators. A study of the endemic pigeons of the Atlantic islands in relation to their natural food supply would be of great interest.

The Fruit-eaters 175

Seed-predators and pulp-predators

We have already mentioned the differences between 'legitimate' fruit-eaters, which act as seed-dispersers, and two other kinds of fruit-eaters: seed-predators that eat the seeds, discarding the pulp, and pulp-predators that eat fruit-pulp without dispersing the seed. Functionally the distinction between the different categories is crucial to the plant, and there are important evolutionary implications. Thus a plant should be adapted to encourage legitimate fruit-eaters and to prevent or discourage exploitation by seed-predators. But it is not always possible to fit each bird species neatly into one of the categories. In our area Woodpigeons are either seed-predators or dispersers, depending on what fruits they are eating; and they may on occasion be involuntary dispersers even of seeds that they normally destroy, if they are killed with a full crop by a predator. Tits may be either seed-predators or pulp-predators, or occasionally dispersers, depending on the fruit and on the behaviour of the feeding bird. Some legitimate fruit-eaters may on occasion be pulp-predators, for example Starling when feeding on cherries. The finches, however, are relatively straightforward: when they feed on fleshy fruits they – or at least, the species in our study area – are always seed-predators.

The impact of seed-predators on a fruit crop may be severe, resulting in the destruction of a high percentage of all seeds. In this section we describe the feeding behaviour of the seed-predators and pulp-predators in our area and, later, give a general discussion of the ecological and evolutionary implications of their activities.

176 The Fruit-eaters

The finches

Four species of finch are seed-predators of fleshy fruits in our area: Bullfinch, Greenfinch, Chaffinch and Brambling. (The Hawfinch, an important seed-predator of several kinds of fleshy fruits, is unfortunately rare or absent in the area.) Other finches appear not to exploit fleshy fruits or, if they do, only to an insignificant extent. All the finches, whether or not they exploit fleshy fruits, are of course adapted for seed-eating; their various specialisations for tackling seeds of different kinds and their methods of dealing with them are well described by Newton (1978). For convenience we include our records for House and Tree Sparrows in this section.

Bullfinch *Pyrrhula pyrrhula*

Of the four finches which we recorded taking the seeds of fleshy fruits, the Bullfinch took the widest variety of native species, 12 in all (Table 67). It was also the most persistent of the seed-predator finches, as indicated by the total of 549 records compared with 318 for the Greenfinch. We saw Bullfinches taking seeds from ripe fleshy fruits from July to February inclusive; a few January and February records were of birds taking seeds from dried-up fruits still on the bush. In contrast to the variety of native fruits exploited, a single Bullfinch taking the seeds of *Pyracantha coccinea* was our only record for an introduced or cultivated plant.

We recorded Bullfinches feeding on the seeds of all the native species of Caprifoliaceae that occur in our area; no other finch showed such a close association with this plant family. They took a substantial proportion of the seed crops of both honeysuckles and guelder rose (Table 67), but little from the wayfaring tree, probably because the latter ripens at the same time as the rowan. The total number of records for the Rosaceae was almost exactly the same as for the Caprifoliaceae, but this is almost entirely due to the strong preference shown for the three species of *Sorbus*, rowan, whitebeam and wild service tree. Although many different seed-predators feed on whitebeam seeds, Bullfinches were the only ones that we saw feeding on rowan, which has smaller seeds than whitebeam. They take a substantial part of the rowan seed crop; in fact some trees in shady sites, whose fruits ripen slightly later than the others, appear to lose most of their seeds to Bullfinches.

Newton (1967a) pointed out that although Bullfinches and Greenfinches are of about the same size, not only is the Bullfinch's bill smaller than the Greenfinch's but its jaws are surprisingly weak compared with the Greenfinch's. This accounts for the fact that the seeds taken by Bullfinches have thinner seed-coats and are generally smaller than those taken by Greenfinches. The Bullfinch's bill, however, has a very sharp cutting edge, which Newton interprets as an adaptation for breaking into buds and fleshy fruits. For the nine native fruits for which we have ten or more Bullfinch feeding records, the average seed weight is 1.9 mg; for the Greenfinch, using the same criteria, the average seed weight is 3.7 mg. Consequently there is little overlap in their diets, at least as far as the seeds of fleshy fruits are concerned; only whitebeam and blackberry are commonly taken by both species.

Bullfinches use two techniques to get at seeds within a fleshy fruit. The most commonly used method is to open the fruit in situ and remove the seeds, leaving the skin and some pulp hanging; this is the method mainly used when feeding on rowan, whitebeam, guelder rose, buckthorn, privet and woody nightshade.

TABLE 67 *Bullfinch: monthly summary of feeding records*

	Jul	Aug	Sep	Oct	Nov	Dec	Jan	Feb	Totals
Rosaceae									
Rowan	18	66	21						105
Whitebeam			4	24	43	26			97
Wild service tree					5	14			19
Blackberry		1	5	1				1★	8
Caprifoliaceae									
Elder		2	14	1	12				29
Guelder rose						26	27	4	57
Wayfaring trees		1					2★		3
Perfoliate honeysuckle	21	44							65
Common honeysuckle	40	6	6					52	
Others									
Buckthorn				2	22	23	3	3★	53
Privet				9	17	9	11	2	48
Woody nightshade		2	1	1	1	8			13
No. different fruits	2	7	6	7	6	6	4	4	
Total monthly records	39	156	51	44	100	106	43	10	549

Note: Records of seeds taken from dried-up fruits attached to the plant are indicated by an asterisk

178 The Fruit-eaters

The other method is to pluck the whole fruit and mandibulate it to obtain the seeds, dropping the skin and pulp. This is most often used when feeding on elder, the two honeysuckles, and blackberry. When feeding on blackberries a Bullfinch plucks one drupelet at a time. Bullfinches quite frequently hover to pluck a fruit, which they then mandibulate on a perch near by; 7% of our records include some hovering, at seven different kinds of fruit. Hovering was most frequent at common honeysuckle (46% of all records), followed by guelder rose (17%), and mostly occurred late in the season of the plant, when the last remaining fruit can only be reached in this way.

Whichever method is used, it is characteristic of the Bullfinch to spend a very brief period bending to pluck a fruit or extract a seed before its head is up again and alert for possible danger. Bullfinches when feeding are at risk from predators such as Sparrowhawks, because much of the fruit is presented peripherally on the plant; also, what are apparently the same individuals may feed at a particular fruit source for several weeks. Defence against predators is probably also the reason why Bullfinches usually feed gregariously at the same fruiting plant with conspecifics or, less often, with birds of other species. Of our 549 feeding records, only 15% were of birds feeding solitarily. We had records of gregarious feeding with Greenfinches and Chaffinches at whitebeam, where Blackbirds and Redwings also were often feeding. On such occasions the Bullfinches respond to the alarm calls of Chaffinches and Blackbirds, and all leave the tree for a period.

Mistle Thrushes defending buckthorn trees (p. 128) recognise Bullfinches as food competitors and drive them away. On four occasions when defended buckthorns were being watched, in December, a single Bullfinch managed to slip in for a feed when we, or a passing pedestrian, frightened the Mistle Thrush away.

BULLFINCHES ELSEWHERE

Newton (1967b), who studied their feeding habits in Marley Wood near Oxford, did not find the seeds of fleshy fruits of great importance for Bullfinches except in September when 16% of his feeding observations were at blackberry, 12% at privet (unripe fruit), and 1% at elder. (In his area honeysuckle was scarce, and rowan and whitebeam apparently absent.) Our records, collected by a different method, indicated a peak in August and a smaller winter peak in November–December, and are more akin to the findings from eastern Europe (Erkamo 1948) where Bullfinches appear to prefer seeds from fleshy fruits as long as crops last.

Radford (1968) recorded a pair of Bullfinches tearing off and swallowing pieces of pulp of the introduced snowberry (*Symphoricarpus rivularis*) in winter. Commenting on this, Newton mentioned that he had three records of Bullfinches eating the pulp of haws. We had no record of Bullfinches feeding on fruit pulp.

Greenfinch *Carduelis chloris*

We recorded Greenfinches taking the seeds of fewer kinds of native fleshy fruits than Bullfinches (Table 68). A marked difference from the Bullfinch was that, with the exception of a single perfoliate honeysuckle record, there were no records for any of the Caprifoliaceae. As mentioned above, Greenfinches tend to take larger seeds than do Bullfinches, a difference doubtless related to their bigger, more powerful bills (Newton 1972). Greenfinches seem to be more apt to learn new feeding habits than Bullfinches; thus we had 68 records of their taking seeds

TABLE 68 Greenfinch: monthly summary of feeding records

	Jan	Feb	Mar	Apr	May	Jun	Jul	Aug	Sep	Oct	Nov	Dec	Totals
Native plants													
Yew	4							15	42	31	87	32	211
Mistletoe											2		2
Spurge laurel						8							8
Dog rose	4	7										11	22
Field rose												16	16
Whitebeam									2	1	39	11	53
Blackberry										1	4		5
Perfoliate honeysuckle							1						1
													318
Introduced/cultivated plants													
Rosa rubrifolia										5	41	5	51
Amelanchier laevis						4							4
Cotoneaster bullatus									4				4
C. cornubia			3							1		3	7
C. franchetii											1		1
Apple (cult.)												1	1
													68

of six introduced or cultivated fruits compared with only a single record for the Bullfinch. Pettersson (1956, 1961) has described how Greenfinches began feeding on unripe seeds of *Daphne mezereum* in gardens and the habit rapidly spread, apparently as the birds learnt from one another (but see p. 95).

The habit of feeding on unripe seeds of spurge laurel *Daphne laureola* in our area is probably also a habit learnt locally. We recorded it only in one area, where 65 spurge laurel bushes formed part of two frequently trimmed hedges on either side of a road through open farmland, a most unusual site for this plant which usually grows in the deep shade of beech woods. Here over 98% of the seeds were taken while still unripe, probably all by Greenfinches.

Two trees, yew and whitebeam, make up 83% of our records of Greenfinches taking seeds of native fleshy fruits, and most of these records are for September–December. Greenfinches begin to take yew seeds in August, when some fruit is still unripe; at times fledged young follow their parents into yew trees, to be fed on opened ripe and unripe seeds. Greenfinches show a marked tendency, not noted in other finches, to extract whitebeam seeds from brown or blackish dried-up fruits on the tree, even when plenty of ripe fruit in good condition remains. Of our November records 31%, a minimum figure, were of birds choosing these dried-up fruits.

Greenfinches normally exploit fleshy fruits from a perched position. On only five occasions did we record a bird hovering to pluck fruit, in all cases at yew late in the season when the easily plucked fruits had been taken but some less accessible fruits remained on branchlets below the main branches. When feeding on yew seeds, Greenfinches hold the whole fruit in the bill and mandibulate the seed so that the fleshy part falls away. They treat blackberries in the same way, taking one drupelet at a time. On the other hand when feeding on whitebeam, rose hips and cotoneaster, they usually open the fruit on the tree or bush and remove the seeds one at a time, leaving the skin and pulp in place. A Greenfinch feeding on rose hips will perch for up to 30 minutes, moving only when bending to open a fruit or extract a seed. At *Rosa rubrifolia* we recorded male Greenfinches dealing with seeds at a rate of 3.7 per minute, with feeds lasting for 15–20 minutes.

Greenfinches usually feed gregariously (68% of all our records), mostly with conspecifics but also with dispersers or, when feeding on whitebeam, with other seed-predators. In spite of this precaution against predation, we saw a Greenfinch taken by a male Sparrowhawk while feeding on yew. At the time, in late November, the fruit was sparse and scattered, as were the six other Greenfinches feeding in the same yew tree.

GREENFINCHES ELSEWHERE

The most thorough information on the food of Greenfinches in Britain is given by Newton (1967a), based on field work near Oxford. His data indicate that Greenfinches took the seeds of the same fleshy fruits as we found in our study area, *ie* yew, *Rosa* spp. (especially *R. canina*), blackberry and *Cotoneaster* spp. He did not record mistletoe, spurge laurel, or whitebeam. Radford (1980) observed Greenfinches eating the seeds of snowberry (*Symphoricarpus rivularis*), tearing the pulp to get at the seeds.

In Holstein in northern Germany, Eber (1956) recorded snowberry seeds as an important winter food of Greenfinches. Turček (1961) included *Crataegus* spp. and guelder rose among the seeds eaten by Greenfinches in Czechoslovakia, unfortunately without giving any information about their frequency in the diet. Though both these latter were available in our study areas, neither Newton nor we had evidence of Greenfinches taking them.

Chaffinch *Fringilla coelebs* and Brambling *F. montifringilla*

Chaffinch and Brambling take few seeds from fleshy fruits. We recorded them taking seeds of whitebeam (Table 69), opening the fruits on the tree and extracting and eating the seeds there, at all heights where the fruit was available. Chaffinches also take the seeds of crab-apples and domestic apples. When feeding at crab-apples they may peck away some of the pulp to get at the seeds, but mostly they wait for seeds to be exposed by Blackbirds or other birds that eat the pulp. Apple seeds are taken mainly from fallen fruit on the ground, but they may also be taken from fruits still on the tree after they have been opened by other birds.

We often saw Chaffinches in or below yew trees where Greenfinches or Grey

TABLE 69 *Chaffinch and Brambling: monthly summary of feeding records*

	Feb	Apr	Sep	Oct	Nov	Dec	Totals
Chaffinch							
Whitebeam			3	6	42	4	55
Crab-apple	13	1					14
Woody Nightshade	1						1
Brambling							
Whitebeam					21		21

Note. The woody nightshade record was of a bird taking seeds from dried-up fruits in situ.

Squirrels were feeding, pecking small grains of endosperm from the seed-husks cracked in half by the two seed-predators. These half seed-husks quite frequently lodge in the trees. We have not included these records in Table 69, as they constitute scavenging rather than seed-predation.

It is possible that Chaffinches take the seeds of mistletoe berries. They are repeatedly chased by Mistle Thrushes guarding mistletoe clumps, and persistently return to the same bunches of mistletoe. They may, however, only be taking seeds that have been defaecated and are adhering to the branches, as do Blue Tits also.

CHAFFINCHES AND BRAMBLINGS ELSEWHERE

In his thorough study of the food of finches in southern England, Newton (1967a) recorded no seeds of fleshy fruits for either Chaffinch or Brambling. In northern Germany, however, Eber (1956) recorded seeds of *Sorbus* sp. in the diet of Chaffinches and Bramblings, and Turček (1961) listed some fleshy fruits for both species in Czechoslovakia, but without any details of how, or how regularly, they are taken.

The sparrows

We recorded House Sparrows as true seed-predators only at ornamental plants, with 53 records at *Pyracantha coccinea* and one at *Cotoneaster cornubia*. When feeding on these seeds they open the fruit in situ, discarding the pulp and eating the seeds.

The House Sparrow has a special method of exploiting the fruits of elder, which spoils them but does not destroy the seeds. They pluck the berries, squeeze out the juice, and then drop the skin and seeds below the elder bush. We had 73 records of House Sparrows feeding on elder in this way, all but four of them in August and September. We also saw two Tree Sparrows *Passer montanus* feeding on elder fruits in the same way. Without giving details, Zedler (1954) and Schlegel & Schlegel (1965) report that House Sparrows feed on elderberries in Germany.

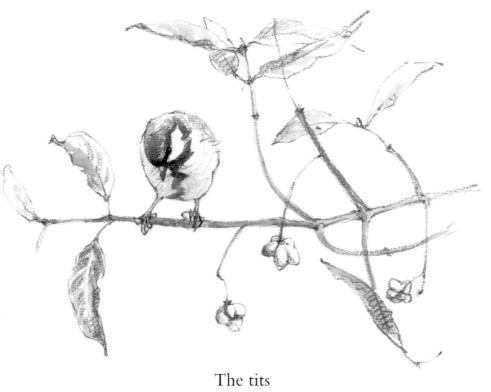

The tits

As the tits use various methods when feeding on fleshy fruits, and may act either as seed-predators, pulp-predators, or occasionally dispersers, we describe their feeding methods in some detail. Casual references to fruit-eating by tits in the literature do not usually give enough detail for a proper assessment of their role as predators or dispersers; consequently we have not included sections on records from elsewhere, but at the end of each account we refer to some of the more informative published observations that we have come across.

Great Tit *Parus major*

Our 54 records of Great Tits feeding on soft fruits indicate little dependence on this food source (Table 70). Most records are from spindle, for which the Great Tit is a pulp-predator. Its technique is to pluck a whole fruit (seed plus aril) while perched, take it to a near-by perch (nearly always, in our experience, in the same tree or bush), hold it under a foot while removing the aril, and then drop the seed, often with some of the aril still adhering. Up to six fruits may be dealt with in this way in the course of a feed. Only once did we see the aril removed at a neighbouring bush, so that dispersal over a short distance was achieved.

 The Great Tit is a seed-predator for the four other fruits. At yew, whitebeam and elder it plucks the whole fruit and then holds it under a foot to extract the

184 The Fruit-eaters

TABLE 70 Great Tit: monthly summary of feeding records

	Oct	Nov	Dec	Jan	Feb	Totals
Yew			6			6
Whitebeam		2				2
Crab-apple					3	3
Elder	1					1
Spindle	1	28	11		2	42
						54

seed, discarding the pulp. The large yew seeds are broken up by hammering as they are held underfoot. For crab-apple, our observations are of Great Tits taking seeds already exposed by other birds, and holding them under a foot to open them. Fruits that are plucked are nearly always taken by a perched bird; only 7% of our records included hovering to pluck a fruit.

Elsewhere in Europe Great Tits are known to be both pulp-predators and seed-predators. In Germany they have been recorded pecking flesh from wild cherry fruits and feeding it to fledged young (Heymer 1966), and in Dalmatia taking the flesh from olives (Tutman 1969). In the Vercors district of south-eastern France we have seen them feeding on the pulp of dogwood fruits in the same way as they exploit spindle. Oakes (1942) recorded them as seed-predators of whitebeam in Lancashire.

Blue Tit *Parus caeruleus*

Blue Tits were observed feeding at 12 different native fruits and four introduced or cultivated fruits (Table 71). In nearly three-quarters of the records – those for

TABLE 71 Blue Tit: monthly summary of feeding records

	Aug	Sep	Oct	Nov	Dec	Jan	Feb	Totals
Yew	1	1						2
Hip					2	5		7
Haw			6	20	2	1	17	46
Rowan		2						2
Whitebeam		20	5	7		2		34
Blackberry	4	24	2					30
Elder	6	99	20	1				126
Perfoliate honeysuckle	3							3
Common honeysuckle		1		7				8
Dogwood		4	6			1		11
Privet			3			1	5	9
Woody nightshade					1			1
								279

Note. For methods of dealing with the different fruits, see text. The two January whitebeam records were of birds taking seeds from dried-up fruits.

elder, the two honeysuckles, whitebeam and blackberry – they were acting as seed-predators. Their methods of taking seeds from these fruits differ. When feeding at elder, which has 2–4 seeds per fruit, a Blue Tit usually plucks a whole fruit, holds it under its foot, removes the seeds and drops the remainder. The fruit is usually opened where it is plucked, but sometimes the last fruit in a feed is carried 2–3 m before being opened, and on these occasions a few seeds may be dispersed, as seed removal is sometimes incomplete in multi-seeded fruits. When feeding at the honeysuckles, it opens and takes the seeds in situ; all our records for common honeysuckle were of birds taking seeds from dried-up fruits. Blue Tits feeding on whitebeam seeds open the fruits in situ, and possibly eat some of the pulp while getting at the seeds. We never saw them taking seeds of rowan, but twice saw them taking pulp remains left on the tree by Bullfinches. When feeding on blackberries, a Blue Tit may pluck one drupelet at a time and hold it underfoot to extract the seed, or may peck into the blackberry in situ, presumably also to extract the seeds.

When feeding on haw, hip and privet, Blue Tits act as pulp-predators. They usually peck a few beakfuls of pulp from a fruit in situ, leaving much of it, including the seeds, intact on the plant. At dogwood they also act as pulp-predators, but usually pluck the whole fruit and hold it under the foot, removing the pulp and dropping the seed. Occasionally this is done away from the parent plant, so that the seed is dispersed. Haws are also occasionally treated in this way. Yew berries are plucked and held under the foot, then dropped after a little of the pulp has been removed; but they seem to be only rarely eaten. We were surprised that, in many hours of watching, we never saw a Blue Tit feeding on the aril of spindle, one of the most important fruits eaten by the other tits. Possibly its very short beak is inefficient at extracting the arillate seed from its capsule.

Once a Blue Tit was seen to pluck an elder fruit on the wing, but all other records were of perched feeding, in marked contrast to the Marsh Tit.

There are records from elsewhere of Blue Tits acting as seed-predators of fleshy fruits. Oakes (1942) observed them taking the seeds of whitebeam in Lancashire. According to Heim de Balsac & Mayaud (1930), who made detailed observations on seed-dispersal of mistletoe in France, Blue Tits eat mistletoe seeds adhering to branches after being defaecated by Mistle Thrushes. We also suspected this as we often saw Blue Tits persistently foraging high in trees near mistletoe clumps that were defended by Mistle Thrushes.

Marsh Tit *Parus palustris*

Taking into account their much lower density compared with Blue Tits, our 48 records of Marsh Tits at eight fruit species indicate a similar, or perhaps somewhat higher, tendency to exploit fleshy fruits (Table 72). In 56% of the records the Marsh Tits were acting as pulp-predators, and as seed-predators in the rest. Unlike the Blue Tit, none of the Marsh Tit pulp-feeding involved pecking pulp from the fruit in situ; it always plucked the whole fruit and then removed the pulp or aril while holding the fruit under its foot. Marsh Tits frequently hover to pluck a fruit; 27% of our records included some hovering.

Both as a pulp- and as a seed-predator Marsh Tits may disperse seeds. On seven occasions we noted them moving spindle, dogwood and privet seeds for distances of 1–8 m before removing the aril or pulp and then dropping the seed. While this is not far, it is likely that the seed will be deposited at a site suitable for the plant. During 26 minutes when a Marsh Tit fed on the seeds of 15 honeysuckle

TABLE 72 *Marsh Tit: monthly summary of feeding records*

	Aug	Sep	Oct	Nov	Dec	Jan	Feb	Totals
Crab-apple							2	2
Wild service tree					3			3
Elder	2							2
Honeysuckle		6	3					9
Common spindle				2	12	1	4	19
Dogwood				1	6			7
Privet						1		1
White bryony		2		3				5
								48

Note. Marsh Tits may be either seed-predators, pulp-predators or occasionally dispersers. for details see text.

fruits, three were taken about 4 m from the parent plant. After it had finished feeding, we searched below the perches that it had used when opening the fruit and found 15 fruit skins, four of which still contained a seed. The irregular number of seeds in a honeysuckle fruit (1–8) may help to ensure that seed-predators miss some of them.

It is well known that Marsh Tits store much food in autumn and winter (Perrins 1979). Some white bryony fruits which we saw Marsh Tits pluck in September were taken about 20 m away for storage. Presumably some stored seeds are never retrieved and may germinate. As late in the winter as February, when fallen crab-apples are opened up by other birds, a Marsh Tit was seen carrying a seed for more than 50 m, presumably also for storage.

MARSH TITS ELSEWHERE

In Marley Wood near Oxford, Marsh Tits have been recorded feeding on elderberries in September and spindle fruits in December and January (Gibb 1954, Perrins 1979); the method of dealing with the fruits is not described. Sorensen (1981) listed seven fruits for 'Marsh/Willow Tit' from observations in the same woodland, including bramble, black bryony and woody nightshade. Also without giving details, Oakes (1942) reported from Lancashire that Marsh Tits are very fond of guelder rose fruits. W. D. Campbell's record of a Marsh Tit storing whole crab-apples is quoted on page 64.

Willow Tit *Parus montanus*

Willow Tits seem to exploit fleshy fruits very little. As Perrins (1979) suggests, their slightly more slender beaks are probably less suited to dealing with seeds than the Marsh Tit's stouter beak. We had a single record of a Willow Tit taking seeds from dried-up white bryony fruits in late October. It opened up the fruits in situ, removed seeds and hid them 2–3 m away, repeating the process four times.

Long-tailed Tit *Aegithalos caudatus*

We had 22 records of Long-tailed Tits exploiting fleshy fruits, all in January and in three different years: eight records for spindle and 14 for dogwood. At spindle, a small flock came twice during a 25-minute watch and removed the arils from the seeds while hanging onto the capsule upside down. With the removal of the aril, some seeds fell to the ground below. By January any uneaten dogwood fruits remain on the bush with the pulp partly dried up. The flocks of Long-tailed Tits pulled off strips of the pulp, leaving the seed on the bush.

Gibb (1954) observed Long-tailed Tits exploiting spindle fruits in Marley Wood near Oxford in January–March. In his thorough account of the ecology and behaviour of the Long-tailed Tit in Germany, Riehm (1970) mentions two fruits in his discussion of their diet, sloe and spindle. He gives no details of their method of exploiting these fruits, but there can be no doubt that it is the pulp and not the seed that is taken. Sandring (1944) recorded Long-tailed Tits taking spindle 'seeds', also in Germany; but it must surely have been the pulp that they were eating.

3: INTERPRETATION

Evolutionary ecologists have built up a considerable body of theory over the last 15 years, in an attempt to explain the ways in which birds and plants have interacted and have become adapted to one another. Two relationships are of particular importance: that between nectar-eating birds and flowers (well reviewed by F. G. Stiles (1980)), and that between fruit-eating birds and fruits, the subject of this book. It is generally accepted that most fleshy fruits are adapted for dispersal by birds, and it is equally well established that many birds are adapted for a diet of fruit. Typical bird-dispersed fruits are visually conspicuous, of a size that birds can swallow, presented in such a way that birds can reach and pluck them, and lacking a scent. In some of these characters they contrast with fruits adapted for dispersal by other animals. But these very general statements are only a starting point; they lead to further, and in some cases more controversial hypotheses involving adaptation and evolution. The purpose of this part of the book is to discuss our observations on fruit-eating, in a small area of southern England, in the light of these hypotheses and in a wider geographical context.

The subject is complex and the many different aspects, some ornithological and some botanical, are all interrelated. Our treatment is not exhaustive but aims to give an adequate (we hope not over-simplified) idea of ways in which the facts presented in Parts 1 and 2 may be given an evolutionary interpretation. It consists of a series of short sections, each dealing with a more or less discrete aspect of the subject. The first sections deal with aspects that primarily concern the plants; then follow some sections on aspects primarily concerning the birds; and a final section attempts a synthesis of the evolutionary picture. Readers who wish to go more thoroughly into the subject will need to consult other works; we recommend the recent discussions and reviews by Herrera (1985a, 1985b, 1986), Howe (1986) and Moermond & Denslow (1985).

Physical characteristics of fruits: size and seed burden

Botanists have an elaborate system for classifying fruits, based on anatomy and development. Use of this system is essential in any analysis of the relationships and evolution of plants, but fortunately for our purposes we can generally ignore it, because we are concerned primarily with a functional division of the fruit into the 'pulp' (the part offered by the plant to the fruit-eating bird, and digested), and the 'seed' (the part that is regurgitated or defaecated). We may simply note that the pulp may be formed from part of the pericarp (wall of the ovary), or from the receptacle; or in the case of arillate fruit may be an outgrowth of the seed itself. Also, that what we here call the seed may be botanically either an unprotected seed or a seed enclosed within a hard endocarp (the inner layer of the pericarp). Fruits whose seeds are not so enclosed are berries; those with enclosed, protected seeds are drupes. Appendix 1, in which all the wild fruits in our study area are listed, gives the correct botanical term for each fruit.

FRUIT SIZE

In tropical forest, relatively huge fruits up to 30 mm or more in diameter are plucked and swallowed whole by hornbills, toucans, and other large frugivorous birds. Other specialised frugivores, though smaller, have extremely wide gapes which enable them to swallow fruits that are huge for the size of the bird. No British birds that eat fruit are either very large or have very wide gapes, and the fruits that they eat are correspondingly small.

Fig. 7 shows the size distribution of the wild bird-dispersed fruits found in our study area. There is a peak in the size range, 8–9 mm (diameter), and over 50% of the fruits are in the 7–10 mm range. Few exceed 12 mm and none averages more than 15 mm. This is approximately the upper size limit of fruits that the larger thrushes can swallow whole. Only one fleshy fruit is much larger, the crab-apple (about 25 mm) and, as discussed elsewhere, it is probably not adapted primarily for dispersal by birds. One other fruit, deadly nightshade, is very variable in size but often larger than the mean value of 13.5 mm used in the figure, fruits up to 16 mm in diameter being common. This is a puzzling fruit and is probably bird-dispersed though there is little evidence (p. 89).

Fig. 7 also shows the gape widths of the main avian dispersers of fruits in southern England, based on measurements from skeletons. The measurement given is the diameter of the largest round or ovoid object that can pass between the rami of the lower jaw. The general correspondence between gape widths and fruit diameters is good, but birds can swallow fruits somewhat larger than indicated by the measurements from skeletons. Thus the larger thrushes (gape widths 10.7–11.9 mm) can swallow fruits up to about 14 mm diameter, and the Blackcap (gape width 7.4 mm) can swallow fruits in the 8–9 mm range. This is presumably made possible by the flexibility of the rami of the lower jaw, which must bow outwards to some extent during the swallowing of large fruit.

The only comprehensive data available for comparison from elsewhere in Europe are those given by Herrera (1984a) for two scrubland areas in southern Spain. There, most fruits are in the 5–10 mm size range, few being more than 10 mm. Herrera notes that this corresponds closely to the gape size of the main frugivorous birds of the area. Warblers are much the most important dispersers. Thrushes are much fewer: the Blackbird is the only resident species, and the Song Thrush a not very numerous winter visitor.

190 Interpretation

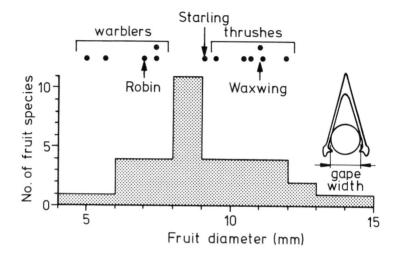

Figure 7 Diameters of bird-dispersed fruits, southern England, and gape-widths of the main dispersers.
Diameters are for fruits listed in Appendix 2, with the exception of *Rubia peregrina* (not found in study area), *Rubus* spp. ('polydrupes' not amenable to a single measurement), and wild strawberry and crab-apple (probably not primarily bird-dispersed). Gape-width is defined as the maximum diameter of a round object that can pass between the rami of the lower jaw – see inset drawing.

How is this correspondence between fruit size and gape width of the dispersers to be explained? Theoretically one can envisage two processes that may be involved, which are not mutually exclusive.

(1) The correspondence might be the result of coevolution between plant and bird, mutually adapting each to the other. Here we must anticipate arguments that are more fully developed later. Although it seems that such a coevolutionary process must have taken place, there may not necessarily have been any tight coevolution between particular plants and particular bird species; but instead a much 'looser' co-evolution, adapting plants for dispersal by birds of some general kind and size. This seems likely in view of the fact, which Herrera (1985a) has stressed, that fruit characters have been much more evolutionarily conservative than bird faunas have been. Thus the fruits of living species of *Taxus* (yew) seem 'essentially identical to those of *Palaeotaxus*, an Upper Triassic close relative living about 175 Myr [million years] ago'. Obviously, one cannot explain the size of yew berries as the result of selection for dispersal by present-day thrushes. Other plant genera, of which *Ilex* (holly) is one, show striking uniformity of fruit characters among species occurring in widely different parts of the world and in very different habitats and climates, species which must have had a long history of dispersal by very different assemblages of frugivorous birds. In view of this stability of fruit characters, in attempting to explain the correspondence in size between fruits and birds, in any particular area, one is led to attribute a major

role to a second process that may be involved.

(2) The assemblage of different fruits found in our study area, and in Europe generally, may to a large extent be the result of the dispersal by birds, from a larger 'pool' of plant species, of those whose fruits were of the right size for birds to exploit. At the height of the last glaciation few if any of the plants that now produce fruits for frugivorous birds were present in southern England; they must have retreated to milder refuges to the south and, in view of the slow rate of evolutionary change of fruit characters, most of them must have had fruit similar to what we see today. As the climate ameliorated, birds would have been responsible for the northward spread of these plants, and it is reasonable to suppose that one of the factors affecting their dispersal was the size of their fruits. Thus the dispersers may have been responsible for selecting the plants which spread north as the ice retreated, and thereby helped to create a flora suited to the birds' needs.

SEED SIZE, AND THE SEED BURDEN

Seeds vary tremendously in size. From the range of wild bird-dispersed fruits in our study area the heaviest seed (wild cherry, 0.29 g) is 260 times as heavy as the lightest (black nightshade, 0.0011 g). If wild strawberry is counted as a bird-fruit (p. 100), the heaviest seed is over 700 times as heavy as the lightest, both being in the family Rosaceae. Trees have, on average, the largest seeds, and herbs and subshrubs the smallest. There is, as would be expected, an inverse correlation between seed-size and the number of seeds per fruit (Fig. 8). Thus a plant has a 'choice' between a range of possibilities, the extremes being to produce a fruit containing one large seed or a fruit containing very many small seeds. It is a general rule (though, like all such rules, there are plenty of exceptions) that forest trees have large seeds, as the seedling needs food reserves in order to survive the

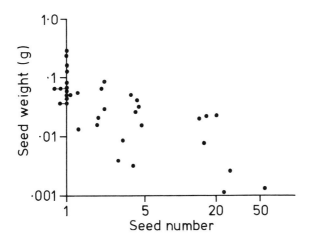

Figure 8 Relationship between seed number and individual seed weight in 36 native bird-dispersed fruits, southern England (data from Appendix 2). There is a highly significant inverse correlation (Spearman Rank Correlation, one-tailed $r_s = 0.733$, $P < 0.01$).

192 Interpretation

early stages of growth in poor light under the woodland canopy, while plants that colonise temporarily open ground, wood-edges and such places do best if they produce as many seeds as possible, even if there are little or no food reserves in each seed.

From the bird's point of view, the profitability of a fruit is affected not only by the size of individual seeds (this may be important in determining whether they are regurgitated or defaecated, and hence the rate at which the fruit is processed – see p. 217), but also by the total seed burden, which we have expressed as a percentage of the fresh weight of the fruit. This also varies tremendously between different fruits, from as little as 3.4% in rowan and 5.8% in wild currant to 59% in spindle. In the sample of plants occurring in our study area, there is no clear relationship between the seed burden and the growth form of the plant.

Within the same species of plant, also, there may be considerable variation in the seed burden of different fruits, and a bird should benefit by taking, if possible, fruits with small seed burdens. We do not know whether any birds in our study area do this. A few examples will illustrate the potential opportunities for the bird. Ivy berries may contain 1–5 seeds. For the fruit of any individual ivy plant, the seed burden is closely correlated with the number of seeds per fruit (Fig. 9); the smaller fruits have fewer seeds and a lower seed burden. A bird feeding on ivy berries would do best, in terms of minimising its seed intake, to select small fruits. It seems to be a common rule for multi-seeded fruits, in which the number of seeds varies, that smaller fruits have fewer seeds and a lower seed burden. By contrast, when feeding on single-seeded fruits, the best strategy would be to select the largest fruits, as these tend to have high pulp/seed ratios; in other words, a large size is achieved more by an increase in the amount of pulp than by an increase in seed size. Figure 10 illustrates this point in two species of *Prunus*.

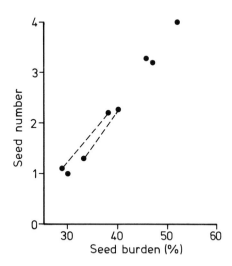

Figure 9 Relationship between mean seed number and seed burden (seeds as a percentage of the fresh weight of the whole fruit) in samples of ivy berries from six different plants. For two plants, large fruits and small fruits are shown separately, linked by a dotted line. To maximise the amount of nutritive matter eaten, a bird should select small fruits (see text).

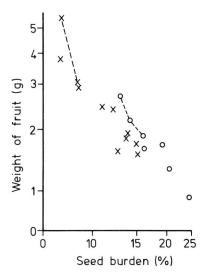

Figure 10 Relationship between fruit size and seed burden (seed as a percentage of the fresh weight of the whole fruit) in ten samples of sloe (crosses) and five samples of wild cherry (circles). For one sloe sample, large and small fruits are shown separately. To maximise the amount of nutritive matter eaten, a bird should select large fruits (see text).

Our ignorance on this point does not exclude the possibility that there may be subtle discrimination by birds of fruits according to their size and shape, and that this may be related to their seed burden. Herrera (1981c), studying the dispersal of fruits of *Smilax aspera*, a spiny climber, in southern Spain, found that birds tend to select one-seeded fruits in preference to those with two or three seeds, presumably because they were the most profitable. Plants growing in different areas differed in the proportions of one-, two- and three-seeded fruits that they produced, and there was evidence that such differences had a genetic basis. Herrera also found that in areas where there were many other kinds of fruits available at the same time as *Smilax*, the *Smilax* fruits tended to have a low mean seed number; where there were few other kinds of fruit present, the *Smilax* fruits had more seeds. Herrera tentatively interpreted these facts as showing a subtle adaptive response by *Smilax* to local differences in competition for dispersers. Where there are many other kinds of fruits available, the *Smilax* has to make its fruits more attractive to dispersers, hence fruits with few seeds are favoured; where there are few other kinds fruits, interspecific competition for dispersers is reduced and multi-seeded fruits are favoured.

Several of the fruits in our study area have the same kind of variability in seed number as *Smilax*, for example ivy, black bryony and privet. It would be of great interest to carry out studies similar to Herrera's on such plants, to see whether the same kinds of selective factors are at work.

The colour and taste of fruits

That a ripe strawberry is as pleasing to the eye as to the palate, – that the gaily coloured fruit of the spindle-wood tree and scarlet berries of the holly are beautiful objects, – will be admitted by every one. But this beauty serves merely as a guide to birds and beasts, in order that the fruit may be devoured and the manured seeds disseminated: I infer that this is the case from having as yet found no exception to the rule that seeds are always thus disseminated when embedded within a fruit of any kind (that is within a fleshy or pulpy envelope), if it be coloured of any brilliant tint, or rendered conspicuous by being black or white.

Charles Darwin: The Origin of Species, 1859

Nearly all the bird-dispersed fruits in our study area are red or black when ripe. All but two of them change colour on ripening from green to their final colour (in some cases, when changing from green to black, through an intermediate purple stage, or when changing to red, through intermediate yellow or orange stages). The striking exception is the wayfaring tree, whose fruits change from green to bright red while still unripe and then, on ripening, change rapidly from red to black; blackberry makes a similar but more gradual change through dull red to black. One fruit, mistletoe, is white when ripe; and two, gooseberry and wild currant, are yellowish and translucent (elsewhere, wild currant is generally red-fruited). The fruits of sloe and juniper have a bluish-mauve bloom when they first ripen, which wears off to leave a blackish colour. A few fruits have bicoloured displays: spindle, with bright pink capsule and orange aril; elder and dogwood, with black or blue-black fruits borne on red stalks.

Undoubtedly the colours of these fruits serve to advertise them to birds, which have good colour vision. Indeed it is not easy to make any more valid or wide-ranging generalisation than Darwin's, though he ought to have said that their colours are 'a guide to birds' rather than 'birds and beasts', mammal-dispersed fruits being typically dull-coloured and scented (Pijl 1972).

Several attempts have been made to explain why bird-dispersed fruits are one colour rather than another, but with small success. Willson & Thompson (1982) investigated the colours of temperate North American bird-dispersed fruits and tested a number of hypotheses, without any very clear-cut results. Stiles (1982) made the suggestion, again on the basis of North American data, that autumnal colour changes of the leaves of some deciduous plants may act as a signal to dispersal agents, and that such 'foliar flags' should be best developed in plants whose fruits are otherwise not very conspicuous; but later analysis (Willson & Hoppes 1986) failed to confirm such a relationship. Willson & Melampy (1983) investigated bicoloured fruit displays, which are apparently commoner in the North American flora than in the European. They tested the hypothesis that red and black fruit displays (as, for example, in our elder) enhance the rate of fruit removal by dispersers. The hypothesis was generally supported, but the manipulations necessary for the test must have resulted in very unnatural fruit displays, and it is uncertain what effect this may have had on the birds. One result of the experiments was to show that fruit removal was more rapid in light-gaps in woodland than in the interior of woodland, as other studies have shown; but again, whether this is due to the brighter colour display in light-gaps or to a general preference for birds to forage in light-gaps in uncertain. Recent research has amply confirmed the broad generalisations that were apparent to Darwin, but the differences in colour between the fruits of different bird-dispersed plants still await explanation. What, if anything, can we say of the colours of fruits in our study area?

Most of our local fruits are red when ripe; black is the next commonest colour. Almost certainly red is the most conspicuous colour for birds, as it is for us. Black is also fairly conspicuous, but not nearly as much as red against a green background, and this may explain why the conspicuousness of black elderberries and dogwood fruit is enhanced by crimson-red stalks, and why wayfaring tree fruits go through a red phase before they ripen, thus advertising the place where ripe black ones will be found. But it is difficult to see why a particular fruit is red or black, or some other colour. For fruits in our study area there does not seem to be any clear connection with the season of ripening, though one may suspect that there is a tendency, which analysis of more complete data on whole floras would confirm, for early-ripening fruits to be more often red than later-ripening fruits. Thus, taking related pairs of fruits, the wild cherry (red) ripens before the sloe (black), the raspberry before the blackberry; and, to go beyond our area, the red elder (*Sambucus racemosa*) before our black-fruited elder. But the number of related pairs for which one can make this kind of comparison is small, and this hint of a difference connected with the season of ripening may be illusory. Our two species of *Daphne* ripen at almost the same time (July), and one has red and the other black fruits. The two viburnums go the opposite way: guelder rose (red) ripens later than wayfaring tree (black). Willson & Thompson's (1982) analysis of the colours of bird-dispersed fruits in woodland in Illinois shows that fruits that are black when ripe greatly outnumber red fruits (in contrast to our area) and also ripen on average earlier.

An important consideration which is usually omitted from discussions of the significance of fruit colours is the sensitivity of birds to different parts of the spectrum. Recently it has been shown that birds of several families are sensitive to ultra-violet rays, and it seems probable that this is true of most birds. Burkhardt (1982) has shown that the bloom on several kinds of fruits, including sloe and juniper, has quite strong ultra-violet reflectance; if the bloom is rubbed off, the reflectance disappears. The bloom is due to waxes of various kinds exuded from the fruits. Burkhardt concludes: 'While for us from some distance berries of the wild plum [sloe] or of the juniper are hardly detectable among leaves and needles, they must contrast fairly well for birds due to the UV reflection which is nearly lacking in the case of the leaves.' Not all the fruits found by Burkhardt to show UV reflectance are blue to our eyes; among them was the green grape *Vitis vinifera*, and the orange-brown fruit of the tree tomato *Cyphomandra betacea*, a tropical American plant.

The colours of ripe fruits certainly attract fruit-eating birds; but do the birds show any preferences for one colour over another? Turček (1963) made a statistical examination of supposed colour-preferences of fruit- and seed-eating birds, but he lumped dispersers together with seed-predators in his analyses, and the results are unusable. All that he seems to have shown is that most European fruits adapted for dispersal by birds are red or black, and fewer of them blue or other colours. A more significant observation was published by Diesselhorst (1972). He reported that Blackbirds, faced with a choice between (cultivated) red and white currants which were, apparently, similar in size, taste and softness, chose the red ones, ignoring the white ones until the red were finished. He concluded that the choice depended on colour alone. Not surprisingly Diesselhorst had not come across Gilbert White's (1789) observation on the same subject, made nearly 200 years earlier. 'Birds are much influenced in their choice of food by colour; for though white currants are much sweeter fruit than red, yet they seldom touch the former till they have devoured every branch of the latter.'

Where cultivated fruits occur in white and coloured forms, it is probably common for the coloured ones to be preferred by birds. In addition to the currants mentioned above, Feare (1984) records that Starlings prefer red cherries to white, and blue grapes to white ones. These preferences need not indicate any innate tendency to prefer one colour to another; they may be learnt. There seems to be no evidence for innate colour preferences in fruit-eaters, but appropriate experiments have not been carried out. In places where there are no native white fruits, it would be expected that white might be ineffective as an advertisement, and this could explain the apparent failure of birds to recognise mistletoe berries as food, as mentioned under that plant, and the fact that snowberries tend to be ignored, though their peculiar texture may also be involved, and that introduced white-fruited species of sorbus may be left uneaten for a long time.

Generalisations about fruit colours that may be valid for temperate regions may not be so for tropical forest. In particular, the very specialised tropical fruit-eaters must know intimately the varied fruit resources of the areas in which they live, and the qualities of the different fruits, and they probably do not need to be attracted by such conspicuous visual signals as are used by northern fruits to attract the migratory birds that feed on them. This may be the reason why many bird-dispersed tropical fruits, especially those that are fed on by specialised frugivores, produce fruits that are not only very nutritious but also relatively inconspicuous.

We know of only one attempt to find out whether taste is a factor in the preferences shown by fruit-eating birds. Sorensen (1983) tested thrushes with 'dough-fruits' (fruit-shaped dough pellets) flavoured with the juice of various wild fruits and presented in pairs to the experimental birds (four wild-caught Blackbirds and one Song Thrush kept temporarily in captivity). Apart from the fruit-juice flavour all other factors that might have affected the choice were eliminated. The birds showed distinct preferences for fruit flavours, and the 11 fruits tested could be arranged in an order of preference:

ivy	significantly preferred to	sloe, hip, white bryony, woody nightshade, buckthorn
black bryony	,, ,,	hip, blackberry, woody nightshade, buckthorn
sloe	,, ,,	hip, white bryony, buckthorn
hip	,, ,,	elder, haw, woody nightshade, buckthorn
elder	,, ,,	haw, white bryony, buckthorn
haw	,, ,,	guelder rose, white bryony, buckthorn
guelder rose	,, ,,	ivy, white bryony, buckthorn
white bryony	,, ,,	hip, woody nightshade, buckthorn
blackberry	,, ,,	woody nightshade, buckthorn
woody nightshade	,, ,,	haw, buckthorn
buckthorn	,, ,,	—

Sorensen noted that only three out of 32 preferences in the above table (guelder rose–ivy, white bryony–hip, woody nightshade–haw) are out of place, and that the consistency of the order suggests that the preferences indicated are 'transitive' (eg, if a is preferred to b and b is preferred to c, then a is preferred to c).

It is far from clear whether these taste preferences have much influence on the choice of different fruits by thrushes in the wild. They do not fit the observed preferences well (pp. 219–23). Thus Blackbirds prefer haw to sloe, hip, elder and buckthorn, but three of these four were preferred to haw by the experimental birds on the basis of taste. Sorensen's interpretation is invalidated by a lack of appreciation of the extent to which the various fruits are eaten in the wild. Thus she states that four fruits with the less preferred flavours (guelder rose, white bryony, woody nightshade, buckthorn) are avoided by thrushes in the wild; but in fact all of them are taken freely, only one (woody nightshade) being avoided when other, more highly preferred fruits are available (p. 87). Buckthorn, in particular, is eaten in great quantities, and is probably preferred by Song thrushes to haw (Table 80). Based on these supposed avoidances, Sorensen put forward the hypothesis that thrushes have taste aversions for certain fruits. She speculated that taste aversions 'are related to toxic compounds evolved by fruits to prevent thrushes from consuming fruits, and so dispersing seeds to unsuitable habitats'; and suggested that tits are the proper dispersers of the thrush-avoided fruits, and more likely to disperse the seeds to suitable sites. Tits, however, are generally not dispersers but seed- and pulp-predators.

Levey (1987) carried out experiments with six species of tropical American fruit-eating birds, testing their ability in captivity to detect differences in diets containing different sugar concentrations, within the range of concentrations found in the wild fruits that they eat. He found that three species of tanagers were able to discriminate between concentrations of 8%, 10% and 12% sugar, consistently choosing the higher concentration when offered a choice of two, but two species of manakins failed to discriminate between these diets, as did a fourth tanager species which is probably more insectivorous than the other three. The tanagers, an exclusively New World family, crush fruits in the bill, swallowing the pulp and letting the seeds and skin drop, a method of fruit-eating not used by European frugivores (except by sparrows, which are however only occasional fruit-eaters). The tanagers' method gives them the opportunity to taste the fruit pulp before swallowing it. Manakins, which swallow fruit whole, like our thrushes and warblers, do not normally have this opportunity and so would not be expected to be so discriminating. The role of taste in the selection of fruits by birds in the wild remains an open question, awaiting further research of the kind begun by Sorensen and Levey.

Competition between plants for dispersers

If a plant depends on birds for dispersal of its seeds it is likely to face competition from other plants with bird-dispersed fruits, unless the demand from dispersers matches, or is in excess of, the available fruit supply. Fruit is often in excess of the needs of the dispersers in our area, and it is clear that there is a similar excess in other areas too. One might therefore expect plants either to vie with one another, in an evolutionary sense, in ability to attract dispersers by offering more nutritious, more abundant or more conspicuous fruit; or to avoid competition by evolving different fruiting seasons. Clearly there will be limits to either strategy. A plant might put more of its resources into producing more nutritious or more attractive fruits, or into fruiting more abundantly, but there are other calls on its limited resources, too, which cannot be neglected without making the plant less efficient in other ways. The alternative strategy, of evolving earlier or later fruiting seasons, might fail for want of sufficient suitable pollinators for the flowers; or because the season is too late to permit the fruit to ripen and the seed to be dispersed with a good chance of germination.

What evidence is there that competition for dispersers has had either or both of these effects? Intuitively it seems likely that competition between different plant species has been a factor in the evolution of their fruit characters, but it is not easy to prove. The fact that fruits are conspicuous and attractive to birds and provide them with nourishment does not prove the operation of *inter*specific competition; *intra*specific competition in the absence of competing species would be expected to produce the same result. One would have to show that in a part of the plant's range where competitors were present its fruits were better, in quality or quantity, than in a part where there were no competitors; or one would need to show that, within a local population, individuals producing poorer-quality fruits had poorer dispersal when competing with other species, than those producing higher-quality fruits. Very detailed research would be needed to demonstrate such effects, and so far it seems we have only Herrera's investigation of the fruits of the common climber *Smilax aspera* in southern Spain, already mentioned (p. 193). To recapitulate the points in his study relevant to the present discussion, the berries of *Smilax aspera* may contain one, two or three seeds. One-seeded fruits have the highest pulp/seed ratio, and three-seeded fruits the lowest pulp/seed ratio; consequently the fewer the seeds the higher the fruit's food value to a bird eating it. Herrera found that the mean number of seeds per fruit in 12 different populations was correlated with the number of other plant species bearing fruit at the same time and equally or more abundant than *Smilax*. Other environmental variables had much smaller or insignificant effects. To quote Herrera's summary: 'Assuming that the intensity of competition among plant species for dispersers must be roughly proportional to the number and population sizes of competitor species, these results suggest that inter-population variation in the number of seeds per fruit is actually related to variation in competitive pressures, provided that (1) dispersers are in short supply, (2) dispersers select to some extent the most profitable fruits, and (3) number of seeds has some genetic component.' He was able to give evidence supporting all three assumptions, and so concluded that the local variation in seed numbers was attributable to inter-specific competition for dispersers. It seems likely that similar, thorough studies in other plant species will confirm the importance of interspecific competition.

Evidence that interspecific competition for dispersers may have an effect on fruiting seasons is still scanty. In Trinidad, in the humid tropics, where a seasonal but relatively equable climate makes it possible for fruits of the same general kind to ripen throughout the year, the many species of trees and shrubs of the genus *Miconia* produce a succession of fruits covering the whole year. In one valley 18 species, each with fruiting seasons generally of 2–3 months, had their seasons staggered in such a way that they were more or less evenly distributed through the year and at any time at least two species, more usually 3–5, had ripe fruit (Fig. 11). Some time ago (Snow 1965) we suggested that the staggered fruit seasons of *Miconia* species were best interpreted as the result of interspecific competition for dispersers, in this case mainly tanagers and other small fruit-eating forest birds. More recently Hilty (1980) has documented a similar case from the very humid tropics, in western Colombia, and has proposed the same explanation. It has been doubted, however, whether these staggered fruiting seasons are significantly different from random distributions, in which case our interpretation would be unwarranted (Poole & Rathcke 1979). The statistical analysis of such 'circular' distributions is complex and somewhat controversial, and we are not competent to give an opinion. Nevertheless we still think it likely that competition for dispersers has been a factor in the fruiting seasons of *Miconia*, for the following reason. Although *Miconia* species evidently can produce ripe fruit in all months, Trinidad has a markedly seasonal climate, with a pronounced dry season and a wet season that is interrupted by a minor dry period. The 18 species of *Miconia* shown in Fig. 11 all occur in a single valley; many are very similar to one another and grow side by side in apparently identical conditions of soil and climate. If there were no other factors involved, one would expect them to show some more or less common phenological response to the rainfall regime. The fact that their fruiting seasons are spread over the year, to such an extent as to be arguably random, indicates that some other factor is involved other than a response to physical environmental factors.

How then may we interpret the seasonal succession of fruits in our study area, shown graphically in Fig. 1? Climatic conditions in Britain prevent a plant modifying its fruiting season in the manner of plants in Trinidad. Flowering seasons too are likewise restricted. Of our 29 native species shown in Fig. 1, 69%

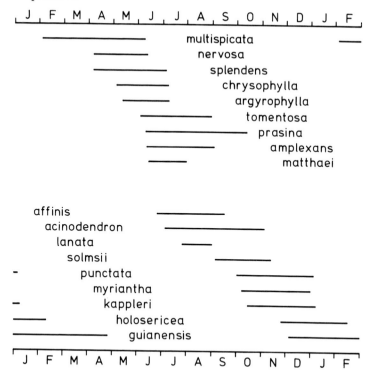

Figure 11 Fruiting seasons of 18 species of *Miconia* in the Arima Valley, Trinidad (from Snow 1965).

ripen their fruits in the four months July–October. If interspecific competition has modified fruiting seasons, it would not be expected to have led to the staggering of the seasons of closely related species throughout the year, as in *Miconia*. In fact, we have no plant genus with anywhere near the same number of species as *Miconia* in Trinidad. But might there be some staggering of the fruit seasons of such closely related species as there are? There are several genera in which this can be examined; the ripening seasons of the species involved are set out below. (Species that are probably not truly native to our study area, such as bird cherry *Prunus padus*, are omitted.)

Prunus avium	late June (N)
Prunus spinosa	September (S)
Rubus idaeus	late July, early August (N)
Rubus fruticosus	August on, mainly September (S)
Rosa arvensis	late September
Rosa rubiginosa	late September, October
Rosa canina	late November, December

Crataegus monogyna	late August, September
Crataegus laevigata	apparently earlier than monogyna (p. 45)
Solanum dulcamara	late July onwards (N)
Solanum nigrum	September (S)
Sorbus aucuparia	second half of July (N)
Sorbus aria	September (S)
Sorbus torminalis	November (S)
Viburnum lantana	late July, August
Viburnum opulus	December

This list shows that, with the possible exception of *Crataegus*, there are differences in the season of ripening of the fruit of congeneric species, and in most cases the differences are quite marked. It should be noted, however, that where the congeneric species have different overall geographical ranges those with the more northerly range (indicated by (N) in the list) ripen their fruit earlier than those with a more southerly range (S). This is part of a geographical trend in fruiting seasons, discussed in the following chapter. Clearly, in an area such as ours where there are few congeneric species this trend invalidates any attempt to relate differences in fruiting seasons between closely related species to the effects of interspecific competition for dispersers.

Although competition is usually expected to be most intense between closely related species, plants that are quite distantly related may compete for dispersers. In our study area, several plants (*eg* woody nightshade, wayfaring tree) compete rather unsuccessfully with elder, whose fruit season overlaps theirs. These plants would achieve better seed dispersal if they fruited earlier than elderberry; hence there may well be a selective pressure favouring earlier fruiting. But there must be many other constraints on their flowering and fruiting seasons, and the temporary situation in one locality, where the whole flora has been much distrubed by man and where natural plant associations hardly exist, is not likely to have had any long-term evolutionary effect on plant phenologies.

To sum up, there is some evidence that competition between plant species for avian dispersers may influence fruit characters and ripening seasons; but none of it is from Britain. Very detailed research may, in particular cases, show an effect on fruit characters similar to that found for *Smilax aspera* in Spain by Herrera (1981c); but it seems unlikely that, in the comparatively impoverished and highly disturbed British flora, interspecific competition for dispersers has had any detectable influence on fruiting seasons.

Adaptation of fruiting seasons and fruit quality to changing bird populations

In areas where there are marked seasonal shifts in bird populations, it might be supposed that plants should produce ripe fruit when most fruit-eating birds are present. There is ample evidence that this is the case. For example, in the eastern forests of North America, where winters are very severe and all fruit-eating birds are migratory, the seasonal production of bird-dispersed fruits, and the latitudinal patterns of fruit abundance, coincide exactly with the southward passage of frugivorous migrants. Whether one compares different plant species with different latitudinal ranges, or different individuals of the same plant species at different latitudes, one finds that fruit-ripening generally occurs earlier at higher latitudes (Stiles 1980). Geographical trends in fruit seasons thus run counter to what one could expect on climatic grounds; if there were no counter-selection exerted by the dispersers one would expect plants to fruit earlier in the south, where spring is earlier and the weather warmer.

Essentially the same thing is found in the less extreme conditions of western Europe. If we divide the fruits occurring in our study area into species that are widespread in Europe from north to south, and species with mainly southern distributions, we find that the southern species have on average later fruiting seasons than the more northern or more widespread species (Fig. 12). A few examples were given in the previous chapter, in the discussion of the possible role of interspecific competition between plants in determining fruit seasons. It might be that plants with more southern distributions are adapted to a warmer climate than plants with more northern distributions, so that, where they occur together in southern England, they simply flower later and take longer to ripen their fruits than the plants adapted to more northern conditions. If this were so there would be no need to suppose that there has been any adaptation to the changing populations of their dispersal agents. But the plants with southern distributions also fruit late in southern Europe. Thus whitebeam ripens its fruit in September in southern Spain just as it does in England, and ivy in winter, as in England (Herrera 1982a). The data on fruiting seasons in southern Spain, presented by Herrera (1982a), include 18 plants that occur in our English study area. For 11 of them the fruiting seasons are essentially the same in Spain and England, for four the season is earlier in Spain, and for three it is earlier in England.

Jordano (1985) investigated seasonal changes in the numbers of fruit-eating birds in Mediterranean scrubland (the dominant habitat) in southern Spain. He found that variations in the abundance and number of species of frugivorous passerines (mostly *Sylvia* warblers, Robins, thrushes and starlings) are extremely seasonal and are closely associated with the annual cycle of fruit production by the native plants. Jordano's and earlier studies by Herrera in southern Spain lead to the conclusion that the fruiting seasons of Mediterranean plants are adapted to the wintering of huge numbers of migrant frugivores from northern Europe. If detailed studies were carried out of the fruiting seasons of all European plants

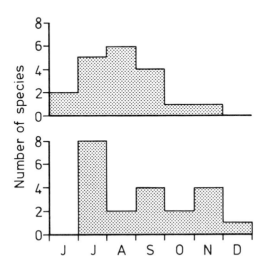

Figure 12 Seasons of ripening of 40 European fruit species in southern England: upper histogram, 19 species with widespread or northern distributions; lower histogram, 21 species with primarily southern distributions.

from north to south at different latitudes they would probably show the same kind of seasonal shift in the peak of fruit abundance that Stiles found in eastern North American forests, but it would be more extreme. In northern Scandinavia the peak would probably be in July–August, in southern England it is in September, and in southern Spain it is in November–December.

Of course, in the far north climatic constraints on the phenology of plants are so extreme that plants could hardly do otherwise than produce fruit at the end of the very short summer, and so no adaptation to the presence of populations of frugivores need be involved – unless the ability to remain in good condition beneath the snow through the winter, and so be available to migrants arriving in spring, is such an adaptation. But further south, where plant growing seasons are longer, there is scope for adaptations to the seasonal presence of fruit-eaters, and the evidence for an adaptive basis for fruiting seasons is strong, at least when Mediterranean plants are compared with plants from the latitude of Britain. A consideration of flowering seasons strengthens the argument. Herrera (1982a) gives the flowering seasons in southern Spain of 15 species of bird-dispersed fruit-bearing plants common to southern Spain and our English study area. Although, as mentioned above, fruiting seasons in the two areas are not very different, the same plants flower on average ten weeks earlier in southern Spain than in England. This implies an adaptive delay in the ripening of their fruits in southern Spain, and the likely basis for such an adaptation is the seasonal presence of their dispersers.

In addition to adaptation of fruiting seasons to seasonal changes in populations of fruit-eating birds, it might be supposed that natural selection would favour the evolution of fruits that meet the nutritional needs of their dispersers, and there is convincing evidence that this is so.

Fruits ripening at different times of year tend to have different nutritive qualities. As summarised in Table 73, which gives data for fruits from our study area, early-ripening fruits tend to have low seed weights and very watery pulp, and later-ripening fruits a higher seed weight and less watery pulp. There are similar differences between early- and late-ripening fruits in southern Spain (Herrera 1982a). To some extent the low seed weight of early-ripening fruits compensates for the wateriness of their pulp, but by no means entirely; thus the later-ripening fruits yield more nutritive matter and less water per gram of fruit eaten than the early-ripening fruits.

TABLE 73 *Seed burden and water content of pulp of native fruits ripening at different seasons, southern England*

Season of ripening	No. fruit spp.	Seed burden	Water content of pulp (%) (ranges and means)	Relative yield (%)
Summer (Jun–Aug)	18	3.4–33, 14.0	76–93, 84.9	6.5–22.5, 13.0
Autumn (Sep–Oct)	13	7.1–41, 23.3	60–90, 74.1	7.8–37.2, 19.6
Winter (Nov–Jan)	5	14–59, 34.8	56–84, 70.2	12.9–31.7, 18.4

Notes. Seed burden is the weight of the seed(s) expressed as a percentage of the fresh weight of the whole fruit. Relative yield is the amount of potentially nutritive dry matter expressed as a percentage of the fresh weight of the whole fruit, derived from seed burden and water content of pulp. Table based on data in Appendix 2.

TABLE 74 *Lipid content of dry pulp of native fruits ripening at different seasons, southern England*

Season of ripening	No. fruit spp.	% lipid content (range and mean)
Summer (Jun–Aug)	10	0.2–6.8, 2.8
Autumn (Sep–Oct)	11	1.3–24.9, 5.3
Winter (Nov–Jan)	5	2.8–35.8, 15.6

In addition, the later-ripening fruits have, on average, pulp with higher lipid (fat) contents than early-ripening fruits. Our data show this (Table 74), and it has been demonstrated for fruits in southern Spain (Herrera 1982a). Lipids when metabolised produce much more energy per gram of dry weight than carbohydrates or protein, the two other main nutritive components of fruit pulp (see Appendix 4).

It would seem that these seasonal differences in fruit quality are an adaptation to the needs of birds that eat the fruit. In summer a higher water intake is required, especially in hot dry weather; at the same time less energy is needed for the maintenance of body temperature, and a longer feeding period is available. In winter the opposite conditions prevail and a drier, more energy-rich food should be more suitable. Before accepting this interpretation, however, one must consider a possible alternative explanation. A plant that ripens its fruit relatively rapidly in summer might well produce a less nutritious, more watery pulp demanding less resources of the parent plant. Conversely, if a fruit is to ripen in late autumn or winter, maturing over a longer period, it should be less demanding for the plant to provide it with more concentrated nutrients. Also, a less watery pulp might make the fruit less vulnerable to frost, so that it should last longer. Thus, even without any adaptation to the needs of the dispersers one might expect to find the kind of differences between early and late-ripening fruits that in fact are found. On the basis of our data alone it would be hard to refute this argument, but, as Herrera has argued, the situation in southern Spain (and indeed in the whole Mediterranean region) strongly suggests that there has been adaptation of the nutritional qualities of fruits to the needs of their avian dispersers.

The crucial difference between southern Spain and England, in this connection, is that in southern Spain winter is the period when water is most available to plants and the weather is also mild enough for photosynthetic activity to continue. The summer, by contrast, is very dry and hot. 'Soil water deficit coupled with high temperatures impose serious limitations on plant life, and these factors have been responsible for the evolution of some characteristic features of mediterranean plant communities' (Herrera 1982a). In these conditions a bird's water requirements increase, as evaporative water loss is an essential means of reducing body temperature. Thus the most watery fruits are produced at the time when water is scarcest and the bird's need for water is greatest. Furthermore, the plants not only produce the driest fruits at the season when it should be easier for them to produce watery fruits, they are also investing more of their energy in the pulp (in the form of lipids) than at other times of year; and it is hard to see why they should do so if it were not that their dispersers need more nutritious food in winter. Hence we conclude that the quality of fruits ripening at different seasons has evolved in response to the needs of the dispersers.

206 *Interpretation*

Seed-predators and the plants' defences

In our study area we have 14 regularly occurring species of dispersers, and seven species which exploit fleshy fruits either wholly or to some extent as seed-predators – three finches (exclusively seed-predators), three tits (pulp-predators and seed-predators) and the Woodpigeon (a disperser of some seeds, and thus included in both categories). Seed-predators were recorded taking seeds of 19 of the 34 wild fruits on which we have observations, and we calculate that significant proportions of the seed crop of several plants are destroyed by them. A summary is given in Table 75.

A plant should have fundamentally different strategies in relation to dispersers and seed-predators. The more dispersers it can attract, in numbers and in diversity of species, the more efficient its dispersal is likely to be. Conversely, all seed-predators are disadvantageous, and throughout its evolution the plant, anthropomorphically speaking, should have tried to outwit seed-predators by evolving impenetrable seed-coats, poisons or other forms of defence. This must be one of the reasons for the fact that, whereas bird-dispersed fruits of different kinds are in many cases very similar to one another in size and shape, and can be eaten by many different dispersers, the seeds inside them are extremely diverse. As a consequence most of the fruits in our study area have many dispersers but only one or two seed-predators. The latter are much more specialised feeders than are the dispersers with their general ability to swallow whole fruits. Thus Greenfinches feed on the seeds of yew and hips, which Bullfinches, with their smaller and weaker bills of different shape, apparently cannot tackle. Bullfinches, on the other hand, take several kinds of seeds (*eg* elder, guelder rose, privet, buckthorn) which we never recorded for Greenfinches. Chaffinches, with more pointed, much less robust bills than either Bullfinch or Greenfinch, seem to be limited mainly to the soft seeds of whitebeam and crab-apple.

Seed-defences may be mechanical or chemical; that is, the predator may be prevented from getting at the seed, usually by some kind of hard casing, or the seed may be distasteful or poisonous. Mechanical defences may be fairly easily recognised, but little is yet known about chemical defences, which involve many different substances, and almost nothing about their effectiveness against different seed-predators. Substances which may be poisonous for mammals may not be for birds, and not all birds may be equally susceptible to a particular poison. To generalise from the little available information, it seems probable that different

TABLE 75 *Impact of seed-predators on some fruits*

	No. feeding visits	Seed-predators Species	% of total	Estimate of % of seed crop destroyed
Yew	2363	GF, GT	9	9
Ivy	2457	WP	13	78
Field rose	17	GF	94	94
Rowan	756	BF, BT	14	14
Whitebeam	855	BF, GF, CH, BR, GT, WP	29	53
Blackberry	563	BF, GF, BT	7	[7]
Crab-apple	216	CH, GT, MT	10	[10]
Elder	1565	BF, BT, GT, MT, WP	14	[84]
Guelder rose	175	BF	33	33
Common honeysuckle	110	BF, MT	63	63
Perfoliate honeysuckle	221	BF, GF	29	29
Privet	217	BF, WP	24	47
Buckthorn	601	BF	8	8
Woody nightshade	125	BF, BT, WP	12	26

Notes. Fruits listed are those for which seed-predator visits totalled at least 5% of all feeding visits recorded, except spurge laurel, for which seed-predators and dispersers were recorded in separate areas. Estimates of percentages of seed-crops destroyed are very tentative, made on assumption that seed-predators' meal sizes are the same as dispersers' meal sizes with the exception of the Woodpigeon, whose meal sizes are assumed to be 25 times as great. For ivy, calculations are, however, based on recorded meal sizes of all species concerned. Square-bracketed estimates are too high for the following reasons: blackberry, because a significant part of the fruit crop remains uneaten; crab-apple, because locally the fruit crop may be eaten by mammals; elder, because Woodpigeons were probably over-recorded by comparison with dispersers, being much more conspicuous at a distance, and locally some of the fruit crop may remain uneaten. A general caveat in assessing the estimates is that seed-predation is often very patchy, varying according to habitat.

Abbreviations of seed-predators as follows: BF, Bullfinch; BR, Brambling; BT, Blue Tit; CH, Chaffinch; GF, Greenfinch; GT, Great Tit; MT, Marsh Tit; WP, Woodpigeon.

plant families have tended to adopt either mechanical or chemical defence, but not both. Thus palms have seeds lacking poisons, but many have very hard cases (botanically endocarps, the fruits being drupes), which are nevertheless opened by a few very specialised seed-predators such as macaws; while the huge, mainly tropical laurel family (Lauraceae) has seeds that lack physical protection, and are often notably soft, but are apparently toxic. We think it probable that the reason why parrots seem never to have been recorded eating the seeds of lauraceous trees, although accessible and nutritious, is that they are poisonous to parrots. Dispersers which specialise in lauraceous fruits, such as Quetzals (Wheelwright 1983) and some cotingas (Snow 1982), need to be able to process these fruits in the stomach without subjecting the seeds to mechanical damage that might release toxins; hence it is probably significant that lauraceous seeds are easily separated from the enclosing pulp. The avocado pear is an example; its seed, which is poisonous, comes cleanly away from the pulp; but it is far larger than the fruit of other members of the family, having been subjected to selection by man.

None of the plants in our study area with fleshy, bird-dispersed fruits have very large, hard seeds like palms, or large, rather soft seeds like the Lauraceae; but they probably have adaptations against seed-predators essentially similar to those shown by tropical plants. Some of the fruits that are, technically, drupes have the seed protected by a hard endocarp; such are holly, the three *Prunus* species (cherries and sloe), the two hawthorns and dogwood. These are immune, or almost immune, to avian seed-predators in our study area – in the absence of Hawfinches which, like macaws, are specialised to break into the defences of the most well-protected seeds.*

Several of the fruits are known to contain poisons, which in some cases are probably concentrated in the seed, but we can say little about their effectiveness. We suggested (pp. 30 and 32) that the poisonous seeds, which are very soft, may deter corvids from feeding on ivy berries and, possibly, the harder yew; but Woodpigeons are clearly unaffected by ivy poison. Greenfinches may be immune to the poisonous seeds of spurge laurel and mezereon; but the former are taken when unripe, when possibly the poisons have not developed, and the cultivated mezereons which Greenfinches have been reported to exploit may have seeds of reduced toxicity as a result of artificial selection. Spindle seeds are probably poisonous; they are fairly large and soft, easily opened by a finger-nail, yet no seed-predators attack them although potential seed-predators such as Marsh and Great Tits pluck the fruit for the aril and would be capable of opening them.

In spite of many uncertainties, one thing that seems significant is that several genera of the family Rosaceae, and all the locally occurring members of the Caprifoliaceae, are especially vulnerable to seed-predators. The five species of Caprifoliaceae, in three genera, show few signs of seed defence; all have soft seeds, easily opened, which are much exploited by the Bullfinch and even by birds as small as the Blue Tit. There is no evidence that any of their seeds contain poisons. Seed-predators accounted for 25% of our combined (disperser + seed-predator) records for the Caprifoliaceae.

Seeds of the species of Rosaceae occurring in our area are not known to contain toxins, but some (especially the *Prunus* species and haws) have thick woody protective layers which are effective defences against seed-predators. We did not record seed-predators at any of these fruits, but the Hawfinch, with its massive, powerful beak, is able to feed on their seeds (Mountfort 1957). Taking the

* Another specialised seed-predator, the Rose-ringed Parakeet *Psittacula krameri*, has recently become feral in southern England. The only time that we have seen it in our area, it was extracting and eating the seeds of holly berries.

Rosaceae as a whole there is great diversity in the incidence of seed-predation; seed-predators accounted for only 5% of our combined (disperser + seed-predator) records, but they accounted for 86% (wild service tree), 29% (whitebeam) and 14% (rowan) of our records for *Sorbus* species, and for nearly all (94%) of those for field rose.

By contrast, the seeds of most fleshy, bird-dispersed fruits in the tropics seem to be immune to avian seed-predators. In our studies of frugivory in Trinidad (Snow & Snow 1971, and earlier papers) we had no evidence that any fruits eaten by dispersers were also exploited by seed-predators. Published research from elsewhere in the New World tropics suggests that there are few avian seed-predators of fruits adapted for dispersal by birds; and it seems that this may also be true of the Old World tropics. One of the main families of seed-predators that take fruit from the parent plant (*ie* excluding those that take the seeds of fallen fruit), the parrots, feed very largely on fruits not adapted for dispersal by birds.

A tentative general conclusion is that, in the tropics, where plants have had vast periods of time evolving under relatively constant conditions and perfecting their adaptations to their environments, including the animals that exploit them, seed-defences have reached a higher degree of pefection than in north-temperate regions, where faunas and floras have undergone drastic changes during the last two million years (the approximate extent of the ice-ages). In this connection it seems significant that the two northern families most vulnerable to seed-predators, the Rosaceae and Caprifoliaceae, are the only ones with primarily north-temperate distributions, and they give evidence of recent evolution and speciation. Thus they contain genera which consist of large numbers of closely related species (*eg Rosa, Crataegus, Sorbus, Lonicera*), and in consequence several genera are botanically 'difficult', and there is much disagreement as to how many species there are and how they should be subdivided. Perhaps those plant groups that have been most successful in exploiting the ecological opportunities in the shifting northern scene have not been able to evolve defences against what must have been a constantly changing array of seed-predators. On the other hand, the plants whose seeds apparently rely on defence by toxins – ivy (Araliaceae), yew (Taxaceae), spindle (Celastraceae), lords and ladies (Araceae), probably black bryony (Dioscoreaceae) – are members of very old and/or primarily tropical families which must have had very long histories of interaction with potential seed-predators. Like the other interpretations offered in this chapter, this is highly tentative and offered mainly as a stimulus to further research.

210 Interpretation

Physical limitations to fruit-eating

As has been mentioned in the various species accounts, fruit-eater species differ in their methods of taking fruit, in their success at plucking it, and ability to swallow it. Their physical characteristics determine the range of fruits that they can successfully feed on. This is perhaps rather an obvious point to make; all birds are limited by their physique to certain methods of foraging and particular kinds of food. (We use the word 'limited' deliberately, as it is neutral and factual; to say 'adapted' begs the question which we are about to discuss.) But the limitations are not always obvious; quite subtle differences between species may be important in determining their feeding habits.

Fruit-eaters have two main methods of taking fruit: from a perch, or in flight. Most of the very specialised frugivores of the tropics use mainly, or almost exclusively, one or other of these methods; even so, a specialised 'sallier' will take fruit from a perched position if it can do so, and a specialised perched feeder may sally to take a particularly inviting fruit. In this connection Moermond & Denslow (1985) have studied the very varied array of neotropical forest fruit-eaters and have found that both main feeding methods can be subdivided to reveal further differences among the fruit-eaters.

Birds that take fruits from a perch may do so by three methods, which Moermond & Denslow term 'picking', 'reaching' and 'hanging'. Picking refers to cases where a bird takes fruit close to the perch without extending the body or assuming special positions. Reaching is where the bird extends its body well away from the perch, outwards or downwards (sometimes with wing-fluttering). Hanging means that the bird's entire body and legs are below the perch. Among the neotropical frugivores examined by Moermond & Denslow, only woodpeckers regularly use the hanging method; the others, when feeding perched, use the picking and reaching methods, but there are great differences between species in their ability to reach. For instance, the Collared Araçari *Pteroglossus torquatus*, a small toucan, can reach a full body length below the perch, whereas the Slaty-tailed Trogon *Trogon massena* cannot reach at all, and picks fruits slightly below the perch only with difficulty. The neotropical thrushes are also poor reachers. These differences are associated with differences in leg structure. Trogons have very weak feet and their leg musculature makes up only about 3% of the body weight. Toucans have much larger leg muscles. Thrushes have long strong legs, typical of ground feeders, which prevent them performing feats of reaching that are possible for birds such as tanagers that are primarily arboreal feeders. Tanagers and other efficient reachers have legs that are strong but short.

Birds that take fruit in flight may do so, according to Moermond & Denslow, in four ways: by 'hovering', 'stalling', 'swooping' or 'snatching'. 'A hovering bird pauses in front of the fruit while flapping its wings so as to maintain zero air speed ... A stalling bird pauses briefly in front of the fruit by using a very steep wing attack angle allowing the bird to slow down and stall just in front of or below the fruit ... Swooping and snatching both involve continuous movement past the fruit as it is taken. In swooping, the wings are held out and the bird glides up to the fruit, whereas in snatching, the wings are flapped throughout.' Moermond & Denslow found that, among neotropical birds, the first three methods are used by those species that take most of their fruit in flight – in other words, by the most specialised flight-feeders. Snatching is apparently a less specialised technique, and at times is probably used by nearly all the frugivores.

Just as methods of perched feeding are associated with differences in leg structure, the methods of flight-feeding used by different species are associated with differences in wing morphology. Long wings and low wing-loading reduce the energetic costs of flight and favour the ability to hover. Short wings increase manoeuvrability and the ability to accelerate. Wing-tip slotting reduces stalling speed. Size is also very important: a decrease in size brings improvements in most aspects of aeial performance. In general, constraints on flight are less critical for birds weighing below about 100 g than for heavier birds. But bird flight is very complex, and any analysis by single factors is unlikely to explain all the differences between species.

Among neotropical fruit-eating birds, hovering is commonly used by small birds such as manakins, small flycatchers and small tanagers weighing less than 20 g. Stalling is the main method used by the Slaty-tailed Trogon and probably other trogons with deeply slotted wing-tips. Swooping is a rather specialised method, best exemplified by the Purple-throated Fruit-crow *Querula purpurata*, a cotinga with long broad wings and consequently very low wing-loading. Snatching, as already mentioned, is apparently a less specialised technique and is probably used by nearly all fruit-eaters.

Moermond and Denslow's study of the very varied fruit-eaters of the New World tropics helps towards an understanding of the methods used by fruit-eating birds in southern England; but a full understanding will not be possible until much more detailed observations have been made, including high-speed filming of birds feeding under controlled conditions.

The five thrush species in our study area take fruit mainly from a perch, but also in flight. They are not good at reaching, and when they try to reach a fruit that is to the side or below the perch they beat their wings to maintain and recover their position. Almost certainly they cannot reach far below the perch and then recover their position by the use of legs and feet alone, as good reachers such as some tanagers can. As mentioned above, these limitations are a consequence of their leg morphology, which is primarily adapted for ground-foraging. When they take fruit in flight, the thrushes usually fly a short distance upwards, more or less vertically or obliquely, taking the fruit quickly at the top of the sally, then drop back to a lower perch (or to the ground, if the sally has been made from the ground). At the moment of plucking the fruit they presumably hover briefly. If a fruit is at more or less the same level as their perch, they may 'snatch' it in the way described by Moermond & Denslow, by flying past and plucking it without any perceptible pause in their flight.

To judge from their size and wing-loading, *Turdus* thrushes should not be efficient at feeding in flight. Moermond and Denslow's analysis of a wide range of neotropical fruit-eaters showed that, among those heavier than 100 g, birds with wing-loadings greater than 0.42 g/cm^2 take fruits primarily from a perch. British thrushes are around this critical 100 g weight, and their wing-loadings are in the range 0.35–0.47 g/cm^2 (Table 76). Brief flight sallies to take fruit are

TABLE 76 *Wing-loadings of thrushes and other fruit-eaters*

	No.	Mean weight (g)	Wing-loading ($g/cm^2 \pm S.D.$)
Blackbird (Nov–Jan)	10	113.9	0.46 (±0.03)
(Apr–Jul)	7	95.9	0.40 (±0.03)
Song Thrush (Nov–Feb)	17	86.2	0.47 (±0.04)
Mistle Thrush (Nov–Feb)	10	144.0	0.46 (±0.03)
(Apr–Jun)	9	128.7	0.40 (±0.02)
Fieldfare (Sep–Jan)	10	117.9	0.42 (±0.02)
Redwing (Sep–Nov)	10	64.9	0.35 (±0.02)
Robin (Nov–Feb)	10	22.1	0.26 (±0.03)
(Apr–Jun)	10	19.6	0.21 (±0.01)
Blackcap	11	19.3	0.25 (±0.02)
Garden Warbler	10	23.7	0.29 (±0.02)
Common Whitethroat	20	14.6	0.21 (±0.02)
Lesser Whitethroat	5	12.5	0.20 (±0.01)
Starling (Nov–Feb)	10	92.1	0.47 (±0.04)

Notes. In the samples measured there are no significant differences between males and females, except for egg-laying females which for obvious reasons have higher wing-loadings than usual. In order to exclude the latter, data are given for males only, all from the British Isles.

All data from unpublished MSS by R. Meinertzhagen, deposited in the Sub-department of Ornithology, British Museum (Natural History), Tring. Wing-loadings are based on fully extended wings (excluding the body between them). Agreement with other available data is good: Robin, 0.26 (Tatner & Bryant 1986); Fieldfare, Finland, breeding season, 0.36 (Pulliainen *et al* 1983); Blackbird, 0.47 (pers. obs.); Starling, 0.47 (pers. obs.).

TABLE 77 *Incidence of flight-sallies by thrushes feeding on ivy berries*

	No. feeds	% of feeds which included flight sallies
Blackbird	869	25($\pm$3)
Song Thrush	379	31($\pm$5)
Mistle Thrush	119	28($\pm$8)
Fieldfare	110	14($\pm$7)
Redwing	168	49($\pm$8)

Note. Figures in parentheses are 95% confidence limits.

presumably profitable for them, but one would expect perched feeding to be the primary method. A factor that may be important for them, but less so for tropical fruit-eaters, is that they undergo marked seasonal changes in weight and are markedly heavier, on average, in winter than in summer, with consequent effects on wing-loading. For birds built on the same general plan, wing-loadings are higher in the larger species, as indicated for the four warblers in Table 76. In the five thrushes, however, wing-loading is not correlated with size: the three resident species have similar wing loadings, while the the two winter visitors, with relatively longer and more pointed wings, have lower wing-loadings. One would predict that the Redwing, with much the lowest wing-loading, should take fruit by aerial sallies more than the other four, and this is confirmed by records for ivy berries, the only fruit for which we have adequate records for a comparison between all five species (Table 77).

With their long slender legs, primarily for locomotion on the ground, Robins are much less efficient at 'reaching' than the thrushes. Their small size, however, must make them more efficient at flight-feeding than the thrushes, and they take almost all fruit in flight. Even so, it is an energetically demanding process, as Tatner & Bryant (1986) have shown. Using a sophisticated technique (doubly labelled water) and Robins in captivity, they found that flight costs were very high (23 times the basal metabolic rate) but that the average flight duration was only 0.78 sec. The very short flights recorded for their captive birds (from one side of the cage to the other) were of much the same length as the very rapid flight sallies with which Robins take fruit, which suggests that they must need to economise on this very energy-demanding activity.

The Blackcap has much shorter legs than the Robin. Its wing-loading is similar, but its wings are longer and considerably more pointed. Like the other *Sylvia* warblers, it is adapted primarily for foraging in vegetation, not on the ground. Probably as a consequence of its leg morphology, the Blackcap is much better at reaching than the Robin. With its longer, narrower wings it must also be less efficient at rapid acceleration, and it makes little use of flight sallies, taking fruit almost always from a perch.

Starlings are primarily adapted for ground-feeding, using their special technique of probing the ground and opening the bill to look into the hole thus made, a method of foraging for which the head morphology is specially adapted (Feare 1984). They are also adapted for rapid flight, not for short-range manoeuvrability. Their wing-loading is similar to that of thrushes of equivalent size, but their wings are a quite different shape from a thrush's, with pointed tips and hardly any wing-tip slotting. Their legs, although quite long, are strong, and they are better at reaching than the thrushes, being able to reach down to fruit well below the perch. The pointed, almost triangular wing shape probably makes them inefficient at making short flight-sallies to take fruit, and in our experience they never do so. They may, however, make upwards flight-sallies and will even hover briefly to take very energy-rich food. They regularly take fat in this way from a hanging bell in our garden. No doubt for such food the energy expended is worth while; whereas for fruit the expenditure is probably greater than the reward.

Adaptations for a fruit diet

If fruit is, at least at times, an essential part of a bird's diet one might expect to find structural and physiological adaptations enhancing its ability to obtain fruit, swallow it, and process it in the alimentary canal. Such adaptations are well known in some tropical and subtropical birds. Thus specialised fruit-eaters often have very wide gapes, enabling them to swallow relatively large fruits whole. In some the alimentary canal is highly modified for a fruit diet; the mistletoe-berry specialists and fruit-pigeons are perhaps the best-known examples. Features common to all specialised frugivores include a thin-walled, relatively non-muscular stomach, a short intestine, and an ability to process fruit pulp very rapidly and to evacuate the seeds efficiently and quickly. But with the exception of the Waxwing, European fruit-eating birds are dietary generalists. To what extent are they structurally or physiologically adapted to the fruit component of their diet?

We have seen in the last chapter that the size and structure of the fruit-eaters affect their efficiency in reaching and plucking fruit. Robins are efficient at sallying to pluck fruit on the wing; Blackcaps are not. The thrushes are quite good at plucking fruit in flight sallies whereas Starlings, with very differently shaped wings, are apparently unable to do so. But this in itself is not evidence that the structural characteristics producing these abilities, or disabilities, have evolved in relation to fruit-eating. It might be that size and structure have been influenced by other selective factors, and that as a result some species happen to be more efficient than others at exploiting fruit. The Starling, with its adaptations for a special method of ground-foraging and for rapid flight, is a case in point. Thus a bird might be to a greater or lesser extent 'pre-adapted' to frugivory (Bock 1959). Although all physical aspects of a species are interrelated, to produce an organism well fitted to survive in its complex environment, it is best to consider the possible adaptations (or pre-adaptations) for fruit-eating one by one.

SIZE

In an analysis of the passerine birds of Mediterranean scrubland, Herrera (1984c) showed that fruit-eaters tend to be larger than birds of other feeding types, but only slightly so. Jordano (1987) made a more detailed analysis of six species of

Sylvia warblers occurring in southern Spain, and found that the degree of frugivory is closely related to body size: the largest species (Blackcap, Garden Warbler) are the most frugivorous, and the smallest species (Dartford Warbler *S. undata*) least frugivorous. Common Whitethroat, Subalpine Warbler *S. cantillans* and Sardinian Warbler *S. melanocephala* are intermediate in size and in degree of frugivory. Correlated with the size differences are subtle differences in bill shape and leg structure, which make the larger species more efficient at foraging for fruit. Jordano suggests that size is important primarily in determining the range of different fruits that a warbler can exploit. As he points out, 'if only a small fraction of the fruit species available can be used, sustained frugivory is difficult to achieve'. This is because any single fruit species is likely to be an inadequate diet; for a full complement of nutrients, a varied fruit diet is necessary (see also p. 233). Thus Jordano found only two kinds of fruits in the diet of Dartford Warblers, compared with 14 and 12 respectively in those of Blackcaps and Garden Warblers.

Considering the genus *Sylvia* as a whole, Jordano points out that frugivory seems to be best developed in the larger species that are long-distance migrants, which breed in a variety of northern and central European habitats and winter mainly south of the Sahara (Barred Warbler *S. nisoria* and Orphean Warbler *S. hortensis*, in addition to the species that he studied); frugivory is least developed in the smaller species (probably also Spectacled *S. conspicillata* and Marmora's Warblers *S. sarda*, in addition to those studied) which breed in Mediterranean scrub and winter mainly between 35° and 10°N. Body size is presumably part of a complex of adaptations to one or the other of these contrasting 'life-styles', its relationship to fruit consumption being perhaps a secondary consequence rather than a primary dietary adaptation.

A relationship, whether adaptive or not, between body size and frugivory is most likely in a group of small birds such as warblers, for which fruit-size will be an important limiting factor. Whether there is any correlation between size and frugivory in birds above the warbler size range is uncertain. Herrera's (1984c) analysis of the passerines of Mediterranean scrubland lumped together such diverse birds as pipits, chats, warblers and tits; with such a heterogeneous sample there is a danger that irrelevant factors and sample bias (unequal representation of different families, with basically different ecologies) may have undue influence. There would be scope to test the idea in the large thrush genus, *Turdus*, with some 60 species that are similar in structure but cover a wide size range, from about 40–150 g. It is not at first sight apparent that the larger thrushes are more frugivorous than the smaller ones; a proper investigation would be a major piece of major research, involving tropical as well as temperate species, and might throw light on the factors promoting specialised frugivory.

BILL SIZE AND SHAPE

A bill of generalised insectivorous type – more or less straight, moderately broad, and with a strong, slightly hooked tip – is also useful for plucking fruit. A wide gape, typical of flycatchers, also enables a bird to swallow relatively large fruits. Thus birds that are primarily adapted for insect-eating may be pre-adapted for frugivory; and of course the opposite might be the case. Without the detailed evolutionary history of a species it may not be possible to distinguish adaptations from pre-adaptations.

In extreme examples it may be possible to identify with some certainty adaptations fitting a bird's bill for fruit-eating. Thus the neotropical cotinga family includes a variety of species, with rather wide bills, that are partly frugivorous and partly insectivorous. There are also a few genera that have become extreme fruit specialists, with relatively very wide bills and gapes; among these the bellbirds (*Procnias*) are the most extreme. They feed mainly on large fruits, which in many cases are also very nutritious, and there seems no doubt that the wide bill and gape in such species is an adaptation to frugivory. Is an increase in bill and gape width an adaptation for fruit-eating in the less specialised north-temperate fruit-eaters also?

In his analysis of Spanish Mediterranean scrubland passerines Herrera (1984c) showed that the fruit-eaters tend to have flatter and broader bills than birds in other feeding categories. For instance, among the warblers, relative bill-width (width/length) is in the range 0.43–0.53 for the fruit-eaters, and 0.36–0.46 for the others. The division between the two groups is in fact generic, the fruit-eaters consisting of five *Sylvia* species and one *Hippolais*, and the non-frugivores belonging to three other genera (*Phylloscopus, Locustella* and *Cisticola*, the last two associated with habitats that are poor in fruits). Analysing the *Sylvia* warblers in more detail, Jordano (1987) showed that within the genus the more frugivorous species have broader bills and wider gapes, relative to their body-size, than the less frugivorous species.

The genus *Sylvia* is centred on Mediterranean scrublands, a habitat very rich in bird-dispersed fruits small enough for *Sylvia* warblers to eat (Herrera 1984a), and it seems probable that the evolution of the genus has been closely bound up with this kind of habitat. It may be concluded from Herrera's analysis that the rather wide bills of *Sylvia* warblers, compared with other warbler genera, are an adaptation for fruit-eating. If this is so, it follows that the small species, whose size is presumably related to other aspects of their ecology (foraging in dense vegetation in the case of the Dartford Warbler), have had to sacrifice some of their ability to exploit fruits.

In the *Turdus* thrushes, too, it seems probable that bill shape is adapted to a diet containing a significant proportion of fruit. Again, the evidence is indirect, based on comparisons with other woodland or forest-living turdine genera containing species with a size range similar to *Turdus*. The southern Asian whistling thrushes (*Myiophonus*) consist of seven species which feed on the ground, mainly near streams, apparently exclusively on invertebrates and other animal food. Compared with *Turdus* they are relatively long-legged (indicating strongly terrestrial habits) and their bills are very narrow (laterally compressed) with a markedly hooked tip. At the other extreme are the three species of cochoas (*Cochoa*), also from southeast Asia. They are short-legged (indicating arboreal habits) and have bills that are laterally much expanded, approaching those of the specialised fruit-eating cotingas. Though not much is known about them in life, it seems that they are specialised frugivores. There are several other genera with bills that are like *Turdus* and which, like *Turdus*, include a proportion of fruit in their diets; *Zoothera* is the largest of such genera. We may conclude that the typical thrush bill represents a compromise solution for birds with an unspecialised, partly frugivorous diet.

DIGESTIVE SYSTEM

Some specialised fruit-eating birds have highly modified digestive systems. Though different groups have their own peculiarities, they share as common features a thin-walled, non-muscular stomach and short intestine. (The typical bird's stomach is in two parts, a glandular proventriculus, and a muscular ventriculus or gizzard.) A muscular gizzard is unnecessary because their food, consisting of fruit pulp, does not need to be ground, and a short alimentary canal allows food to be passed rapidly, presumably because, being relatively unnutritious compared with animal food, a large amount has to be processed daily. Extreme examples of these specialisations are found in the fruit-pigeons (Wood 1924); the euphonias, which live exclusively on mistletoe berries (Wetmore 1914); and some of the flower-peckers, which are also mistletoe-berry specialists (Docters van Leeuwen 1954). There is also some evidence that frugivores tend to have a relatively large liver, possibly because it is needed as a detoxification organ, as toxins are more often present in fruits than in animal matter (Pulliainen *et al* 1983).

Do the less specialised European frugivores show any adaptations of the digestive system to fit them for a fruit diet? They would, of course, be expected to be much less extreme than those of the tropical fruit specialists. Pulliainen *et al* (1983) examined the digestive sytems of six species of birds that feed on rowan fruit in Finland. Three of them – Bullfinch, Pine Grosbreak *Pinicola enucleator* and Parrot crossbill *Loxia pytyopsittacus* – are seed-predators; the other three – Fieldfare, Starling and Waxwing – are seed-dispersers. They are a very heterogeneous lot, from four different families, and perhaps not much could be expected from the limited comparison between them, which was restricted to liver size, gut length, and the presence or absence of a gall bladder. The only clear difference that emerged was that the Waxwing, the most specialised of the seed-dispersers, proved to have the largest liver (an example of the tendency for frugivores to have large livers, mentioned above), and lacked a gall-bladder, which was tentatively related by Pulliainen *et al* to a low-fat diet.

Herrera (1984c) examined the gizzard mass (muscular gizzard, disregarding the proventriculus), intestine length and liver mass of 25 species of Spanish passerines, and found no signficant differences between the fruit-eaters and the others. He pointed out that most of the birds examined are strongly frugivorous for part of the year and strongly insectivorous at other seasons, and that the conflicting demands of a seasonally changing diet would be expected to preclude major structural adaptations to frugivory★. Jordano (1987) similarly found no significant differences in gizzard weight, intestine length or liver weight between the more frugivorous and less frugivorous species of *Sylvia*.

Functional adaptations, however, are also possible. One of the most likely of such adaptations, and the only one that has been analysed so far, is the speed with which fruit is processed in the digestive canal. Fruit pulp is bulky and, for its mass, relatively unnutritious; so a specialised fruit-eater should need to eat a lot and process it rapidly in order to obtain adequate nourishment, and should also be able to get rid of the seeds quickly to make way for more food. Some specialised frugivores have been shown to pass food through the alimentary canal, and evacuate the seeds, very rapidly. Thus the Phainopepla *Phainopepla nitens*, a mistletoe-berry specialist, may defaecate the seed only 12 minutes after eating a fruit (Walsberg 1975); the Olive-backed Euphonia *Euphonia gouldi*, another mistletoe specialist, after only 9.2 minutes and manakins after 10–12 minutes (Levey 1986). Seeds that are regurgitated are got rid of even more rapidly; thus manakins can regurgitate a seed in about half the time they need to defaecate a seed of the same kind (Levey 1986). For the less specialised European fruit-eaters the only

---

★ Starlings have been shown to undergo seasonal changes in gut length associated with changes in diet (Feare 1984). The gut is longer in winter than in summer, and the lengthening is associated with a change from mainly animal to mainly vegetable food. This suggests an adaptation opposite to that shown by the specialised frugivores, whose guts are very short; but the winter food of the Starlings investigated consisted mainly of seeds, not fruit pulp.

relevant study has been made by Herrera (1984c), who investigated 'gut passage time' in 30 species of Mediterranean scrubland passerines. Size for size, he found a marked difference in gut passage time between the seed-dispersers and the others. Thus the gut passage time (about 50 minutes) of an 80 g disperser was similar to that of a 9 g non-disperser. (The non-dispersers were divided into fruit-predators and non-frugivores, and there was little difference between them.) The difference persisted, but was less, when the analysis was restricted to the muscicapids, thus indicating that it was not merely a fortuitous consequence of the diverse taxonomic composition of the birds in the two groups.

Herrera's measurements of gut passage time were obtained from birds in very unnatural conditions. It was the time taken for a small dose of barium sulphate, stained with a harmless dye and introduced into the gizzard by catheter, to appear in the faeces of a captured bird held in a cloth bag. While admitting the artificial nature of the experiment, he argued that the consistency of the results indicated a real difference between the species examined, and concluded that 'Significantly shorter GPTs [gut passage times] of scrubland seed dispersers should be interpreted as an indispensable feature for sustained subsistence on a strongly frugivorous diet. In addition, this characteristic is an adaptive evolutionary response to fruit-feeding.' It would be of great interest to obtain good field data on this point. One would need not only to investigate the time between ingestion of food and defaecation (and in the case of seeds, regurgitation where appropriate) and to compare frugivores with non-frugivores, but also to compare gut passage times of frugivores when processing fruit and animal food.

To conclude, structural adaptations of the digestive system are found in specialised frugivores, but with the exception of the Waxwing have not been demonstrated, and indeed are unlikely to be very marked, in frugivores that also eat insect and other animal food, as all European frugivores do. Specialised frugivores process fruit in the alimentary canal very rapidly. There is experimental evidence that less specialised European frugivores, too, process their food more rapidly than non-frugivores, but confirmation is needed from studies of free-living birds.

Fruit preferences

In dealing with the different dispersers, in Part 2, we have mentioned some outstanding cases of fruit preference, and have briefly discussed reasons why a species feeds largely on some kinds of fruit and apparently avoids others. This chapter attempts a more thorough treatment of the question of preferences. It is easy to get a superficial, perhaps inaccurate idea of why, for example, Song Thrushes are seldom seen feeding on holly, but less easy to know how best to investigate the question. To supplement our field observations we shall make some use of analyses of the nutritive contents of different fruits, and of some experimental results obtained by Sorensen (1984).

It is essential to investigate preferences in a way that is free from bias and the influence of extraneous factors. Merely to compare the total numbers of records obtained for different bird species feeding on different kinds of fruits may be misleading. A larger, dominant species may prevent access by other species to fruit which they would otherwise eat. An obvious example is the virtual lack of records for any species except the Mistle Thrush feeding on mistletoe berries. In such cases the number of feeding records for the various species is no guide to their preferences. Or a fruit that a larger, more powerful species can pluck when it is first ripe may be too securely attached for a smaller species to pluck until some time later, so that apparent preferences will be different at different stages of the fruit season. In the case of fruits of variable size, whereas a larger species may always be able to pluck and swallow them, a smaller species may only be able to deal with fruits towards the lower end of the size range, so that, unless the size of the fruits being watched is known, observations will be difficult to interpret. Or a fruit may be easy for one species to reach, but another may find it difficult and so tend to avoid it, although both may like it equally if it is presented to them experimentally in a feeding test. Nevertheless, the total numbers of records obtained over a long period, under a variety of conditions, may be useful in indicating some of the more marked preferences, and we have made some comparisons between the thrushes on this basis (Table 78).

The results of experiments with captive birds cannot be uncritically accepted as indicating preferences under natural conditions, especially if the different fruits being tested are made equally easy to obtain by being presented on a dish. One may conduct an experiment which shows that under the experimental conditions one kind of fruit is preferred to another; but under natural conditions the preference may be suppressed or reversed by other factors such as those mentioned above. To sum up, it is meaningless to talk of preferences without taking into account the circumstances in which they are shown.

PREFERENCES INDICATED BY OBSERVATION OF CHOICE SITUATIONS

In addition to the indications provided by the overall numbers of records of different bird species feeding on different fruits, we studied fruit preferences by field observation in special circumstances. We considered that a preference could potentially be shown if two different kinds of fruit, growing close together, were kept under observation at the same time, provided that both were present in comparable quantities, neither of them was being defended by a dominant bird, and we had previous evidence that both fruits were ripe and being eaten. We had significant observations only for thrushes and the Starling; they are summarised in Tables 79–82.

220 *Interpretation*

TABLE 78 *Records of the five thrushes feeding on some important fruits*

	Total thrush records	Blackbird	% contribution by			
			Song Thrush	Mistle Thrush	Field-fare	Redwing
Holly	1705	36	3	22	5	34
Yew	1278	38	40	18	0	4
Ivy	1643	53	23	7	7	10
Hip	1057	57	1	1	41	+
Haw	3345	47	3	4	30	16
Sloe	724	37	43	7	5	8
Elder	362	47	52	1	+	+
Guelder rose	114	12	83	4	0	1
Dogwood	233	31	25	+	0	43
Buckthorn	449	51	8	9	2	29

Notes. Contributions of less than 0.5% are shown as +. In interpreting the figures as indications of preference for, or avoidance of, particular fruits, taking the Blackbird (a catholic feeder) as standard, account must be taken of the total number of feeding records obtained for each species (all fruits): Blackbird 6,869, Song Thrush 1,984, Mistle Thrush 1,450, Fieldfare 1,676, Redwing 1,819. Thus the almost equal contribution of Blackbird and Song Thrush to the yew records indicates a strong preference by the Song Thrushes, and the 83% contribution of Song Thrush to guelder rose indicates a very strong preference. For further discussion, see text.

The figures for holly are affected by the fact that some hollies are protected by Mistle Thrushes, which prevent the other thrushes from feeding; but the very small percentage for Song Thrush, compared with Blackbird and Redwing, indicates a degree of avoidance of holly by Song Thrushes.

PREFERENCES SHOWN BY BLACKBIRDS

Table 78 shows that Blackbirds are catholic in their choice of fruits, their percent contributions to records for the fruits listed being much more uniform than is the case for the four other thrushes. Nevertheless, observation of choice situations (Table 79) showed that Blackbirds prefer haws over hip, sloe, dogwood, buckthorn and elder. In the choice between haw and sloe they show a seasonal change, with haw strongly preferred at first (October) and less strongly later. This change probably reflects the fact that it becomes easier for Blackbirds to pluck sloes as winter advances. As Table 37 shows, based on observations combined from all months, Blackbirds find it considerably easier to pluck haws than sloes.

Haws are probably preferred to hips even more strongly than Table 79 suggests. In November, when most observations were made, we had no cases of hips being eaten when there was a choice between hips and haws. The few instances of preference for hips were in December and January, when the haw crop is nearly finished and the remaining fruits tend to dry up and are sometimes hard to pluck. The preference for haws was strikingly seen on several occasions when we watched Blackbirds taking the last, terminal haws from hawthorn bushes surrounded by dog-roses laden with hips; but these are excluded from Table 79 because the two fruits were not present in comparable quantities. Also excluded from the table are records showing that Blackbirds strongly prefer haws to guelder rose fruit: when Song Thrushes in numbers were visiting a guelder rose, for whose fruit they show a strong preference (see below), Blackbirds were going exlcusively to haws near by to take the very sparse fruits that were left, ignoring the guelder rose.

Table 79 shows that holly is preferred to yew, but not very strongly, and yew to elder. This is in the opposite order to the success shown by Blackbirds in taking these fruits (76%, 89% and 93% respectively; Table 37), which points to a preference based on nutritive qualities or ease of internal processing, as discussed below.

PREFERENCES SHOWN BY SONG THRUSHES

We have already seen that Song Thrushes show some clear fruit preferences, based on comparisons with fruits taken by Blackbirds. Table 78 brings out these differences very clearly. The Song Thrush's most striking preferences are for yew, sloe, elder and guelder rose, for which Song Thrush records outnumber Blackbird records, in spite of the fact that we had altogether more than three times as many Blackbird as Song Thrush feeding records, and Song Thrushes are often more secretive feeders. The clearest avoidances, of the major fruits, are for hip, haw and holly. Table 80 agrees with and extends the list of preferences, showing that Song Thrushes also prefer buckthorn to haw, and yew to both elder and holly.

TABLE 79 Fruit preferences shown in choice situations by Blackbirds

	No. obs.	Results of choice: % of feeding visits to:			
Haw/hip	90	haw	96($\pm$4)	hip	4
Haw/sloe, Oct.	84	haw	95($\pm$5)	sloe	5
Nov.	236	haw	88($\pm$4)	sloe	12
Dec.	119	haw	55($\pm$9)	sloe	45
Haw/dogwood	44	haw	89($\pm$9)	dogwood	11
Haw/buckthorn	209	haw	70($\pm$6)	buckthorn	30
Haw/elder	227	haw	76($\pm$6)	elder	24
Yew/elder	141	yew	84($\pm$6)	elder	16
Yew/holly	44	yew	30($\pm$14)	holly	70

Note. All data refer to feeding visits (not numbers of fruit taken) to one or the other fruit source. Figures in parentheses are 95% confidence limits, which apply equally to the percentages for the other fruit.

TABLE 80 Fruit preferences shown in choice situations by Song Thrushes

	No. obs.	Result of choice: % of feeding visits to:			
Haw/sloe	237	haw	13($\pm$4)	sloe	87
Haw/elder	49	haw	14($\pm$10)	elder	86
Haw/buckthorn	24	haw	29($\pm$19)	buckthorn	71
Yew/elder	152	yew	61($\pm$4)	elder	39
Yew/holly	27	yew	93($\pm$10)	holly	7
Haw/guelder rose	41	haw	2($\pm$4)	guelder rose	98

See footnote to Table 79

222 Interpretation

These preferences are only partly in line with the Song Thrush's ability to handle the different fruits. Thus Song Thrushes are more successful at plucking and swallowing yew ($88\pm4\%$) and elder ($97\pm2\%$), which they prefer, than haw ($73\pm9\%$); but they are much more successful at plucking and swallowing haw than sloe ($37\pm8\%$), but nevertheless prefer sloe. As with Blackbird preferences, differences in the ease of internal processing of fruits may be critical. The fruits that Song Thrushes prefer have soft pulp and/or seeds that are easily detached from the pulp, and it may be that Song Thrushes are less efficient than Blackbirds at processing fruits with dense pulp adhering to the seeds, such as haw. Hips, however, are probably avoided mainly because of their large size.

PREFERENCES SHOWN BY OTHER THRUSHES

Mistle Thrushes strongly prefer sloes to haws (Table 81). Presumably because of their large size and greater strength, they are more successful at taking sloes than the smaller thrushes (success rate of 68%, compared with 54% for Blackbird and 37% for Song Thrush); but they are even more successful at taking haws (success rate 92%), hence their preference for sloes presumably depends on some quality of the fruit. In the other choice situation shown in Table 81, Mistle Thrushes strongly preferred yew to elder.

Redwings much prefer haw to sloe (Table 81). Sloes are mostly too large for Redwings to swallow, and they feed on them mainly by picking out pieces in situ. The high percentage of dogwood fruits taken by Redwings (Table 78) suggests a preference, but we cannot demonstrate this from observations of choice situations. Dogwood fruits are never too large for Redwings to swallow, and they are easily detached, whereas haws may be at or near the upper size limit for Redwings and may be hard to pluck. Haw is usually preferred to buckthorn, though locally we have recorded the opposite preference, probably because the haws were less easy to pluck than the smaller, more easily detached buckthorn fruits. We have no data on choice situations involving holly, but general observations (pp. 138–41) indicate that holly fruit is much favoured by Redwings.

TABLE 81 *Fruit preferences shown in choice situations by Mistle Thrushes, Redwings and Fieldfares*

	No. obs.	Result of choice: % of feeding visits to:			
Mistle Thrush					
Haw/sloe	35	haw	$11(\pm 11)$	sloe	89
Yew/elder	38	yew	100	elder	0
Redwing					
Haw/sloe	201	haw	$90(\pm 4)$	sloe	10
Haw/buckthorn	75	haw	$75(\pm 10)$	buckthorn	25
Fieldfare					
Haw/sloe	368	haw	$98(\pm 1)$	sloe	2
Haw/buckthorn	99	haw	$97(\pm 3)$	buckthorn	3
Haw/hip	80	haw	100	hip	0

See footnote to Table 79

TABLE 82   *Fruit preferences shown in choice situations by Starlings*

	No. obs.	Result of choice: % of feeding visits to:			
Haw/sloe	91	haw	90(±6)	sloe	10
Haw/dogwood	1300	haw	0	dogwood	100
Haw/buckthorn	58	haw	2(±4)	buckthorn	98
Haw/elder	178	haw	1(±1)	elder	99
Haw/yew	21	haw	0	yew	100
Yew/elder	224	yew	50(±7)	elder	50
Yew/holly	79	yew	100	holly	0

Fieldfares strongly prefer haw to sloe and buckthorn. We lack adequate data to compare their success at taking these fruits, but it is unlikely that the preference for haw is influenced by their ability to take them, since Song Thrushes greatly prefer sloe though they are smaller and weaker than Fieldfares, and buckthorn fruits are easy to pluck. Possibly because haw is much the most abundant fruit available in the open country where Fieldfare flocks mainly feed, they develop a preference for haw independent of its quality relative to other less common fruits.

PREFERENCES SHOWN BY STARLINGS

Compared with the thrushes, Starlings show very marked preferences for certain fruits and avoidance of others (Table 82). Most obviously they prefer dogwood, elder, yew and buckthorn to haw; but haw is preferred to sloe, which are often too large for Starlings to swallow. Between yew and elder there seems to be no clear preference.

These preferences are reflected in the total numbers of feeding records for different fruits summarised in Table 60, in which dogwood scores highest, followed by yew and elder. In spite of the preference for buckthorn we have altogether more records for Starlings eating haw than buckthorn (198 and 90, respectively), as buckthorn is a much less abundant plant in our study area.

The orders of preference can easily be accounted for by the Starling's ability to take the various fruits (Table 61). We discuss elsewhere the Starling's physical limitations as a fruit-eater. They are most successful at taking dogwood and elder fruits (97% and 93%), rather less successful at taking yew (61%), mainly because they drop many that they pluck rather than being unable to pluck them, and least successful at taking haw (40%; we have inadequate data for buckthorn). The rate for haws should be even lower than this, as Starlings feed only at certain hawthorns, apparently those whose fruits they find they can pluck; so the figures are largely based on a biased sample.

SUMMARY OF PREFERENCES SHOWN BY FIELD OBSERVATION

The fruits that have been compared cannot be arranged in any simple order of preference that is valid for all the thrushes and the Starling, nor would this be expected, as the reasons for the preferences are not the same for each bird. Thus Blackbirds, Redwings, Fieldfares and Starlings prefer haw to sloe, while Song Thrushes and Mistle Thrushes prefer sloe to haw. In the case of Redwing and Starling, the preference for haw probably reflects their physical inability to pluck and swallow sloes successfully. In the case of the Song Thrush, the preference for sloes, in spite of the fact that they are near the upper size limit, may reflect

224 Interpretation

differential ability to process haws and sloes internally. But the data indicate some consistent preferences:

All the thrushes prefer haws to hips. Because of their large size, hips can be tackled successfully only by Blackbird, Mistle Thrush and Fieldfare. Song Thrushes and Redwings rarely attempt to eat them.

Yew is preferred to elder by Blackbird, Song Thrush and Mistle Thrush (Redwing and Fieldfare are not present during the main period of elderberry availability).

Ivy is strongly preferred to holly, the only other major fruit available during the ivy season.

The Blackbird is the only species for which we have data adequate for comparing any other fruit with holly; it prefers holly to yew, but not very strongly.

FACTORS AFFECTING FRUIT PREFERENCES

It is evident that in some cases a preference by a particular bird species for one fruit over another is based on its ability to pluck and swallow the fruits. In other cases a preference may run counter to the bird's ability to 'handle' the fruits, and it seems that its ability to process them internally is involved. Another factor that may be important is the nutritive quality of the fruits concerned. Possible additional factors are the colour and taste of the fruits, but it is probably justified to exclude these from consideration here; they have been discussed in a previous section and there is little evidence that they play a significant part in preferences by birds for native fruits. A bird that is behaving 'rationally' would be expected to base its preferences on a comparison of the net energy gain from eating each kind of fruit, taking into account the bird's handling proficiency, its speed and efficiency at ejecting the seeds and digesting the nutritive matter, and the nutritive content of the pulp. But 'nutritive content' is too general a term. A further complication, suggested by Herrera (1985a), is the possibility that different fruits may provide specific nutriments not provided by other locally occurring fruits; if this is so, a bird's choice may be influenced by subtle biochemical differences between fruits. Clearly, we are a long way from understanding the basis of the fruit preferences shown by birds in the wild.

Experiments can provide some leads, and it is in this content that it is relevant to consider the results of Sorensen's (1984) experimental investigation of the Blackbird's preference for, and efficiency in processing and digesting the pulp of, six kinds of wild fruits (haw, sloe, hip, blackberry, elder, ivy), all but one of which (blackberry) are among those discussed above.

On the basis of earlier field observation in woodland near Oxford, Sorensen had concluded that Blackbirds prefer haw to elder and blackberry in autumn, sloe to hip in winter, and ivy to hip in early spring. Supporting these conclusions, based on feeding records, she noted that the fruit crops of haw, sloe and elder in her study area were all depleted, whereas many fruits of elder, blackberry and hips rotted on the plants uneaten. All of this is broadly consistent with findings in our study area.

From laboratory analysis of the six different fruits, Sorensen found no overall relationship between their nutritive properties (based on the water, protein, fat and carbohydrate content of the pulp, and the proportion of pulp to seed) and the Blackbird's preference as indicated by field observation. But for the winter and spring fruits there was a suggestion of a relationship: sloe and ivy fruits both had significantly more energy per gram of pulp than hip, and they were preferred to hip.

But the relative 'profitability' of different fruits as indicated by laboratory analysis is not the same as their actual profitability to a bird eating them, as other constraints are involved, affecting the bird's speed and efficiency in processing each fruit. Using seven wild-caught Blackbirds temporarily held in captivity, Sorensen investigated their efficiency at extracting energy from the six different fruits. The results showed that, based on the potential energy content of the pulp, Blackbirds were more efficient at extracting energy from sloe and ivy fruits than from hips. Thus sloe and ivy were found not only to be more nutritious than hips, but to be more efficiently digested. For the other, autumn fruits there was no relationship between preferences observed in the field and the Blackbird's efficiency at digesting the pulp.

It should be noted that other analyses (Appendix 4) have shown that the pulp of hips has a higher energy content than that of sloes; hence the Blackbird's efficiency at digesting the different pulps is likely to be more important than their potential energy contents. Hip pulp has a high tannin content, which may inhibit digestion by birds (Herrera 1984b).

The clearest result of Sorensen's experiments was the relationship between observed fruit preferences and 'seed passage times' – that is, the times taken for the seeds to be regurgitated or defaecated. The captive Blackbirds invariably regurgitated the comparatively large seeds of sloe, haw and ivy (weights 0.24 g, 0.13 g and 0.053 g respectively), and defaecated the smaller seeds of hip, elder and blackberry (weights 0.024 g, 0.0039 g and 0.0028 g respectively). All sloe, haw and ivy seeds were regurgitated 6.5–8.9 minutes after the fruits had been eaten; the first hip, elder and blackberry seeds were defaecated 25.5–39.1 minutes after the fruits had been eaten, and the last seeds 33.8–50.4 minutes afterwards.

Summarising her findings, Sorensen concluded that 'seed passage rates play an important role in determining preferences, particularly if nutritional and other properties (e.g. accessibility) of fruit species are similar. Calculations show that birds obtain a high rate of energy gain by consuming fruits whose seeds are regurgitated. ... Seed regurgitation results in a rapid elimination of non-nutritional seed "ballast" and creates space in the gut for additional food. Fruit species containing seeds which are defaecated have lower rates of energy gain because the seeds remain (and take up space) in the gut for much longer periods of time.'

Sorensen's conclusions help to explain some of the preferences that we found. Thus, Blackbirds in our area strongly preferred haw to elder (as Sorensen also found) and to hip; they also preferred haw to buckthorn, whose seeds are small (0.027 g) and probably defaecated. In other cases, however, where there is a preference between two fruits, both of whose seeds are regurgitated, other factors must be involved. The ease with which pulp can be separated from seed in the bird's stomach is probably important. Thus ivy seeds slip easily out of the pulp, and this should enhance the fruit's attractiveness in addition to the high nutritive quality of the pulp. It is interesting to note from Sorensen's experiments that there was no significant differences in the times taken for haw and sloe seeds to be regurgitated (means of 8.6 and 7.8 minutes respectively), but sloe pulp was passed through the gut much more rapidly than haw pulp (19.2–25.2 and 32.2–40.0 minutes, respectively, for passage of first and last pulp). Hence in this case seed passage time is probably not important, but the difference in the speed of processing of the pulp in the gut, if valid for other species, may help to explain the strong preference for sloe shown by the Song Thrush and Mistle Thrush.

An important general conclusion from these experiments is that the rate of energy gain from fruit-eating depends primarily on the speed and efficiency with which the fruit can be processed in the alimentary canal. From this it follows that it does not usually depend (unless fruit is very sparse) on the rate at which the fruit can be plucked and swallowed. This has important implications for the time-budget of frugivores, as discussed in another chapter.

The importance of regurgitation of seeds for efficient processing of fruits is also of wider evolutionary interest. It suggests that in the coevolution of birds and bird-fruits, as birds became more and more specialised for fruit-eating they will have tended to prefer fruits with single large seeds to fruits with many small seeds, because the more dependent a bird is on fruit the more important it must be to process the fruit efficiently. For the same reason, they will have tended to prefer fruits with seeds easily separated from the pulp to fruits with pulp closely adhering to the seed. A specialised fruit-eater must also tend to prefer a fruit with a more nutritious pulp to one that is less nutritious. Fruits with these qualities are prominent in the diet of very specialised tropical fruit-eating birds, and very rapid regurgitation of the seeds is one of the striking characteristics of the feeding behaviour of such specialists.

Foraging strategies of fruit-eaters

Two contrasting foraging strategies are employed by the fruit-eaters in our study area: social feeding without aggression, and aggressive defence of a fruit source. Each method doubtless confers an advantage on the bird using it; to speak of a 'strategy' implies an adaptive behavioural response, presumably evolved and maintained by natural selection. Both strategies are employed by other birds too, not only fruit-eaters, and the conditions and circumstances which favour the adoption of one or the other strategy are likely to be essentially the same for them as for the fruit-eaters.

SOCIAL FORAGING

Social foraging in a flock, whether of a single species or of several species, is a common strategy of tropical forest birds, and there has been much discussion of its function. It is an impressive sight when, after wandering through the forest for some time with hardly a bird to be seen, one is suddenly surrounded by a mixed flock of many different species, all assiduously foraging as they move through the trees. Two main functions have been proposed: that foraging in a group in some way enhances the efficiency with which each individual obtains food; and that it confers safety against predators, since many pairs of eyes are better than one, and one bird's warning calls can alert the whole group. The two functions are not mutually exclusive, and may be related; enhanced safety may also tend to improve foraging efficiency, as each individual may need to spend less time on the look-out.

There is some evidence, mostly indirect, for both main functions, but on balance it seems likely that enhanced safety is the primary function, at least in tropical forest (Powell 1985). Most research has been carried out on mixed-species flocks of insectivores, and it is possible that one bird may make prey available to other individuals; for instance, one bird may flush an insect which another may take, or it may see where another is finding food and try the same or a similar site. But for fruit-eaters the situation is different. For them, social foraging may help initially in locating sources of food, since the combined knowledge available to a flock is likely to be greater than that of an individual; but once a fruit source has been found, the principal benefit is likely to be the group vigilance against predators, which may enable a bird to take its meal more quickly. Help in locating new fruit sources is probably of minor importance, except perhaps at times of migration; most frugivores are, in our experience, thoroughly familiar with their local fruit sources. Hence one would suppose enhanced safety to be the primary function of social feeding by fruit-eaters.

In the case of fruit-eaters in our study area we are confident that this is so, because the threat from predators, even in our man-altered habitat, is surprisingly high. In winter, when fruit eating is the main activity of most frugivores, the likelihood of attack by a predator is so great that efficient anti-predator responses are vitally important. During October–February, in 1983–85, in 113 hours at five of our main observation sites, we had 33 records of hawks either actively hunting birds that were visiting fruit or, more often, flying close enough in a manner suggesting hunting to elicit a reaction from them. Of these 33 records, eight were instances of a Sparrowhawk *Accipiter nisus* attacking a group of fruit-eating birds, and there was one case of a Kestrel *Falco tinnunculus* catching a thrush. The overall rate of one hawk per 3.4 hours of observation was influenced by the data from one site where only one hawk was recorded in 10 hours; at the four other sites the rates were one hawk per 1.5, 2.4, 2.7 and 3.3 hours of observation. These observed rates are perhaps considerably lower than the actual rates, as it is easy to miss a passing hawk while concentrating on watching a fruit source. Moreover, we had a few cases of a hawk perching for several minutes in cover near a visited fruit source, and occasionally sallying to attempt a capture, so we may well have missed concealed hawks on other occasions.

When several birds are visiting the same fruit source from some distance away, the timing of their visits is usually synchronised; they come as a loose group and leave at about the same time. It is common to see a solitary Blackbird, ready for another feed, spend some time giving anxiety calls – the familiar *chook* – and beginning tentatively to approach the fruit tree, until it is joined by others and flies to feed. In the case of some late-summer and autumn fruiting shrubs, it is often very striking to see one individual of several species of warblers, thrushes and Robin come to feed within half a minute, after a 15–20-minute interval when no birds have been present. Further details are given in the account of the Blackcap. It is clear that it is important for a fruit-eater to feed, if possible, in company with others.

Exposure to predation while feeding also has a conspicuous effect on the order in which different fruiting plants are exploited by fruit-eaters, and in which different parts of an individual plant are exploited. This may make it hard to obtain accurate data on the beginning of the fruit seasons of some plants, such as white bryony, in which the first fruits to be taken are those that are best concealed within thick vegetation. The general rule, as would be expected, is that fruit in a safe position is taken first. One observation, though 'anecdotal' in the sense that it has no statistical significance, is surely meaningful. Privet is an important late-winter fruit on the Chiltern slopes. The last remaining fruits are always terminal ones, exposed at the ends of slender bare twigs. One of our few records of a bird being caught while eating fruit was a Blackbird, caught by a Sparrowhawk while taking the sparse terminal fruits of a privet in hard weather in January.

In the case of trees and large shrubs, the last fruits to be taken are usually those that occupy terminal positions on lower branches. Probably this is the position in which a feeding bird is most vulnerable to attack by Sparrowhawks, which tend to fly low when hunting. As already mentioned, a practical result of this is that marking and counting fruits as a method of studying the time of their removal by frugivores is liable to give biased data, as the only easily accessible fruits are often those that are going to be taken last.

Social feeding may also have another important advantage in very hard weather. If a fruit source is being defended by a larger, dominant species, for example a holly tree by a pair of Mistle Thrushes, the defenders can easily keep single intruders away. But a flock of birds that persistently intrudes can eventually overwhelm the defenders, even if individuals are repeatedly chased away. The intruders keep on returning, until their numbers are too much for the defenders. We saw Mistle Thrushes overwhelmed many times by the smaller thrushes (p. 130), and Creutz (1952) has similarly described Waxwing flocks overwhelming Mistle Thrushes attempting to defend mistletoe in Germany. We had one case of a Fieldfare defending a fruit source and being swamped by smaller thrushes during a very cold spell.

This advantage of social foraging for fruit, though important for the individuals that benefit from it, is probably secondary, a consequence of the primary function of increasing safety against predators, since the antipredator function presumably works for all species in nearly all circumstances, whereas the overwhelming of a dominant species' defences is an advantage accruing to certain species only, in special circumstances.

DEFENCE OF A FRUIT SOURCE

In a neat series of feeding experiments with White Wagtails *Motacilla alba* wintering in Israel, Zahavi (1971) showed that, by altering the distribution of food, a bird could be switched from flock feeding to defence of a food supply. When food was well scattered the wagtails fed in flocks without aggression; if food was put into discrete piles, individuals began to defend a pile against their fellows. Similar alternation of behaviour has been shown experimentally by Tye (1986), working with Fieldfares feeding on apples laid out on an open field. Defence gives the advantage of a more assured food supply but it involves energy costs in driving away intruders. Tye made some reasonable assumptions about energy intake and expenditure, then compared the energy gain of individuals defending apples with that of 'non-territorial' individuals (*ie* those not defending apples), and concluded that defence was maintained as long as it improved net energy gain. This general conclusion probably is relevant to other cases involving a choice between alternative feeding strategies, but in more natural conditions other factors, such as the impact of predators, must also play a part.

Turning from experimental to more natural conditions, food resources that are defendable include, in addition to concentrated fruit sources, concentrated nectar sources and some rather special food supplies such as the honeycombs defended by the Orange-rumped Honeyguide *Indicator xanthonotus* in India (Cronin & Sherman 1976). Nectar sources provide a close parallel with fruit resources, as their defence by hummingbirds depends on their spatial concentration much as is the case for fruit; only concentrated nectar sources, such as a profusely flowering bush, are defended. (There is, however, an important difference: hummingbirds not in possession of a defended nectar source forage alone, not socially.) The ability to switch from unaggressive foraging to defence of a food source may be widespread in birds, but to date this has been investigated mainly in nectar-eaters and fruit-eaters.

Even so, not many instances have been recorded of birds defending fruit sources. It has rarely been recorded in tropical forest, although fruit supplies and frugivores are abundant and varied, and those cases that have been found involve short-term defence over periods of a few days (Pratt 1984). Most tropical fruits do not last long in good condition and, in any case, with no winter and a wide variety of fruits, the possession of an exclusive fruit supply may be less important than in more northern latitudes, where fruit supplies are less varied and winter survival is critical. Hence it is not surprising that the few known cases of long-term defence of fruit resources concern north-temperate birds. Mockingbirds *Mimus polyglottos* in North America defend fruit supplies in winter (Moore 1978); in Europe the Mistle Thrush is the prime example, with the Fieldfare and Blackbird showing less developed forms of the same behaviour. Probably further study will bring a few other cases to light, particularly among thrushes.

The circumstances that allow Mistle Thrushes to defend fruit supplies successfully throughout the winter have been fully discussed earlier and need not be repeated in detail. As regards the fruit, it is clear that the size of the crop, its spatial concentration, and its lasting qualities are all important. As regards the bird, dominance in encounters with likely intruders seems to be the key factor, and dominance depends primarily on size. The Mistle Thrush (mean weight c 130 g) is the largest European thrush. Next in size is the Fieldfare (mean weight c 115 g), some individuals of which defend fruit supplies temporarily, followed by the Blackbird (mean weight c 100 g), which may defend fruit supplies if circumstances allow. Of these three, only the Mistle Thrush can be regarded as a specialist in the defence of a winter fruit supply, as indicated by the following characteristics in which it differs from Fieldfare and Blackbird. (1) The attempt to defend a fruit supply, by pairs or single birds, is a major activity in autumn, apparently involving all adults in our study area, where they are resident. (2) The main fruits defended are long-lasting ones (holly, mistletoe, haw). (3) They feed on alternative fruit supplies for as long as they are available, which thus conserves the defended fruit as a long-term resource. (4) All food competitors, including seed-predators, are chased away from the defended fruit supply.

One is left to wonder if success in defending a fruit supply has been the main selective factor responsible for the Mistle Thrush's large size; or whether other selective factors have been responsible, and the Mistle Thrush's size has secondarily enabled it to develop fruit-defence as an important part of its overall strategy for survival. Further questions are raised. Is it significant that the Mistle Thrush's breast is the most heavily spotted of all the thrushes, thus enhancing its effectiveness as a visual signal in attack? Is it significant that it and the Fieldfare (a notably aggressive species, in its breeding habitat as well as when defending fruit) are the only *Turdus* species with white under-wing coverts, and that these may be conspicuous in aggressive encounters? Finally, does it seem to other birds as it does to the human observer, that an attacking Mistle Thrush looks quite like a Sparrowhawk as it bears down on them in rapid flight and swerves past and round its defended tree? These can only be tentative suggestions, drawing attention to the possibility that the Mistle Thrush's specialisation for defence of a fruit supply may involve plumage characters and special behaviour with deterrent qualities.

Fruit as a component of the total diet

Some tropical birds live on an exclusive, or almost exclusive, diet of fruit, except that they may need extra calcium for egg production and to supplement the diet of their growing young. Thus the Bare-throated Bell-bird *Procnias averano*, one of the most specialised frugivores, has been recorded with a snail in its stomach, a food known to be eaten by other birds when calcium is needed. The Andean Cock-of-the-rock *Rupicola peruviana*, another specialised frugivore, regularly includes lizards and frogs in the otherwise exclusively fruit diet fed to its young, probably because fruit alone does not provide enough calcium for their growth (Snow 1982). At the other end of the frugivore range are birds that occasionally take fruit, although they can exist perfectly well without it. All the main fruit-eaters in Britain are intermediate, feeding very largely on insects and other invertebrates, at least at some times of year, and regularly also on fruit. How important is fruit in their diet?

Put this way, the question is rather too general. More specific questions need to be asked, and the answers will differ according to species. The main questions are these. What part does fruit play, quantitatively, in the diet at different seasons of the year? Is it an essential part, either as a source of energy or as a source of specific nutrients? Can the bird live solely on fruit? If not, for how long can it survive on a diet of fruit alone; and how is this affected by different fruits or mixtures of fruits? Does fruit-eating, as opposed to foraging for animal food, affect other aspects of the bird's life? To what extent are the various species of fruit-eaters structurally or physiologically adapted to a diet of fruit?

Once one begins to formulate such questions it becomes clear that there are no simple answers, and for some questions no answer of any sort can yet be given. We shall concentrate on a few questions which we can begin to answer. The questions of structural or physiological adaptations to fruit-eating is rather different from the others. Such adaptations are an indication of the importance of fruit-eating in the evolutionary history of a species, as well as throwing light on its present ecology. We have discussed this question in an earlier chapter.

The first of the questions listed above can be answered only in a general way. Fruit forms a large part of the diet of thrushes in autumn and winter, and during periods of drought in summer, and is least important in the months March–June. During spells of very cold weather fruit may, of necessity, constitute the total diet. For the Starling the pattern is different, with fruit forming a large part of the diet in late summer and autumn and much less at other times of year. We have no data enabling us to assess how much of a bird's energy and essential nutrients are supplied by animal food and how much by fruit at all times of year, nor does there seem to be any such information from other areas. Heavy reliance on fruit in winter is certainly not peculiar to southern England; Dieberger (1982) found that for nine weeks after a heavy snowfall, Fieldfares wintering in Austria, in an area where mistletoe *Loranthus europaeus* was abundant, were feeding almost exclusively on its berries. Pénzes's (1972) findings from Hungary are even more striking; from stomach analyses he found that wintering Fieldfares had eaten nothing but fruit from the beginning of December to the beginning of March. In west-central France, Blackcaps may subsist, apparently almost exclusively, on mistletoe berries through the severest winters (Heim de Balsac & Mayaud 1930). Jordano & Herrera (1981) have found that Blackcaps wintering in southern Spain feed almost entirely on fruits during six consecutive months.

232 *Interpretation*

In the German ornithological literature on fruit-eating one quite frequently comes across references to fruit as *Notnahrung* – a word for which there is no English equivalent, meaning food eaten in times of need, when other food is unavailable or scarce. The implication is that fruit is not taken if alternative and presumably more nutritious food (normally animal) is available. This is an impression conveyed by the conspicuous concentrations of birds at fruit in hard weather, when the ground is snow-covered and most other sources of food are inaccessible, but is not the whole story. Birds take fruit at times when alternative foods certainly are available. In order to investigate the matter, and in particular to find out whether birds that eat both insects and fruit have seasonal preferences for one or the other independent of availability, Berthold (1976a) carried out extensive experiments with captive birds in West Germany.

He kept Blackbirds, Robins, Blackcaps and Garden Warblers, and fed them on diets either of insects or of fruit, or gave them a choice between insects and fruit. Several kinds of wild fruits were used, including some that are succulent and watery (raspberry) and some that are drier and more nutritious (ivy, haw). From Berthold's numerous findings we may select those that bear especially on the questions we asked above.

Berthold's Blackbirds, Robins and Garden Warblers did not prefer fruit to animal food when given a choice at any time of year (assessment of preference being the weight of each food eaten). Blackcaps, however, regularly preferred fruit to animal food in late summer and early autumn. Under constant conditions Blackcaps and Garden Warblers, given a choice, showed regular spontaneous changes in the amount of fruit eaten, Blackcaps taking most in late summer and autumn (when they preferred it to animal food) and Garden Warblers taking most in winter (when, however, they still preferred animal food). These last findings were the first demonstration ever made of spontaneous, or 'endogenous', seasonal changes in food preferences of birds.

None of Berthold's experimental birds maintained its weight on an exclusive diet of fruit; all showed a decline from the first day of the experiment, and the decline continued and led to death if no animal food was offered. As soon as animal food was restored, body weight increased. Berthold concluded that weight loss on an exclusively fruit diet was probably due to protein deficiency. If the birds had unlimited access to fruit, only small quantities of insect food were needed for them to maintain their body weight (3 g of *Tenebrio* larvae daily for Robin and Garden Warbler, only 2–3 g for Blackbird).

Berthold also found that his experimental birds decreased their intake of fruit when, normally, they would have been migrating. This finding, combined with the fact that body weight could not be maintained on fruit alone, led him to reject the hypothesis that song birds shift to feeding on fruit in migratory periods in order to accelerate fat deposition (see p. 155).

Berthold's general conclusions were that, for the species studied, fruit 'is a secondary or supplementary food ... This supplementary food is necessary if animal food is scarce or in periods of high nutritive demands if animal food in sufficient quantities can only be obtained with excessive efforts' (English summary). In other words, fruit is essentially a *Notnahrung*. Berthold suggested, however, that the fruit component is important in enabling some birds to winter in higher latitudes than they would otherwise do, and to rear late broods successfully and prepare for migration even though the supply of animal food may be only moderate.

Berthold's work has been widely quoted and his main conclusions accepted, especially the conclusion that unspecialised songbirds cannot maintain their weight on a diet of fruit alone. His experimental findings are unequivocal. We doubt, however, whether the conclusions should be accepted without question. For some reason his birds, when offered a diet of fruit alone, ate far less of it than a bird eats in the wild. As we shall see (p. 236), a Blackbird that we watched for a whole day consumed 337 (c. 88 g) of *Pyracantha coccinea* berries, almost equivalent to its body weight, and this was probably supplemented by some apple. Our observations indicated that a Mistle Thrush, feeding exclusively on holly or haw fruits, eats at least 100 g of fruit per day, and probably a quantity approaching its body weight (Table 83). The daily consumption of *Cotoneaster horizontalis* berries by Waxwings was at least 140% of body weight, and possibly twice the body weight. Neotropical fruit-eating birds of several species kept in captivity by Moermond & Denslow (1985) ate even more per day, mostly 1.8–2.2 g of fruit for every gram of body weight. Berthold's birds by comparison ate very little. His Blackcaps (weights 18–20 g) for example, when fed on fruits of the alder buckthorn *Frangula alnus*, ate on average only about 11 g of fruit on day one, 20 g on day two, and 15 g on day three. Their weight decreased sharply. They increased their fruit intake to about 30 g on days 4–6, and their weight decrease slowed down. The experimental Blackbirds (mean weight about 94 g), when fed on haws alone, ate on average only about 5 g of fruit on day one, 10 g on day two, 15 g on day three, and 22 g on day four. Their weight declined throughout the four days, not surprisingly as their average fruit intake never reached even 25% of their body weight. We cannot suggest why these birds ate so little fruit, but whatever the reason these results surely do not justify the conclusion that thrushes and warblers cannot, even for a day, maintain their body weight on an exclusive fruit diet, especially in view of the field evidence that they may survive well in the wild on such a diet, even through spells of severe winter weather.

There is another reason why the results of experimental feeding of birds in captivity give results that may be misleading if uncritically applied to birds in the wild. In Berthold's experiments fruit and animal foods were equally easily obtained; but in natural conditions much more time and effort must be spent in order to obtain a given weight of animal food than an equivalent weight of fruit. Thus if both foods are present a bird in the wild might be expected to show a greater preference for fruit than was shown by the captive birds. This point is crucial; time saved by feeding on fruit may be used in other ways, and the saving of energy must tend to compensate for the fruit's poorer nutritional quality. These points are discussed in the next section. Secondly, although several wild fruits were used in the experiments, the choice in particular experiments was limited to one, two or three kinds. A bird in the wild may often be able to select from several kinds of fruit so as to make up a more balanced diet. Herrera (1985b) has discussed this point with reference to Blackcaps wintering in southern Spain, where the diet consists almost exclusively of fruit and the birds maintain a constant weight. He found that 'single fecal samples of *Sylvia atricapilla* contain remains of up to nine fruit species, even though the birds face at times a virtually unlimited supply of highly energetic fruits [of one kind, *Pistacia lentiscus*, an abundant shrub in his area]', and suggested that 'the strong nutritional imbalance characterising the pulp of most fruit species forces many frugivores to have mixed-species diets in order to get a balanced input of energy'. This can be no more than a suggestion at present as nothing is known of the detailed nutritional composition of most wild fruits. One should add that, at least in England, in hard weather in winter birds often have only one kind of fruit available.

Jordano & Herrera (1981) pointed out that the fruits eaten by Blackcaps wintering in southern Spain are, on average, more nutritious than northern

European fruits. This may help to explain the apparent anomaly in Berthold's results, that in late summer and autumn Blackcaps prefer fruit to insects, but could not maintain their weight on the fruits which they were offered. The endogenous seasonal change in food preference is probably adapted to the varied and nutritious fruit supply available in the Mediterranean region, the Blackcap's main wintering area, and not to the less nutritious fruits of northern Europe.

During very hard spells in winter, especially if there is thick snow cover, thrushes in our study area may live for many days on fruit alone. Perhaps on such a diet a bird inevitably loses condition; we do not know. Possibly too long a regime of fruit alone would delay its attainment of breeding condition. That the all-fruit diet is inadequate in some way is suggested by the fact that, when a cold spell ends and the ground thaws, there is a conspicuous switch in feeding behaviour, most clearly seen in the flocks of Fieldfares and Redwings which at once leave the fruit sources at which they have been concentrated and are to be seen foraging in the open fields or, if there has been thick snow, on the first patches of open ground from which the snow has thawed. Nevertheless, large numbers of birds survive spells of severe weather on a diet of fruit alone. Hence, even if fruit is nutritionally inadequate in the long term, it may be no less essential than animal food for survival.

Fruit as a diet for young birds is a separate question, which has not been much investigated. Because most fruits are low in proteins, a fruit diet would be expected to be least suitable for nestling birds needing to grow fast. The Oilbird, an exclusively frugivorous bird of tropical America, feeds its young solely on fruits. They are very nutritious fruits, but even so the nestlings grow slowly for their size; for Oilbirds this is not a great disadvantage as the nests are deep in caves where they are safe from most predators (Snow 1962). Birds such as thrushes and warblers, in temperate regions, regularly include a proportion of fruit in the food given to their nestlings, mainly in the latter part of their time in the nest; and fruit may be fed to nestlings in unusual amounts at times when other food is hard to obtain (Glutz von Blotzheim 1973, Berthold 1984). The only thorough study of fruit in the diet of nestlings of a north-temperate bird that is partly frugivorous is that by Breitwisch *et al* (1984), who investigated the Northern Mockingbird in the USA, a bird with rather similar feeding habits to our thrushes. Almost no fruit was fed to the nestlings until they were largely endothermic (could maintain their body temperature); increasing amounts of fruit were then provided, until the amount of fruit fed to older nestlings was sufficient to satisfy their maintenance energy needs. Breitwisch *et al* conclude that 'parent mockingbirds may feed nestlings an optimal mixture of animal protein, for growth, and fruit carbohydrates, for other energetic needs'. This regime is advantageous, as fruit is quickly obtained in comparison with animal food.

Time and energy budgets for fruit-eaters

We have seen in the last chapter that, whether or not fruit alone is an adequate diet for unspecialised frugivores such as thrushes, it is very important in tiding them over periods when other food is unavailable, and it enables them to winter in areas where they could not otherwise survive. Fruit-eating has another important consequence affecting a bird's whole life-style. Once an abundant source of fruit has been located, no time need be spent in searching for food. The limiting factor to the rate of food-intake is, in fact, not the time needed to find and eat but the time needed to process it internally. This means that in hard weather very little energy need be spent in foraging. A Mistle Thrush in unchallenged possession of a protected fruit supply will, in the severest weather, sit motionless between its periodic feeding bouts. Thus, although fruit is of lower quality than animal food, there is a very significant compensation in the much reduced energy expenditure needed to obtain it.

When weather conditions are not severe, the time saved can be put to other uses. In tropical forest, where there are some very specialised frugivores and many of the fruits are highly nutritious, a frugivore's daily food requirements may be quickly met. We have found, by maintaining a continuous watch throughout the daylight hours, that a male Bearded Bellbird *Procnias averano* in Trinidad spent 87% of the entire day-time within his small calling area (where no food was obtained), and a male Black-and-white Manakin *Manacus manacus* spent 90% of the daylight hours at his display court. Similar figures have been obtained for other tropical frugivores. This ability is apparently one of the main preconditions that have led to a social organisation without pair formation, in which males spend most of their time in elaborate courtship display and females carry out all nesting duties single-handed, consorting with the males only for the purposes of mating (Snow 1976). In many species that have adopted this way of life, extreme sexual dimorphism has evolved, leading to extravagantly ornate males and cryptically coloured females; the cotingas of the New World tropics and the birds-of-paradise of the Australasian region are the best examples.

For several good reasons nothing comparable exists in the north temperate region. Nevertheless, when our northern fruit-eaters are living mainly on fruit they certainly have more time for other activities than non-frugivores. In very cold weather the best way to spend this 'spare time' is probably to sit in a safe place sheltered from the elements; the species that form flocks when so resting and digesting, probably reduce the chances of predation.

One would like to be able to bolster these general arguments with actual measurements of the feeding time needed, and the amount of energy obtained from fruits of different kinds. Some measurements are available, but their interpretation is full of uncertainties. In particular, only rough estimates can be made of the energy and nutritional requirements of birds in a wild state, and there is uncertainty about the efficiency with which the energy potentially available in different kinds of food can be used. Carbohydrates and proteins are probably assimilated by birds with about 75% efficiency (Moermond & Denslow 1985). Johnson et al (1985) tested several species of unspecialised frugivorous North American migrant birds, feeding them on a variety of wild fruits, and found that their digestive efficiencies ranged 59–86%; for five thrushes it was 62–85%, with a mean of 73%. We may therefore take a figure of 75% as giving a reasonable idea of the efficiency with which the various constituents of fruit pulp are utilised by thrushes and other unspecialised fruit-eating birds.

An all-day watch on a male Blackbird that was protecting the fruit of a *Pyracantha coccinea*, on 30 November 1980, showed that it ate 337 fruits, and as it was in sight for much of the time, this was almost certainly the bulk of its food for the day (a cold one with a mean temperature of 0°C) though it probably also took some pulp from fallen apples. On the assumption that *Pyracantha* fruit pulp is of about the same nutritive quality as other species of Rosaceae with mealy pulp (*Crataegus, Sorbus*), it can be calculated that 337 fruits should yield some 44 kcal of energy. Assuming a digestive efficiency of 75%, this should have made available 33 kcal towards the Blackbird's energy requirements. On the basis of figures given by Kendeigh (1970), the 'existence energy' of a 90–100 g passerine bird at 0°C should be about 50 kcal. Existence energy may be defined as the energy used by an individual that is maintaining constant body weight and expending negligible amounts of energy in activity or productive processes such as moult or egg-laying (Blem 1985). Kendeigh based his measurements on birds kept in cages, where they had limited opportunity for locomotor activity. A free-living bird would be expected to use more energy, but in cold weather, as we have seen, a bird that is feeding on fruit may be relatively inactive; our bird on 30 November spent long periods sitting near its fruit supply. The agreement between its estimated energy requirements and the amount of energy obtained from the fruit is reasonably good, taking into account the likelihood that some energy was obtained from other food.

We made many observations on Mistle Thrushes feeding exclusively on the fruit of defended hollies and haws in very cold weather, with the results summarised in Table 83. Kendeigh's data indicate that the existence energy requirement of a 130-gram passerine at 0°C should be about 55 kcal per day. This is less than that potentially available to Mistle Thrushes feeding on holly or haw fruits (Table 83), but if a digestive efficiency of 75% is assumed there is a slight shortfall which needs some explanation. The most likely explanation is that the estimates of the total daily fruit intake given in Table 83 are somewhat too low. We made

TABLE 83 *Estimated energy intake by Mistle Thrushes feeding at defended hollies and haws in very cold weather*

	Feeding visits/hr	No feeds per day	Mean no. fruits per feed	No. fruits eaten per day (and weight)	Energy yield of fruit pulp (kcal)
Holly	3.1	28	7.8	218 (96 g)	63
Haw	3.0	27	5.9	159 (100 g)	68

Notes. Feeding visits per hour: see page 129.
Number of feeds per day is based on a 9-hour feeding period (midwinter); but may be an under-estimate, as visits may be more frequent in early morning and late afternoon than at other times of day.
Mean number of fruits per feed: see Table 48.
Energy yield: based on data in Appendices 2 and 4, and on the following figures for calorific content: protein, 5.2 kcal/g; lipid, 9.3 kcal/g; carbohydrate, 4.0 kcal/g (Watt & Muriel 1963).

most observations during the middle part of the day, when light conditions were good, and it may well have been the case that the birds' early morning and late afternoon feeds were more frequent, and meal sizes greater than at other times of day.

Gibb & Gibb (1951) reported that the daily consumption of *Cotoneaster horizontalis* berries by two Waxwings was, respectively, 600–1,000 and about 500 berries. In this case there was no doubt that the birds were feeding exclusively on these fruits; the air temperature at the time is not stated. Assuming a pulp composition similar to that of *Cotoneaster granatensis* (Herrera, C. M. 1987), it can be calculated that the calorific value of 600 fruits is about 43 kcal. According to Kendeigh, the existence energy of a 53 g bird at 0°C should be 30–40 kcal per day. Taking into account the fact that Waxwings established at a fruit supply are relatively inactive, and assuming a digestive efficiency of 75%, the agreement between the two figures is very good.

The rate at which a fruit-eating bird can obtain energy depends, as we have seen, not on the rate at which it can obtain and swallow the fruit but on the rate at which it can process it internally. It is this that must limit its ability to survive on a fruit diet, especially in very cold environments where energy requirements are high. Rough though they are, the estimates and calculations given above indicate that in natural conditions Blackbird, Mistle Thrush and Waxwing – and presumably other fruit-eaters as well – can process fruit rapidly enough, and that available fruits are nutritious enough, to enable them to obtain sufficient energy to survive and to maintain their body weight during severe winter weather in Britain.

To conclude, we may briefly compare and contrast the time and energy budgets of fruit-eaters with those of seed-eaters and insect-eaters, the two other main feeding groups of passerine birds, and speculate about the consequences of the differences between them. (The terms 'insect-eaters' and 'insectivores' are used for convenience for birds whose diet includes other invertebrates, *eg* spiders.) The foods eaten by seed-eaters and insect-eaters are (on the basis of fresh weight) much more nutritious than fruit pulp. But as staple foods, seeds and insects are very different from one another in various ways. A great deal of time must be spent foraging for insects, but once found they can usually be quickly eaten; 'handling time' is short. By contrast, once a source of seeds has been found little or no further time is needed for searching, but handling time for each seed may be comparatively long. Thus seed-eaters habitually spend many minutes at a food source. Sources of seeds, like fruit sources, are often in exposed positions where a feeding bird is vulnerable to predators. Insect-foraging involves more constant movement, typically in thicker cover, which may reduce a bird's exposure to predation, but against this, there is less opportunity to watch out for predators than when feeding on fruit or seeds. We may summarise the differences:

	searching time	handling time	nutritive quality	exposure to predators	main factor limiting food intake
fruit	short	short	low	high, but for short periods	internal processing
seeds	short	long	higher	high, for long periods	handling time?
insects, etc	long	short	v. high	probably lower, but for long periods	success in finding food

These broad differences between the three feeding groups have important consequences. Fruit-eaters have much spare time, which they can devote to other activities or to behaviour that enhances survival in other ways. Though exposed to predation while feeding, they can minimise this by feeding quickly and more or less gregariously. Clumped fruit sources may be defendable, because the owning bird needs to spend little time feeding and can devote the rest of it to driving off intruders and watching for predators. By contrast, seed-eaters, because of the long handling time involved in dealing with seeds, must spend much more time time at a food source. Presumably as a consequence of their greater vulnerability to predators, they are in general much more highly gregarious than fruit-eaters when feeding, especially those species that feed in open habitats or near the ground. Though seed sources are often just as clumped as fruit sources, because of the long handling time a seed-eater will have less time available for driving away intruders. Hence, not surprisingly, it seems that no seed-eating bird maintains a food territory. For insectivores the problems are different again. For most of them food is widely scattered and hard to find, so that a large part of the day must be spent foraging. Success in finding food is paramount, and in many cases solitary foraging is necessary. As a protection against predation, however, gregarious foraging when possible is desirable, and is often adopted, especially in the form of loose mixed-species flocks in which the foraging success of one individual does not reduce the success of others – *eg* winter flocks of tits and other birds in the north, multi-species flocks in tropical forest. If, however, food is sufficiently clumped, an insectivore may rapidly switch behaviour and defend a food source, as has been shown experimentally for White Wagtails (Zahavi 1971). The ability of insectivores to make this behavioural change, whereas seed-eaters seem not to do so, presumably depends on the fact that for each food item handling time is short for insectivores as it is for fruit-eaters.

The question of coevolution

The early evolutionary history of fleshy fruits is controversial – not surprisingly, as the fossil record is very incomplete. Corner (1954) has argued that the primitive fruits of flowering plants were arillate, the edible aril being an adaptation for dispersal by animals. This theory has been challenged, eg by van der Pijl (1966), but there seems to be general agreement that adaptations for dispersal of seeds of flowering plants by animals arose very early in their evolutionary history. Dispersal by reptiles was probably the most primitive condition. It still occurs, and van der Pijl has suggested that reptile-dispersed fruits are characterised by an attractive odour and by being borne basally on the trunk. As birds and mammals evolved and proliferated, they undoubtedly became the main dispersal agents. Fruits adapted for dispersal by mammals tend to be scented, often with a fetid smell, and dull in colour, and in many cases they are very large and fall from the parent plant. Bat-dispersed fruits are a special case, characterised by permanent attachment and exposure away from the foliage, especially hanging below it, or attached to the trunk (van der Pijl 1957).

Corner (1954, 1964) and van der Pijl (1966, 1969), who have been especially concerned with the evolution of fleshy fruits, both agree that their early evolution was essentially tropical. An earlier idea, that wind-dispersal is more primitive than dispersal by animals, was the result of the northern bias of most botanists at the time. Wind-dispersal should rather be seen as a specialisation, an adaptation to less favourable conditions encountered as the early fruiting plants spread to higher latitudes or more severe seasonal environments. Thus there is a tendency for large, mainly tropical families with mainly animal-dispersed fruits to have a few species in temperate regions adapted for wind-dispersal – for example the Oleaceae, represented by many drupe-bearing tropical or subtropical genera and by the wind-dispersed ash *Fraxinus* in the north.

To return to the present time and to the south of England, we have seen that the fruit-eating birds in our study area are well fitted to feed on most of the fleshy fruits that are native to the area, being of adequate size and physically able to reach and pluck the fruits, swallow them, and get rid of the seeds. The Starling is a partial exception, being inefficient at plucking and swallowing some fruits. However, only one species of irregular occurrence, the Waxwing, is specialised for a diet of fruit; the others are generalists for whom fruit provides only a part of the diet, but a part which nevertheless may be essential for survival when other food is scarce. The fruits, for their part, not only show general adaptations for dispersal by birds but also seem to be adapted to the needs of the native fruit-eating birds: they are of the right size for them, they ripen at appropriate times in conformity with movements of bird populations, and according to their seasons of ripening they provide nourishment which matches the needs of birds at different seasons of the year. Does this mean that the birds and fruits have 'coevolved' – and what would such a statement mean?

Coevolution has been defined by Thompson (1982) as 'reciprocal evolutionary change in interacting species'. Thompson goes on to say: 'The key word is reciprocal in the sense of mutual. In an interaction between two species, both species must undergo evolutionary change specifically in response to the interaction to be called a coevolved interaction.' Thus if two species find themselves in the same habitat they may interact because they already have traits which in some way fit them for each other, and the interaction may superficially, but falsely, appear to be the result of coevolution between them. Hence, in considering an interaction between a particular bird and a particular kind of fruit, according to Thompson's definition we can speak of coevolution only if we can show, or give reasons for thinking it highly probable, that the bird has evolved adaptations for eating that fruit, and that the plant has evolved characters promoting the eating of its fruit, and dispersal of its seeds, by that bird. Put like that, the definition appears unduly restrictive, and we should consider a wider concept of what has been called 'diffuse' coevolution, between groups of organisms rather than between single species.

DIFFUSE VERSUS SPECIFIC COEVOLUTION

In the other main interaction between birds and plants, in which the plant provides the bird with food and the bird promotes the reproduction of the plant – the interaction between nectar-eating birds and the flowers which they pollinate – there are many cases where certain flowers are adapted for pollination by one or a few species of birds, and the bird's bill closely matches the flower's corolla tube. There is little doubt that these are cases of coevolution in Thompson's restricted sense. Theoretically it is easy to see that it may be advantageous for a plant to evolve a highly specialised flower, adapted to a single specialised pollinator, because the pollinator will then be more likely to fulfil its part of the 'bargain' by going from one flower to another of the same species and thus effecting cross-pollination. But the bird/fruit interaction is different. When a bird takes the fruit of a certain kind of plant, it performs its part of the bargain by dropping the seeds anywhere away from the parent plant. Whether it goes to another plant of the same species for a later feed is either irrelevant, or even disadvantageous for the first-visited plant, as seeds and seedlings suffer disproportionate mortality near the parent plant (Howe 1986), and presumably also near other mature plants of the same species. Plants therefore gain no advantage by producing fruit adapted for very specialised frugivores; their strategy should be to attract as many species of dispersers as is compatible with other ecological constraints affecting fruit characters. By such a strategy they also avoid the danger inherent in a specialised coevolved interaction, that the extinction or disappearance of one party must lead to the extinction of the other.

These arguments have been put forward by Wheelwright & Orians (1982) and others, and have modified the rather naive ideas proposed when the interactions between birds and fruits were first discussed in an evolutionary context (Snow 1971, McKey 1975). It is now generally accepted that diffuse coevolution between plants and groups of dispersers is to be expected, rather than one-to-one interactions. It has also become clear that the time dimension must be taken into account, as there is evidence that the rate of evolutionary change of plants has been much slower than that of birds.

THE 'ANACHRONISM LOAD'

This useful term (Janzen & Martin 1982) refers to evolutionarily very conservative characters – in this context, of plants – which presumably arose as adaptations to earlier conditions and have remained virtually unchanged from the distant past. Some fleshy fruits in particular are extraordinarily conservative, so that one is forced to reject the idea that such fruits are adapted for dispersal by the bird species that eat them today. The following examples are taken mainly from Herrera (1986).

The fruits of living species of *Taxus* (yew) seem essentially identical with those of *Palaeotaxus*, an Upper Triassic close relative living nearly 200 million years ago, before the appearance of modern birds. Fruits of the important, mainly tropical, family Lauraceae are very similar in the Old World and New World tropics, and those of modern species are very similar to those of fossil species from the Eocene, some 50 million years ago. The frugivorous birds that have almost certainly been the main dispersers of lauraceous fruits throughout this vast span of time must have been very many and diverse; today quite different species eat these fruits in Malaya, Australasia and tropical America, the family's three main centres of distribution. The two species in the small genus *Laurus*, *L. nobilis* of the Mediterranean basin and *L. azorica* of the Azores and Canary Islands, have been isolated from one another probably for about 7 million years. They now live in very different habitats and climatic environments, *nobilis* in Mediterranean forest subject to summer droughts and *azorica* in cloud forest, and their fruits are eaten by different sets of birds which would be expected to exert different selective pressures. But the fruits of the two species are virtually identical, not only in their size and shape but also in the chemical composition of their pulp. In the holly genus *Ilex* there are close resemblances between the fruits of some species occurring in Asia and eastern North America, although they must have been isolated from one another for many millions of years and have undoubtedly been dispersed by quite different arrays of birds.

These are only a few of the examples that could be given, which indicate that conservatism in fruit characters seems to be common. Presumably these fruit characters originally evolved under the influence of selective pressures exerted by dispersers, but what selective pressures, and what dispersers, can now only be guessed at. We are forced to another general conclusion: not only is coevolution between birds and fruits likely to be diffuse, it is also likely to be distinctly one-sided. Birds are more likely to have to adapt themselves to the available fruits than vice versa; or alternatively if there is coevolutionary interaction between birds and fruits, birds may be expected to undergo most of the modification.

However, while evolutionary conservatism in fruit characters seems to be a rather general rule, Herrera (1986) points out that there are some plant genera in which closely related species show dissimilarities in their fruit characters, and he cites the genera *Cornus*, *Rosa* and *Crataegus*. The fact that all these genera have mainly north-temperate distributions may be no coincidence, as we discuss in the following section.

THE RECENT HISTORY OF BIRD/FRUIT INTERACTIONS IN EUROPE

In trying to explain the relationships between birds and fruits which we see today in Britain and the rest of Europe, we must take account of the past. We have drawn on arguments by Herrera (1984a, 1985b) for much of what follows in this section.

Climates of the Mediterranean type apparently first appeared in the late Pliocene, about 3 million years ago (Axelrod 1973). The kind of vegetation found in southern Europe, associated with a Mediterranean climate, is of comparatively recent origin. The woody members of this vegetation are 'survivors of a richer, tropical-margin vegetation that developed ... through the Tertiary' (Raven 1973), a period of about 60 million years. Most bird species which now disperse this vegetation probably arose much later, during the Quaternary (the last 2 million years). Thus the Mediterranean plants with fleshy fruits must originally have been dispersed by birds that are no longer extant, and many of them must have acquired their structural characters in the course of a long evolutionary history in tropical conditions. This, of course, is largely a restatement – in a Mediterranean context – of what was said in the previous section about the 'anachronism load' characteristic of many fleshy fruits.

If there has been any coevolution between these old elements in the European flora and their present-day dispersers, it is probably the timing of flowering and ripening of the fruits, and their nutrient content, that have been modified rather than their gross structure. We have seen in an earlier section that the nutritive content of fruits ripening at different times of year matches the needs of their avian dispersers.

But another kind of interaction between the fruits and their dispersers may have played a significant part, an interaction that does not necessarily involve any evolutionary change in either party. Dispersers can modify their habitat by promoting the reproduction and spread of the plants whose fruits they feed on. Herrera (1985b) has shown how this process has apparently resulted in plant assemblages with differences in mean fruit size in different habitats in Spain, according to whether the dispersers are mainly small birds, especially warblers, or larger birds such as thrushes. On a larger scale, and over a longer time period, such a process may have played an important part in determining which fleshy-fruited plants, from a large tropical assemblage, came to make up the Mediterranean flora.

In northern Europe the course of events must have been very different, as the successive Pleistocene glaciations wiped out or profoundly modified much of the flora. As a temperate flora spread northwards in the interglacials (including the interglacial in which we are now living), birds must have been the chief agents of the spread of fleshy-fruited plants. They must thus have played a large part in determining the composition of the flora, by selecting from a pool of possible plants those whose fruits were of the right size and suited to their needs in other ways. Such a process of selection might not necessarily involve any evolutionary change in the fruits; but there would have been scope for minor adaptive modifications of fruiting seasons, and perhaps of nutritive content of fruits, to match the birds' needs, and we have seen that there is some evidence for such adaptations.

In the doubtless complex interactions between birds and fleshy fruits that took place in Europe during and after the Pleistocene glaciations, two plant families may have been especially affected, the Rosaceae and Caprifoliaceae. These families show evidence of recent speciation and, with the dogwood family (Cornaceae), are the only plant families with primarily north-temperate distributions. Because they have undergone much recent speciation and are perhaps still in the process of speciating, several genera, especially of the Rosaceae, are taxonomically 'difficult' and opinions differ as to how many species there are. *Prunus*, *Rosa*, *Crataegus* and *Sorbus* are outstanding examples. It is perhaps significant that the fruits of many species in these genera are rather large compared with the small fruits of typically Mediterranean species, and this suggests that dispersal by larger frugivores such as thrushes has been a selective agent determining fruit size.

CONCLUSION

Coevolution between plants and the birds that disperse their fruits has helped to mould the characters of both parties to the interaction; some of the details of the process can be glimpsed, but we are far from a full understanding. The process has almost certainly been a 'diffuse' one rather than a matter of mutual adaptation between particular plant and bird species, and an important constraint has been the tendency for plants to undergo much slower evolutionary change than birds. A further complication hindering easy understanding of the process – because it may 'mimic' the results of coevolution – is the ability of birds, without the need for evolutionary change in either party, to select and promote the spread of plants whose fruits suit them, thereby modifying their habitats to suit their needs. The only sure prediction is that future research will lead to a much better understanding of these processes than we have at present.

APPENDIX I
Plants with fleshy fruits native to England and Wales

plant	growth form	fruit	status in study area
Cupressaceae			
Juniperus communis (juniper)	shrub	fleshy cone	local
Taxaceae			
Taxus baccata (yew)	tree	arillate	widespread
Berberidaceae			
Berberis vulgaris (barberry)	shrub	berry	absent
Hypericaceae			
Hypericum androsaemum (tutsan)	shrub	fleshy capsule	absent
Aquifoliaceae			
Ilex aquifolium (holly)	tree	drupe	widespread
Celastraceae			
Euonymus europaeus (spindle)	small tree/shrub	arillate	locally common
Rhamnaceae			
Rhamnus catharticus (buckthorn)	small tree/shrub	berry	locally common
Frangula alnus (alder buckthorn)	small tree/shrub	berry	absent
Rosaceae			
Rubus chamaemorus (cloudberry)	herb	compound drupelets	absent
R. saxatilis (stone bramble)	herb	,, drupelets	absent
R. idaeus (raspberry)	shrub	,, drupelets	local
R. fruticosus agg. (blackberry)	shrub	,, drupelets	common
R. caesius (dewberry)	procumbent shrub	,, drupelets	local
Fragaria vesca (strawberry)	herb	fleshy receptacle with achenes	local
Rosa arvensis (field rose)	shrub	,, ,,	widespread
R. pimpinellifolia (burnet rose)	shrub	,, ,,	absent
R. stylosa (long-styled rose)	shrub	,, ,,	absent
R. canina agg. (dog rose)	shrub	,, ,,	common
R. villosa agg. (downy rose)	shrub	,, ,,	absent
R. rubiginosa agg. (sweet briar)	shrub	,, ,,	local
Prunus spinosa (blackthorn)	small tree/shrub	drupe	common
P. avium (wild cherry)	tree	drupe	widespread
P. padus (bird cherry)	tree	drupe	local, probably introduced
Crataegus laevigata (midland hawthorn)	small tree/shrub	drupe	local
C. monogyna (hawthorn)	small tree/shrub	drupe	common
Sorbus aucuparia (rowan)	tree	pome	local
S. aria (whitebeam)	tree	pome	local
S. rupicola (rock whitebeam)	small tree/shrub	pome	absent
S. torminalis (wild service tree)	tree	pome	rare
Pyrus pyraster (pear)	tree	pome	absent
Malus sylvestris (crab-apple)	tree	pome	widespread

plant	growth form	fruit	status in study area
Grossulariaceae			
Ribes rubrum (red currant)	shrub	berry	local
Ribes nigrum (black currant)	shrub	berry	absent
Ribes uva-crispa (gooseberry)	shrub	berry	local
Thymelaeaceae			
Daphne laureola (spurge laurel)	shrub	drupe	local
D. mezereum (mezereon)	shrub	drupe	very rare
Elaeagnaceae			
Hippophae rhamnoides (sea buckthorn)	small tree/shrub	drupe-like	absent
Loranthaceae			
Viscum album (mistletoe)	parasitic epiphyte	berry	local
Cornaceae			
Cornus sanguinea (dogwood)	shrub	drupe	widespread
Araliaceae			
Hedera helix (ivy)	climber	berry	widespread
Cucurbitaceae			
Bryonia dioica (white bryony)	climber	berry	widespread
Ericaceae			
Arctostaphylos uva-ursi (bearberry)	shrub	drupe	absent
Vaccinium vitis-idaea (cowberry)	shrub	berry	absent
V. myrtillus (bilberry)	shrub	berry	extinct
V. oxycoccos (cranberry)	prostrate shrub	berry	absent
Empetraceae			
Empetrum nigrum (crowberry)	shrub	drupe	absent
Oleaceae			
Ligustrum vulgare (privet)	shrub	berry	local
Solanaceae			
Atropa belladonna (deadly nightshade)	herb	berry	local
Solanum dulcamara (woody nightshade)	climber	berry	common
S. nigrum (black nightshade)	herb	berry	local
Rubiaceae			
Rubia peregrina (madder)	climber	berry	absent
Caprifoliaceae			
Sambucus ebulus (danewort)	herb	drupe	absent
S. nigra (elderberry)	small tree/shrub	drupe	common
Viburnum lantana (wayfaring tree)	shrub	drupe	local
V. opulus (guelder rose)	shrub	drupe	local
Lonicera periclymenum (honeysuckle)	climber	berry	widespread
L. caprifolium (perfoliate honeysuckle)	climber	berry	probably introduced
Liliaceae			
Convallaria majalis (lily-of-the-valley)	herb	berry	absent
Polygonatum multiflorum (Solomon's seal)	herb	berry	very local
Ruscus aculeatus (butcher's broom)	shrub	berry	absent
Trilliaceae			
Paris quadrifolia (herb Paris)	herb	fleshy capsule	absent
Iridaceae			
Iris foetidissima (stinking iris)	herb	arillate	absent
Dioscoreaceae			
Tamus communis (black bryony)	climber	berry	widespread

Appendices 247

plant	growth form	fruit	status in study area
Araceae			
Arum maculatum (lords and ladies)	herb	berry-like	widespread

The following very rare or local species are omitted: *Cotoneaster integerrimus* (Rosaceae), *Asparagus officinalis* (Liliaceae), *Polygonatum odoratum* (Liliaceae).

Some species that are absent as wild plants from the study area may be found planted in gardens, *eg* tutsan, sea buckthorn, lily-of-the-valley, butcher's broom.

APPENDIX 2
'Design components' of native bird-dispersed fruits, southern England

	diam. (mm)	length (mm)	fresh wt. (g)	no. seeds (range and mean)	individual seed wt. (g)	seed burden (%)	water content of pulp (%)	relative yield
Cupressaceae								
Juniperus communis	8.4	–	.70	1–3, 1.9	.016	18	47	43.5
Taxaceae								
Taxus baccata	10.3	10.8	.70	1	0.82	12	79	18.5
Aquifoliaceae								
Ilex aquifolium	9.3	9.8	.44	2–4, 3.7	.040	34	71	19.1
Celastraceae								
Euonymus europaeus[1]	4.5	6.0	.080	1	.037	59	58	17.2
Rhamnaceae								
Rhamnus catharticus	8.6	8.1	.36	3–5, 4.0	.027	25	74	19.5
Rosaceae								
Rubus idaeus[2]	3.0	–	.037	1	.0027	7.4	84	14.8
R. caesius[2]	6.5	–	.12	1	.0079	6.9	93	6.5
R. fruticosus	15.5	–	1.7	*c.* 30	.0028	5.4	87	12.3
Fragaria vesca	9–13	–	.40	*c.* 60	.00041	6.6	88	11.2
Rosa arvensis	11.3	–	.96	5–24, 14.5	.020	28	56	31.7
R. canina agg.	12.8	20.8	1.8	9–28, 20.6	.024	28	61	28.1
R. rubiginosa	11.5	19.5	1.3	7–23, 16.5	.023	25	70	22.5
Prunus spinosa	14.3	–	1.9	1	.24	13	79	18.3
P. avium	11.8	14.6	1.3	1	.29	23	87	10.0
P. padus	6.7	7.2	.20	1	.065	33	83	11.4
Crataegus laevigata	9.9	11.3	.49	2–3, 2.1	.068	31	70	20.7
C. monogyna	10.7	11.5	.63	1	.13	22	73	21.1
Sorbus aucuparia	8.6	9.0	.49	1–4, 2.2	.0082	3.4	76	22.5
S. aria	11.0	12.5	.68	1–4, 1.9	.022	7.1	60	37.2
S. torminalis	9.9	12.6	–	1	0.18	–	–	–
Grossulariaceae								
Ribes rubrum	7.6	–	.21	2–5, 3.9	.0031	5.8	90	9.4
R. uva-crispi	12.5	14.0	1.3	10–25, 16.0	.0080	10	85	13.5
Thymelaeaceae								
Daphne laureola	7.0	10.0	.25	1	.047	19	89	8.9
D. mezereum	8.0	9.5	.30	1	.056	16	90	8.4

248 Appendices

	diam. (mm)	length (mm)	fresh wt. (g)	no. seeds (range and mean)	individual seed wt. (g)	seed burden (%)	water content of pulp (%)	relative yield
Loranthaceae								
Viscum album[3]	8.3	–	.40	1	.16	41	75	14.8
Cornaceae								
Cornus sanguinea	7.3	–	.21	1	.067	32	67	22.4
Araliaceae								
Hedera helix	8.8	7.1	.33	1–5, 2.5	.053	41	72	16.5
Cucurbitaceae								
Bryonia dioica	8.4	–	.31	2–6, 4.6	.016	22	84	12.5
Oleaceae								
Ligustrum vulgare[4]	8.5	–	.30	1–4, 2.2	.030	32	81	12.9
Solanaceae								
Solanum dulcamara	8.6	12.2	.50	19–37, 26.9	.0028	17	85	12.5
S. nigrum	7.0	–	.24	24.8	.0011	12	86	12.3
Atropa belladonna	13.5	10.0	.97	47–63, 55.7	.0014	11	87	11.6
Rubiaceae								
Rubia peregrina[5]	7.3	–	.17	1	.038	21	86	11.1
Caprifoliaceae								
Sambucus nigra	6.2	–	.14	2–4, 2.9	.0039	6.9	88	11.2
Viburnum lantana	6.9	9.1	.28	1	.059	22	77	17.9
V. opulus	8.4	9.0	.38	1	.055	14	84	13.8
Lonicera periclymenon	8.1	9.0	.28	1–8, 3.2	.0087	11	81	16.9
L. caprifolium	6.1	8.3	.19	1–2, 1.3	.013	6.5	81	17.8
Liliaceae								
Polygonatum multiflorum	9.9	–	.42	2–6, 3.7	.041	36	–	–
Iridaceae								
Iris foetidissima[1]	5.0	–	.13	1	.060	45	82	9.9
Dioscoreaceae								
Tamus communis	10.5	11.6	.70	2–6, 4.7	.033	22	90	7.8
Araceae								
Arum maculatum[4]	10.5	13.0	.60	1–4, 1.3	.060	24	89	8.4

Notes: all figures, except ranges, are means of samples, usually of 10 or more fruits in each sample and usually of several samples for each kind of fruit. Lengths are not given for fruits that are essentially globular.
1. Seed plus aril treated as a fruit unit.
2. Individual 'drupelets' treated as fruit units.
3. Seed weight includes mucilaginous coating.
4. Fruits very variable in size; dimensions given for full-sized fruits.
5. Does not occur in or near study area; sample from south Devon.

APPENDIX 3
'Design components' of some introduced and cultivar bird-dispersed fruits, southern England

	diam. (mm)	length (mm)	fresh wt. (g)	no. seeds (range and mean)	individual seed wt. (g)	seed burden (%)	water content of pulp (%)	relative yield
Berberis darwinii	8.9	–	.37	2–7, 4.8	.0086	15	85	12.8
Mahonia aquifolium	7.9	8.4	.28	1–6, 3.5	.018	22	82	14.0
Amelanchier laevis	8.0	–	.36	1–10	.018	11	78	19.5
Rosa rubrifolia	12.5	17.8	1.0	7–19, 13.8	.016	25	77	17.3
Prunus laurocerasus	14.0	15.8	1.9	1	.34	18	80	16.4
Prunus lusitanica	11.0	12.3	.76	1	.13	21	76	19.0
Cotoneaster 'cornubia'	8.7	7.7	.27	2	.016	12	78	19.1
C. horizontalis	6.9	7.7	.15	2–3, 2.9	.0091	17	82	14.9
C. distychus	9.3	11.6	.40	3–5, 4.0	.020	20	83	13.6
Pyracantha coccinea	8.8	–	.26	5	.0047	9.5	84	14.5
P. rogersiana	7.5	–	.16	5	.0036	11	84	14.2
P. atalantoides	7.0	–	.12	5	.0032	13	85	13.1
P. crenatoserrata	5.5	–	.064	4–5, 4.9	.0029	22	72	21.8
Crataegus 'prunifolia'	12–15	–	1.1	2–3	.12	25	74	19.5
C. × lavallei	19	16	2.3	5	.085	19	78	17.8
Lycium barbarum	10	18	.85	11–29, 20.2	.0013	3.5	81	18.3

Note: all figures, except ranges, are means of samples, usually of 10 or more fruits. Lengths are not given for fruits that are essentially globular.

APPENDIX 4
Nutritive values of some bird-dispersed fruits, southern England

Most of the measurements of lipid, protein and carbohydrate content are from Herrera (1987); those for *Euonymus europaeus* and *Arum maculatum* are from E. W. Stiles (*in litt.*). These measurements are percentages, by weight, of the dry pulp. The figures for energy yield per gram of dry pulp are calculated on the basis of caloric contents as follows: lipid, 9.3 kcal/g; protein, 5.2 kcal/g; carbohydrate, 4.0 kcal/g (Watt & Muriel 1963). The figures for energy yield per gram of fresh, whole fruit are derived from the preceding figures and the 'relative yields' of the different fruits (dry pulp as a percentage, by weight, of the whole fruit; Appendix 2).

Note that the energy yields of the dry pulp of different fruits are similar. Nearly all are between 3 and 4 kcal; only two are much higher (dogwood and ivy) and probably a third (spindle, for which the carbohydrate content is not available), these being the three fruits with far the highest lipid contents. Energy yields (per gram) of whole fruits are much more variable, being greatly affected by differences in their seed burdens and the water content of the pulp.

For the remaining constituents (fibre and minerals), and data for other European fruits, see Herrera (1987).

	lipid	protein	soluble carbo-hydrate	energy per g of dry pulp	energy per g of whole fruit
Juniperus communis (juniper)	16.42	4.12	39.81	3.33	1.45
Taxus baccata (yew)	0.21	2.34	93.48	3.88	0.72
Ilex aquifolium (holly)	3.05	4.75	72.52	3.43	0.66
Euonymus europaeus (spindle)	35.8	10.3	—	—	—
Rhamnus catharticus (buckthorn)	5.43	3.81	76.63	3.77	0.74
Rosa canina (dog rose)	2.84	4.31	72.46	3.38	0.95
Prunus spinosa (sloe)	2.23	4.07	60.77	2.85	0.52
P. avium (wild cherry)	1.65	11.81	—	—	—
P. padus (bird cherry)	0.76	5.19	72.40	3.24	0.37
Crataegus monogyna (haw)	2.30	2.45	72.40	3.24	0.68
Sorbus aucuparia (rowan)	3.06	3.16	76.10	3.48	0.78
S. aria (whitebeam)	2.84	3.05	61.56	2.88	1.07
Malus sylvestris (crab-apple)	5.38	2.43	74.32	3.60	—
Daphne laureola (spurge laurel)	1.52	5.20	78.70	3.56	0.32
Viscum album (mistletoe)	8.61	4.12	73.93	3.97	0.59
Cornus sanguinea (dogwood)	24.86	6.43	53.81	4.79	1.07
Hedera helix (ivy)	31.91	5.00	47.36	5.12	0.84
Bryonia dioica (white bryony)	4.90	27.73	37.10	3.38	0.42
Ligustrum vulgare (privet)	3.32	5.88	73.85	3.57	0.46
Atropa belladonna (deadly nightshade))	2.46	7.31	69.46	3.39	0.39
Rubia peregrina (madder)	9.88	5.19	64.51	3.77	0.42
Sambucus nigra (elder)	3.33	17.95	52.72	3.35	0.38
Viburnum lantana (wayfaring tree)	2.61	1.69	66.90	3.01	0.54

Appendices 251

	lipid	protein	soluble carbo-hydrate	energy per g of dry pulp	energy per g of whole fruit
V. opulus (guelder rose)	4.19	0.98	82.88	3.76	0.52
Lonicera periclymenum (honeysuckle)	6.84	4.72	77.35	3.98	0.67
Iris foetidissima (stinking iris)	6.87	4.75	64.42	3.47	0.34
Tamus communis (black bryony)	3.22	4.12	66.96	3.19	0.25
Arum maculatum (lords and ladies)	3.4	7.9	—	—	—

APPENDIX 5

Monthly totals (numbers of hours) of timed watches at wild fruits, southern England 1980–1985

	Jan	Feb	Mar	Apr	May	Jun	Jul	Aug	Sep	Oct	Nov	Dec	totals
Holly (non-MT)	14.3	6.3	16.5	34.3	24.7	39.6	3.8	.5	.5	15.1	16.8	17.5	189.9
(MT-def)	34.8	20.8	11.5							6.4	15.9	14.7	104.1
Ivy (ripe)	15.2	24.5	37.9	57.3	36.6	2.2						3.2	176.9
(unripe)	4.0	10.1	13.5	3.1								2.3	33.0
(M-def)	3.1	2.8	2.4	3.8								.5	12.6
Yew (non-MT)	.8						2.3	21.2	41.8	36.8	28.4	9.1	140.4
(MT-def)	2.3	.5								.2	1.1		4.1
Mistletoe (non-MT)	2.3	1.2	4.3	4.2	2.2	1.3			.5	1.8	2.3		20.1
(MT-def)	11.5	11.2	5.8	3.2	4.0	1.7				2.3	13.7	8.5	61.9
Hip (*canina*)	48.3	26.4	4.4	2.1						1.1	13.8	32.2	128.3
(*arvensis*)										1.6	2.7	3.7	8.0
(*rubiginosa*)									.6	.5	1.3	.5	2.9
Haw (non-MT)	11.8	5.3	6.3					3.4	34.9	69.9	71.2	30.3	233.1
(MT-def)	9.5	6.0									1.3	1.9	18.7
Rowan							3.9	31.2	12.0	1.1	.3		48.5
Whitebeam								7.4	19.3	17.4	3.3		47.4
Cherry					8.4	29.2	.3						37.9
Bird cherry							10.9	1.8					12.7
Sloe	10.2								9.3	13.4	31.1	19.0	83.0
Blackberry								17.6	32.8	5.8	2.8		59.0
Raspberry							9.9	4.4					14.3
Strawberry						6.2	2.3						8.5
Crab-apple		5.8	10.8							2.3	2.4	1.8	23.1
Elder								27.0	101.1	30.9	6.4		165.4
Wayfaring tree							2.5	18.1	3.5				24.1
Guelder rose		11.2	3.3						1.5	5.9	3.2	10.3	35.4

252 Appendices

	Jan	Feb	Mar	Apr	May	Jun	Jul	Aug	Sep	Oct	Nov	Dec	totals
Common honeysuckle						.5	3.8	28.8	7.2	4.8			45.1
Perfoliate honeysuckle							12.6	22.2					34.8
Dogwood	.3	1.0						2.2	25.5	33.9	16.6	4.1	83.6
Spindle	11.8	6.2								3.1	19.9	21.4	62.4
Privet	15.6	6.8								6.5	12.2	11.9	53.0
Buckthorn	3.4							.7	5.3	27.0	31.2	25.3	92.9
Currant						1.1	16.8						17.9
Gooseberry							1.0	2.0	.3				3.3
Woody nightshade	.7						9.6	25.9	19.7	10.3	12.9	11.8	90.9
Deadly nightshade								2.5	2.9	.5			5.9
White bryony							7.4	13.4	4.8	2.2	.5	.4	28.7
Black bryony	5.8								3.2	5.5	14.2	12.1	40.8
Spurge laurel					1.1	6.3	18.4						25.8
Mezereon							2.0						2.0
Juniper (non-MT)	1.2	2.8	2.2	4.8	2.2	1.5	.7	2.6	1.2	1.2	.4		22.2
(MT-def)	5.2	2.0											7.2
Lords and ladies							.4	25.5	3.8				29.7

grand total 2339.5

Notes: in the case of plants sometimes defended by Mistle Thrushes, watches at plants that were undefended (non-MT) and those that were defended (MT-def) are listed separately. 'Sloe' includes a few watches at plants of bullace type. Black nightshade and stinking iris, at which only brief watches were carried out, are omitted.

References

Aplin, O. V. 1910. Daphne-berries eaten by birds. *Zoologist* 14: 394.
Axelrod, D. I. 1970. Mesozoic paleogeography and early angiosperm history. *Bot. Rev.* 36: 277–319.
Banzhaf, W. 1932. Schmetterlings- und Beerennahrung bei Helgoländer Zugvögeln. *Vogelzug* 3: 180.
Bentham, G. & Hooker, J. D. 1943. *Handbook of the British Flora.* 7th ed. L. Reeve & Co., Ltd.
Berthold, P. 1976a. Animalische und vegetabilische Ernährung omnivorer Singvogelarten: Nahrungsvorzugung, Jahresperiodik der Nahrungswahl, physiologische und ökologische Bedeutung. *J. Orn.* 117: 145–209.
Berthold, P. 1976b. Der Seidenschwanz *Bombycilla garrulus* als frugivorer Ernährungsspezialist. *Experientia* 32: 1445.
Berthold, P. 1977. Ueber die kunstliche Aufzucht nestjunger Amseln (*Turdus merula*) mit Beeren des Efeus (*Hedera helix*). *Vogelwarte* 29: 110–113.
Berthold, P. 1984. Beeren des Efeus (*Hedera helix*) als Nestlingsnahrung der Mönchsgrasmücke (*Sylvia atricapilla*). *Vogelwarte* 32: 303–304.
Bettmann, H. 1953. Schneeballbeeren als Vogelnahrung. *Orn. Mitt.* 5: 214.
Blaschke, W. 1976. Nahrungsgäste an den Früchten der Eberesche. *Falke* 23: 424–425.
Blem, C. R. 1985. Energetics. Pp. 186–191 in *A Dictionary of Birds* (eds B. Campbell & E. Lack). T. & A. D. Poyser.
Blondel, J. 1969. Synécologie des passereaux résidents et migrateurs dans le midi méditerranéen français. Centre Régional de Documentation Pédagogique, Marseille.

Bock, W.J. 1959. Preadaptation and multiple evolutionary pathways. *Evolution* 13: 194–211.
Bodenstein, G. 1953. Zur Winternahrung des Buchfinken, des Stars, der Amsel und des Rotkehlchens. *Orn. Mitt.* 5: 134.
Bossema, I. 1979. Jays and oaks: an eco-ethological study of a symbiosis. *Behaviour* 70: 1–117.
Breitwisch, R., Merritt, P. G. & Whitesides, G. H. 1984. Why do Northern Mockingbirds feed fruit to their nestlings? *Condor* 86: 281–287.
Brown, W. H. 1935. *The Plant Kingdom: a Textbook of General Botany.* Ginn & Co.
Bruns, H. & Haberkorn, A. 1960. Beiträge zur Ernährungsbiologie des Stars (*Sturnus vulgaris*). *Orn. Mitt.* 12: 81–103.
Burckhardt, D. 1982. Birds, berries and UV: a note on some consequence of UV vision in birds. *Naturwissenschaften* 69: 153–157.
Butterfield, E. P. 1910. Notes from Yorkshire. *Zoologist* 14: 337–338.
Campbell, W. D. 1980. A country diary. *Guardian*, 10 December 1980.
Campbell, W. D. 1981. A country diary. *Guardian*, 23 December 1981.
Campbell, W. D. 1984. A country diary. *Guardian*, 31 October 1984.
Campbell, W. D. 1985a. A country diary. *Guardian*, 16 January 1985.
Campbell, W. D. 1985b. A country diary. *Guardian*, 20 November 1985.
Castroviejo, J. 1970. Premières données sur l'écologie hivernale des vértébrés de la Cordillière Cantabrique. *Alauda* 38: 126–149.
Christiansen, A. 1914. Taschenbuch einheimischer Pflanzen mit besonderer Berücksichtigung ihrer Lebensverhältnisse. Esslingen & München.
Clapham, A. R., Tutin, T. G. & Warburg, E. F. 1973. *Excursion Flora of the British Isles* (Second Edition). Cambridge University Press.
Colquhoun, M. K. 1951. The Wood Pigeon in Britain. (A.R.C. Report Series No. 10.) HMSO, London.
Cooper, M. R. & Johnson, A. W. 1984. *Poisonous Plants in Britain and their Effects on Animals and Man.* (M.A.F.F. Reference Book 161.) HMSO, London.
Corner, E. J. H. 1954. The evolution of tropical forest. Pp. 34–46 in *Evolution as a Process* (eds J. S. Huxley, A. C. Hardy & E. B. Ford). Allen & Unwin.
Corner, E. J. H. 1964. *The Life of Plants.* Weidenfeld & Nicolson.
Cramp, S. & Simmons, K. E. L. 1983. *Handbook of the Birds of Europe the Middle East and North Africa*, vol. 3. Oxford University Press.
Cramp, S. 1985. *Handbook of the Birds of Europe the Middle East and North Africa*, vol. 4. Oxford University Press.
Creutz, G. 1952. Misteldrossel und Seidenschwanz. *Orn. Mitt.* 4: 67.
Creutz, G. 1953. Beeren und Früchte als Vogelnahrung. *Beitr. Vogelkunde* 3: 91–103.
Cronin, E. W. & Sherman, P. W. 1976. A resource-based mating system: the Orange-rumped Honeyguide. *Living Bird* 15: 5–32.
Cvitanic, A. 1970. [Das gegenseitige Verhältnis der Gedärmelänge, Körperlänge und der Ernährung bei einigen Vogelarten.] *Larus* 21–22: 181–190.
Debussche, M. & Isenmann, P. 1985a. Frugivory of transient and wintering European Robins *Erithacus rubecula* in a Mediterranean region and its relationship with ornithochory. *Holarctic Ecology* 8: 157–163.
Debussche, M. & Isenmann, P. 1985b. Le régime alimentaire de la Grive musicienne (*Turdus philomelos*) en automne et en hiver dans les garrigues de Montpellier (France méditerraneenne) et ses relations avec l'ornithochorie. *Rev. Ecol. (Terre Vie)* 40: 379–388.
Dieberger, J. 1982. Zoologische Komponente bei der Infektion und Verbreitung der Eichenmistel – vorläufige Ergebnisse. *Veröff. Inst. Waldbau Univ. Bodenkultur Wien.*
Diesselhorst, G. 1972. Beeren und Farbenwahl durch Vögel. *J. Orn.* 113: 448–449.
Docters van Leeuwen, W. M. 1954. On the biology of some Javanese Loranthaceae and the role birds play in their life-histories. *Beaufortia* 4: 105–205.
Eber, G. 1956. Vergleichende Untersuchungen über die Ernährung einiger Finkenvögel. *Biol. Abh.* 13–14: 1–60.
Edwards, P. J. 1985. Brood division and transition to independence in Blackbirds *Turdus merula*. *Ibis* 127: 42–59.

Ellis, E. A. 1985. A country diary. *Guardian*, 17 January 1985.
Emmrich, R. 1973–74. Das Nahrungsspektrum der Dornsgrasmücke (*Sylvia communis* Lath.) in einem Gebüsch-Biotop der Insel Hiddensee. *Zool. Abh.* 32: 275–307.
Epprecht, W. 1964. Fütterungsbeobachtungen bei einer Brut des Grauschnäppers. *Orn. Beob.* 61: 137–139.
Erkamo, V. 1948. [On the winter nourishment and biology of the Bullfinch.] *Arch. Soz. Zool-bot. Fenn. Vanamo* 1: 86–101.
Feare, C. 1984. *The Starling*. Oxford University Press.
Frost, P. G. H. 1980. Fruit-frugivore interactions in a South African coastal dune forest. *Acta XVII Congr. Internat. Orn.*: 1179–1184.
Gewalt, W. 1986. Nochmals: Lachmöwen (*Larus ridibundus*) pflücken Baumfrüchte. *Beitr. Vogelkunde* 32: 57–58.
Gibb, J. 1954. The feeding ecology of tits, with notes on Treecreeper and Goldcrest. *Ibis* 96: 513–543.
Gibb, J. & Gibb, C. 1951. Waxwings in the winter of 1949–50. *Brit. Birds* 44: 158–163.
Gilbert, O. L. 1980. Juniper in Upper Teesdale. *J. Ecol.* 68: 1013–1024.
Glutz von Blotzheim, U. 1973. Hartriegelbeeren als Aufzuchtfutter des Grauschnäppers. *Orn. Beob.* 70: 183–184.
Glutz, U. & Bauer, K. M. 1980. *Handbuch der Vögel Mitteleuropas*, vol. 9. Akademische Verlagsgesellschaft.
Glutz, U. 1985. *Handbuch der Vögel Mitteleuropas*, vol. 10/1. Akademische Verlagsgesellschaft.
Gooch, B. 1984. Chiffchaffs eating fruit. *Brit. Birds* 77: 424–425.
Goodacre, M. J. 1960. The origin of winter visitors to the British Isles. 6. Song Thrush (*Turdus philomelos*). *Bird Study* 7: 108–110.
Goodwin, D. 1983. *Pigeons and Doves of the World* (Third Edition). British Museum (Natural History).
Goodwin, D. 1985. Pigeon. Pp. 462–463 in *A Dictionary of Birds* (eds B. Campbell & E. Lack). T. & A. D. Poyser.
Goodwin, H. 1943. Biological flora of the British Isles. Rhamnaceae. *J. Ecol.* 31: 66–92.
Guitian, J. 1984. Sobre la importancia del acerbo (*Ilex aquifolium* L.) en la ecología de la comunidad invernal de Passeriformes en la Cordillera Cantábrica occidental. *Ardeola* 30: 65–76.
Hardy, E. 1969. Mistle Thrushes and mistletoe berries. *Bird Study* 16: 191–192.
Hardy, E. 1978. Winter foods of Blackcaps in Britain. *Bird Study* 25: 60–61.
Hartley, P. H. T. 1954. Wild fruits in the diet of British thrushes. A study in the ecology of closely allied species. *Brit. Birds* 47: 98–107.
Harrison, J. 1966. Invasion by Waxwings. *Birds* 1: 90–92.
Havlin, J. & Folk, C. 1965. [Food and economic importance of the Starling.] *Zool. Listy* 14: 193–203.
Heim de Balsac, H. 1928a. Fragments de bromatologie ornithologique. *Rev. Franç. Orn.* 12: 54–66.
Heim de Balsac, H. 1928b. Contribution à l'étude de la propagation du Gui par les oiseaux. *Rev. Franç. Orn.* 12: 383–390.
Heim de Balsac, H. & Mayaud, N. 1930. Compléments à l'étude de la propagation du Gui (*Viscum album* L.) par les oiseaux. *Alauda* 2: 474–493.
Heinroth, O. 1907. Bericht über die September-Sitzung 1907. *J. Orn.* 55: 622–624.
Herrera, C. M. 1977. Ecología alimenticia del Petirrojo (*Erithacus rubecula*) durante su invernada en encinares del Sur de España. *Doñana, Acta Vertebrata* 4: 35–39.
Herrera, C. M. 1981a. Fruit food of Robins wintering in southern Spanish Mediterranean scrubland. *Bird Study* 28: 115–122.
Herrera, C. M. 1981b. Are tropical fruits more rewarding to dispersers than temperate ones? *Am. Nat.* 118: 896–907.
Herrera, C. M. 1981c. Datos sobre la dieta frugívora del Mirlo (*Turdus merula*) en dos localidades del sur de España. *Doñana, Acta Vertebrata* 8: 306–310.
Herrera, C. M. 1981d. Fruit variation and competition for dispersers in natural populations of *Smilax aspera*. *Oikos* 36: 51–58.
Herrera, C. M. 1982a. Seasonal variation in the quality of fruits and diffuse coevolution

between plants and avian dispersers. *Ecology* 63: 773–785.
Herrera, C. M. 1982b. Defense of ripe fruits from pests: its significance in relation to plant-dispersers interactions. *Am. Nat.* 120: 218–241.
Herrera, C. M. 1984a. A study of avian frugivores, bird-dispersed plants, and their interaction in Mediterranean scrublands. *Ecol. Monogr.* 54: 1–23.
Herrera, C. M. 1984b. Seed dispersal and fitness determinants in wild rose: combined effects of hawthorn, birds, mice, and browsing ungulates. *Oecologia* 63: 386–393.
Herrera, C. M. 1984c. Adaptation to frugivory of Mediterranean avian seed dispersers. *Ecology* 65: 609–617.
Herrera, C. M. 1985a. Determinants of plant-animal coevolution: the case of mutualistic dispersal of seeds by vertebrates. *Oikos* 44: 132–141.
Herrera, C. M. 1985b. Habitat-consumer interactions in frugivorous birds. Pp. 341–365 in *Habitat Selection in Birds* (ed. M. L. Cody). Academic Press.
Herrera, C. M. 1986. Vertebrate-dispersed plants: why they don't behave the way they should. Pp. 5–18 in *Frugivores and Seed Dispersal* (eds A. Estrada & T. H. Fleming). Dr W. Junk, Dordrecht.
Herrera, C. M. 1987. Vertebrate-dispersed plants of the Iberian peninsula: a study of fruit characteristics. *Ecol. Monogr.* 57: 305–311.
Hess, A. 1927. Les baies de la Belladonne et les oixeaux. *Rev. Franç. Orn.* 11: 426–427.
Heymer, A. 1966. Beeren und Früchte als Vogelnahrung. *Beitr. Vogelkunde* 12: 95–102.
Hickling, R. (ed.) 1983. *Enjoying Ornithology.* T. & A. D. Poyser.
Hilty, S. L. 1980. Flowering and fruiting periodicity in a premontane rain forest in Pacific Colombia. *Biotropica* 12: 292–306.
Holyoak, D. 1968. A comparative study of the food of some British Corvidae. *Bird Study* 15: 147–153.
Hope Jones, P. 1962. Mortality and weights of Fieldfares in Anglesey in January 1962. *Brit. Birds* 55: 178–181.
Howe, H. F. 1986. Seed dispersal by fruit-eating birds and mammals. Pp. 123–190 in *Seed Dispersal* (ed. D. R. Murray). Academic Press.
Janzen, D. H. & Martin, P. S. 1982. Neotropical anachronisms: the fruits the Gomphotheres ate. *Science* 215: 19–27.
Jedicke, E. 1980. Auswirkungen des strenges Winters 1978/79 auf die Vogelwelt in Kreis Waldeck-Frankenberg und im Raum Fritzlar-Homberg. *Vogelk. Hefte Edertal* 6: 34–53.
Johnson, R. A., Willson, M. F., Thompson, J. N. & Bertin, R. I. 1985. Nutritional values of wild fruits and consumption by migrant frugivorous birds. *Ecology* 66: 819–827.
Jollet, A. 1984. Variations saisonnières du régime alimentaire de la Corneille noire (*Corvus corone* L.) dans le bocage limousin. *Oiseau* 54: 109–130.
Jordano, P. 1982. Migrant birds are the main seed dispersers of blackberries in southern Spain. *Oikos* 38: 183–193.
Jordano, P. 1985. El ciclo anual de los paseriformes frugívoros en el matorral mediterráneo del sur de España: importancia de su invernada y variaciones interanuales. *Ardeola* 32: 69–94.
Jordano, P. 1987. Frugivory, external morphology and digestive system in mediterranean sylviid warblers *Sylvia* spp. *Ibis* 129: 175–189.
Jordano, P. & Herrera, C. M. 1981. The frugivorous diet of Blackcap *Sylvia atricapilla* populations wintering in southern Spain. *Ibis* 123: 502–507.
Juana, E. de & Santos, T. 1981. Observations sur l'hivernage des oiseaux dans le Haut-Atlas (Maroc). *Alauda* 49: 1–12.
Jurik, T. W. 1983. Reproductive effort and CO_2 dynamics of wild strawberry populations. *Ecology* 64: 1329–1342.
Kammerer, K. 1921. Ueber Beerennahrung. *Orn. Monatsschr.* 46: 174.
Kear, J. 1968. Plant poisons in the diet of wild birds. *Bull. Brit. Orn. Club* 88: 98–102.
Kendeigh, S. C. 1970. Energy requirements for existence in relation to size of birds. *Condor* 72: 60–65.
Kroll, H. 1972. Zur Nahrungsökologie der Gartengrasmücke (*Sylvia borin*) beim Herbstzug 1969 auf Helgoland. *Vogelwarte* 26: 280–285.
Labitte, A. 1952. Notes sur la biologie et la reproduction de *Turdus v. viscivorus* L. 1758. *Alauda* 20; 21–30.

Lack, D. 1948. Notes on the ecology of the Robin. *Ibis* 90: 252–279.
Lack, D. 1976. *Island Biology*. Blackwell.
Langslow, D. R. 1979. Movements of Blackcaps ringed in Britain and Ireland. *Bird Study* 26: 239–252.
Leach, I. H. 1981. Wintering Blackcaps in Britain and Ireland. *Bird Study* 28: 5–14.
Leverton, R. 1983. On the dole. *B.T.O. News* 124: 5.
Levey, D. J. 1986. Methods of seed processing by birds and seed deposition patterns. Pp. 147–158 in *Frugivores and Seed Dispersal* (eds A. Estrada & T. H. Fleming). Dr W. Junk, Dordrecht.
Levey, D. J. 1987. Sugar-tasting ability and fruit selection in tropical fruit-eating birds. *Auk* 104: 173–179.
Link, J. A. 1889. Beobachtungen am Kuckuk. *Orn. Monatsschr.* 14: 439–453.
Löhrl, H. 1957. Kirschen als Futter für nestjunge Stare. *Orn. Mitt.* 9: 23–24.
Lübcke, W. 1980. Mortalität, Masse und Gewicht von Wacholderdrosseln (*Turdus pilaris*) an der deutschen Nordseeküste im Februar 1978. *Beitr. Naturk. Nieders.* 33: 147–152.
Lübcke, W. & Furrer, R. 1985. *Die Wacholderdrossel*. A. Ziemsen Verlag (Die Neue Brehm-Bücherei).
Maas, C. 1950. Beerennahrung der Vögel. *Vogelwelt* 71: 161.
Macleod, H. 1983. Variations in ivy phenology and avian consumption at different altitudes. B.Sc. hons thesis, University of Edinburgh.
Madon, P. 1927. Les deux Sylvies *Sylvia atricapilla* et *S. borin*. *Rev. Franç. Orn.* 11: 12–24.
Mahe-Quinio, M., Rossignol, G. & Foucaud, A. 1975. *Plantes Méd. et Phytothérapie* 9: 182–186.
McCann, C. 1953. The winter food of the Blackbird in New Zealand. *Notornis* 5: 198–199.
McKey, D. 1975. The ecology of coevolved seed dispersal systems. Pp. 159–161 in *Coevolution of Animals and Plants* (eds L. E. Gilbert & P. H. Raven). University of Texas Press.
Meidell, O. 1937. Undersøkelse av mageinnhold hos gråtrost (*Turdus pilaris*). *Nyt. Mag. Naturv.* 76: 59–132.
Mitchell, A. 1974. *A Field Guide to the Trees of Britain and Northern Europe*. Collins.
Moermond, T. C. & Denslow, J. S. 1985. Neotropical avian frugivores: patterns of behavior, morphology, and nutrition, with consequences for fruit selection. Pp. 865–897 in *Neotropical Ornithology* (eds P. A. Buckley *et al.*). American Ornithologists' Union.
Möhring, G. 1957. Zur Beerennahrung der Mönchsgrasmücke. *Falke* 4: 205–208.
Moore, F. R. 1978. Interspecific aggression: toward whom should a Mockingbird be aggressive? *Behav. Ecol. Sociobiol.* 3: 173–176.
Moreau, R. E. 1972. *The Palaearctic-African Bird Migration Systems*. Academic Press.
Mountfort, G. 1957. *The Hawfinch*. Collins.
Müller, A. & Müller, K. 1887. Ueber den europäischen Kuckuk (*Cuculus canorus*). *Orn. Monatsschr.* 12: 59–84 (footnote, p. 63).
Müller-Schneider, P. 1949, Unsère Vögel als Samenverbreiter. *Orn. Beob.* 46: 120–123.
Murton, R. K. 1965. *The Wood-pigeon*. Collins.
Naumann, J. F. 1897–1905. *Naturgeschichte der Vögel Mitteleuropas* (ed. C. R. Hennicke). Gera-Untermhaus.
Newton, I. 1967a. The adaptive radiation and feeding ecology of some British finches. *Ibis* 109: 33–98.
Newton, I. 1967b. The feeding ecology of the Bullfinch (*Pyrrhula pyrrhula* L.) in southern England. *J. Anim. Ecol.* 36: 721–744.
Newton, I. 1972. *Finches*. Collins.
Noll, F. C. 1870. Vögel und Pflanze. *Zool. Garten* 11: 305–311.
Oakes, C. 1942. Birds feeding on berries of service tree and guelder rose. *Brit. Birds* 36: 140.
Olney, P. J. S. 1966. Berries and birds. *Birds* 1: 98–99.
Penhallurick, R. D. 1978. *The Birds of Cornwall and the Isles of Scilly*. Headland Publications.
Pénzes, A. 1972. [Angaben über die Ernährung der Wacholderdrossel in Ungarn.] *Aquila* 78/79: 197–198.
Pérez-Chiscano, J. L. 1983. La ornitocoría en la vegetación de Extremadura. *Studia Botanica* 2: 155–168.

Perrins, C. M. 1979. *British Tits.* Collins.
Petrides, G. A. 1942. *Ilex opaca* as a late winter food for birds. *Auk* 59: 581.
Pettersson, M. 1956. Diffusion of a new habit among Greenfinches. *Nature* 177: 709–710.
Pettersson, M. 1961. The nature and spread of *Daphne*-eating in the Greenfinch, and the spread of some other habits, *Anim. Behaviour* 9: 114.
Pijl, L. van der. 1957. The dispersal of plants by bats (chiropterochory). *Acta Bot. Neerl.* 6: 291–315.
Pijl, L. van der. 1966. Ecological aspects of fruit evolution. *Proc. K. Ned. Akad. Wet. (Ser. C)* 69: 597–640.
Pijl, L. van der. 1969. *Principles of Dispersal in Higher Plants.* Springer-Verlag, Berlin.
Poole, R. W. & Rathcke, B. J. 1979. Regularity, randomness, and aggregation in flowering phenologies. *Science* 203: 470–471.
Powell, G. V. N. 1985. Sociobiology and adaptive significance of interspecific foraging flocks in the neotropics. Pp. 713–732 in *Neotropical Ornithology* (eds P. A. Buckley *et al.*). American Ornithologists' Union.
Pratt, T. K. 1984. Examples of tropical frugivores defending fruit-bearing plants. *Condor* 86: 123–129.
Prime, L. A. 1960. *Lords and Ladies.* Collins.
Pulliainen, E. 1978. The nutritive value of rowan-berries, *Sorbus aucuparia* L., for birds and mammals. *Aquilo, ser. zool.* 18: 28–32.
Pulliainen, E., Helle, P. & Tunkari, P. 1983. Adaptive radiation of the digestive system, heart and wings of *Turdus pilaris, Bombycilla garrulus, Sturnus vulgaris, Pyrrhula pyrrhula, Pinicola enucleator* and *Loxia pytyopsittacus. Orn. Fenn.* 58: 21–28.
Radford, A. P. 1967. Juvenile Song Thrush's method of eating cherry. *Brit. Birds* 60: 372.
Radford, A. P. 1968. Bullfinches eating snowberries. *Brit. Birds* 61: 270.
Radford, A. P. 1978. Blue Tits and Robin eating snowberries. *Brit. Birds* 71: 133.
Radford, A. P. 1980. Birds feeding on snowberries. *Brit. Birds* 73: 361–362.
Raven, P. H. 1973. Plant biogeography. Introduction. Pp. 211–212 in *Mediterranean Type Ecosystems.* (eds F. di Castri & H. A. Mooney). Springer-Verlag, Berlin.
Rey, E. 1910. Mageninhalt einiger Vögel. *Orn. Monatsschr.* 35: 305–313.
Ridley, H. N. 1930. *The Dispersal of Plants throughout the World.* L. Reeve & Co.
Riehm, H. 1970. Oekologie und Verhalten der Schwanzmeise (*Aegithalos caudatus* L.). *Zool. Jhrb., Syst.* 97: 338–400.
Ringleben, H. 1949. Frisst das Rotkehlchen die Früchte des Pfaffenhütchens? *Vogelwelt* 70: 49–52.
Sandring, O. 1944. Zur Nahrung der Schwanzmeise. *Orn. Monatsb.* 52: 116.
Schlegel, S. & Schlegel, J. 1965. Einige Beobachtungen zur Aufnahme pflanzlicher Nahrung durch Vögel im Erzgebirge. *Beitr. Vogelkunde* 10: 448–451.
Schmeil, O. 1911. *Lehrbuch der Botanik.* Leipzig.
Schmidt, E. 1965. Untersuchungen an einigen Holunder fressenden Singvögeln in Ungarn. *Zool. Abh.* 27: 11–28.
Schneider, W. 1957. Einige Beobachtungen über die Ernährung, besonders die Beeren- und Fruchtenahrung unserer Vögel. *Beitr. Vogelkunde* 5: 183–188.
Schuster, L. 1903. Ueber die Beerennahrung unserer Singvögeln. *Orn. Monatsschr.* 28: 114–115.
Schuster, L. 1920. Ueber die Beerennahrung unserer Singvögeln. *Orn. Monatsschr.* 45: 184–187.
Schuster, L. 1930. Ueber die Beerennahrung der Vögel. *J. Orn.* 78: 273–301.
Schüz, E. 1933. Der Massenzug des Seidenschwanzes (*Bombycilla garrula*) in Mitteleuropa 1931/32. *Vogelzug* 4: 1–21.
Siivonen, L. 1939. Zur Oekologie und Verbreitung der Singdrossel (*Turdus ericetorum philomelos* Brehm). *Ann. Zool. Soc. Zool.-Bot. Fenn. Vanamo* 7, no. 1.
Siivonen, L. 1941. Ueber die Kausalzusammenhänge der Wanderungen beim Seidenschwanz, *Bombycilla garrulus* (L.). *Ann. Zool. Soc. Zool.-Bot. Fenn. Vanamo* 8: 1–40.
Simms, E. 1978. *British Thrushes.* Collins.
Snell, H. 1866. Parallele zwischen der Vogelfauna des Taunus und der Wetterau. *Zool. Garten* 7: 201–206.
Snow, B. K. & Snow, D. W. 1971. The feeding ecology of tanagers and honeycreepers in Trinidad. *Auk* 88: 291–322.

Snow, B. K. & Snow, D. W. 1984. Long-term defence of fruit by Mistle Thrushes *Turdus viscivorus*. *Ibis* 126: 39–49.
Snow, D. W. 1958. *A Study of Blackbirds*. Allen & Unwin.
Snow, D. W. 1962. The natural history of the Oilbird, *Steatornis caripensis*, in Trinidad, W.I. Part 2. Population, breeding ecology and food. *Zoologica* 47: 199–221.
Snow, D. W. 1965. A possible selective factor in the evolution of fruiting seasons in tropical forest. *Oikos* 15: 274–281.
Snow, D. W. 1969. Some vital statistics of British Mistle Thrushes. *Bird Study* 16: 34–44.
Snow, D. W. 1971. Evolutionary aspects of fruit-eating by birds. *Ibis* 113: 194–202.
Snow, D. W. 1976. *The Web of Adaptation: bird studies in the American tropics*. Collins.
Snow, D. W. 1982. *The Cotingas*. British Museum (Natural History).
Snow, D. W. & Snow, B. K. 1986. Some aspects of avian frugivory in a north temperate area relevant to tropical forest. Pp. 159–164 in *Frugivores and Seed Dispersal* (eds A. Estrada & T. H. Fleming). Dr W. Junk, Dordrecht.
Sorensen, A. E. 1981. Interactions between birds and fruit in a temperate woodland. *Oecologia* 50: 242–249.
Sorensen, A. E. 1983. Taste aversion and frugivore preference. *Oecologia* 56: 117–120.
Sorensen, A. E. 1984. Nutrition, energy and passage time: experiments with fruit preferences in European Blackbirds (*Turdus merula*). *J. Anim. Ecol.* 53: 545–557.
Stiles, E. W. 1980. Patterns of fruit presentation and seed dispersal in bird-disseminated woody plants in the eastern deciduous forest. *Am. Nat.* 116: 670–688.
Stiles, E. W. 1982. Fruit flags: two hypotheses. *Am. Nat.* 120: 500–509.
Stiles, F. G. 1980. Ecological and evolutionary aspects of bird-flower coadaptations. *Acta XVII Int. Orn. Congr.*: 1173–1178.
Stresemann, E. 1927–34. Aves. *Handbuch der Zoologie* (eds W. Kükenthal & T. Krumbach), vol. 7, pt. 2.
Sunkel, W. 1950. Rotkehlchen und Pfaffenhütchen. *Vogelwelt* 71: 161.
Swann, R. L. 1980. Fieldfare and Blackbird weights during the winter of 1978–79 at Drumnadrochit, Inverness-shire. *Ringing and Migration* 3: 37–40.
Swann, R. L. 1983. Redwings in a highland glen. *Scot. Birds* 12: 260–261.
Szijj, J. 1957. [The food biology of the Starling and its agricultural importance.] *Aquila* 63–64: 71–101.
Thein, R. 1954. Schneeballfrüchte als Vogelnahrung. *Orn. Mitt.* 6: 233.
Tatner, P. & Bryant, D. M. 1986. Flight cost of a small passerine measured using doubly labeled water: implications for energetic studies. *Auk* 103: 169–180.
Thomas, D. K. 1979. Figs as a food source of migrating Garden Warblers in southern Portugal. *Bird Study* 26: 187–191.
Thompson, J. N. 1982. *Interaction and Coevolution*. John Wiley & Sons.
Thompson, J. N. & Willson, M. F. 1979. Evolution of temperate fruit/bird interactions: phenological strategies. *Evolution* 33: 973–982.
Turner, E. L. 1935. *Every Garden a Bird Sanctuary*. H. F. & G. Witherby.
Turček, F. 1948. A contribution to the food habits of the European Magpie (*Pica p. pica*). *Auk* 65: 297.
Turček, F. J. 1961. *Oekologische Beziehungen der Vögel und Gehölze*. Slovak. Akad. Wiss., Bratislava.
Turček, F. J. 1963. Color preference in fruit- and seed-eating birds. *Proc. XIII Int. Orn. Congr.*: 285–292.
Turček, F. J. 1968. Die Verbreitung der Vogelkirsche in den Wäldern durch Vögel. *Waldhygiene* 7: 129–132.
Tutman, I. 1950. [Stomach contents of some birds.] *Larus* 3: 353–360.
Tutman, I. 1962. Beitrag zur Kenntnis der Vogelnahrung in der Umgebung von Dubrovnik. *Larus* 14: 169–179.
Tutman, I. 1969. Beobachtungen an olivenfressenden Vögeln. *Vogelwelt* 90: 1–8.
Tye, A. 1982. Social organisation and feeding in the Wheatear and Fieldfare. Ph.D. thesis, Cambridge University.
Tye, A. 1986. Economics of experimentally-induced territorial defence in a gregarious bird, the Fieldfare *Turdus pilaris*. *Ornis Scand.* 17: 151–164.
Tyrväinen, H. 1975. The winter irruption of the Fieldfare *Turdus pilaris* and the supply of rowan-berries. *Orn. Fenn.* 52: 23–31.

Vertse, A., Zsak, Z. & Kaszab, Z. 1955. [Food and agricultural importance of the Partridge in Hungary.] *Aquila* 59–62: 13–68.
Volkmann, G. 1957. Zu: Stare füttern Junge mit Kirschen. *Orn. Mitt.* 9: 34.
Wachs, Dr. 1922. Experimente zum Vogelzug. *Orn. Monatsschr.* 47: 57–59.
Wahn, R. 1950. Beobachtungen und Gedanken am Nest der Mönchsgrasmücke. *Vogelwelt* 71: 33–39.
Walsberg, G. E. 1975. Digestive adaptations of *Phainopepla nitens* associated with the eating of mistletoe berries. *Condor* 77: 169–174.
Watt, B. K. & Muriel, A. L. 1963. *Composition of Foods.* Agric. Handbook No. 8, U.S. Dept. Agriculture, Washington, D.C.
Wetmore, A. 1914. The development of the stomach in the euphonias. *Auk* 31: 458–461.
Wetzel, R. 1878. Beobachtungen über die Zaungrasmücke (*Sylvia curruca*). *Orn. Monatsschri.* 3: 114–116.
Wheelwright, N. T. 1983. Fruits and the ecology of Resplendent Quetzals. *Auk* 100: 286–301.
Wheelwright, N. T. 1986. The diet of American Robins: an analysis of the U.S. Biological Survey records. *Auk* 103: 710–725.
Wheelwright, N. T. & Orians, G. H. 1982. Seed dispersal by animals: contrasts with pollen dispersal, problems of terminology, and constraints on coevolution. *Am. Nat.* 119: 402–413.
White, G. 1789. *The Natural History of Selborne.*
Wijngaarden, P. 1986. Sociaal gedrag van de Grote Lijster in de winter. *Aythya* 25: 44–46.
Williamson, R. 1978. *The Great Yew Forest: the natural history of Kingley Vale.* Macmillan.
Witherby, H. F., Jourdain, F. C. R., Ticehurst, N. F. & Tucker, B. W. 1938. *The Handbook of British Birds*, vols 1 and 2. H. F. & G. Witherby.
Willson, M. F. & Hoppes, W. G. 1986. Foliar 'flags' for avian frugivores: signal or serendipity? Pp. 55–69 in *Frugivores and Seed Dispersal* (eds A. Estrada & T. H. Fleming). Dr W. Junk, Dordrecht.
Willson, M. F. & Melampy, M. N. 1983. The effect of bicolored fruit displays on fruit removal by avian frugivores. *Oikos* 41: 27–31.
Willson, M. F. & Thompson, J. N. 1982. Phenology and ecology of color in bird-dispersed fruits, or why some fruits are red when they are 'green'. *Can. J. Bot.* 60: 701–713.
Wood, C. A. 1924. The Polynesian fruit pigeon, *Globicera pacifica*, its food and digestive apparatus. *Auk* 41: 433–438.
Zahavi, A. 1971. The social behaviour of the White Wagtail *Motacilla alba* wintering in Israel. *Ibis* 113: 203–211.
Zeddelof, B. zu. 1921. Vom grauen Fliegenschnäpper. *Orn. Monatsschr.* 46: 142.
Zedler, W. 1954. Zur Beerennahrung einiger Vögel im Herbst und Winter. *Orn. Mitt.* 6: 232.

Index

Entries in bold type indicate the main sections for species or topics

Accipter nisus see Sparrowhawk
Acrocephalus schoenobaenus
  eating fruit 167
Acrocephalus scirpaceus
  eating fruit 167
Aegithalos cadatus **187**
Amelanchier laevis
  fruit eaten by birds 104–5
Ampelopsis
  fruit eaten by Starling 162
Apple, crab- **64–5**
  size of fruit 189, 190
Apple, cultivated
  eaten by birds 102

Araçari, Collared
  method of taking fruit 211
Arbutus unedo
  fruit eaten by Song Thrush 124
  fruit eaten by Jay 164–5
Arum maculatum **98–9**
Atropa belladonna see deadly nightshade
Avocado pear
  poison in seed 208

Bellbird
  bill-shape and frugivory 216
  snail-eating 231
  time spent in display 235

Berberis darwinii
  fruit eaten by birds 105, 108
Berry
  terminology 15–16, 189
Bilberry
  eaten by birds 124, 137
Blackberry 60–2
Blackbird **111–19**
  defence of fruit 42, 114–15
  taste preferences 197
  fruit preferences 220–1
  captive experiments 232–3
  daily energy budget 236
Blackcap **147–53**
  defence of mistletoe fruit 39
  method of taking fruit 212, 213
  predominantly frugivorous in winter 231, 232
  captive experiments 232–4
  variety of fruits in diet 233–4
Blackthorn **58–9**
Bombycilla garrulus see Waxwing
Bramble **60–2**
Brambling **180–1**
Byronia alba 91
Byronia dioica **90–1**
Bryony, black **92–3**
  seed defences 209
Bryony, White **90–1**
Buckthorn **79–81**
Buckthorn, alder
  fruit eaten by Cuckoo 167
Buckthorn, sea
  fruit eaten by Fieldfare 137
  fruit eaten by Common Whitethroat 156
  fruit eaten by Magpie 164
Bullace
  size of fruit 58
Bullfinch **176–8**

Capercaillie
  eating whitebeam fruit 53
Caprifoliaceae
  vulnerability to seed-predators 208, 209
Carduelis chloris see Greenfinch
Celtis
  fruit eaten by Starling 162
Chaffinch **180–1**
Cherry, bird **56–7**
Cherry, cultivated
  fruit eaten by Starlings 161
  fruit eaten by Black-headed Gull 56, 167

Cherry, wild **55–6**
  seed size 191
Chiffchaff
  eating fruit 154, 167
Cisticola
  not a frugivore 216
Cloudberry
  eaten by Magpie 164
Cochoa
  bill-shape and frugivory 216
Cock-of-the-rock, Andean
  feeding vertebrates to young 231
Coevolution **239–40**
Columba
  in Canary Islands 174
Columba palumbus see Woodpigeon
Cornus
  fruit characters 241
Cornus sanguinea **75–6**
Cornus stolonifera
  fruit fed to young flycatchers 157
Corvus corone **165–6**
Cotinga
  specialists on Lauraceous fruit 208
Cotoneaster **105–6**
  ornamental species 104, 105–6
Cowberry
  eaten by birds 124, 137
Crab-apple **64–5**
  size of fruit 189
Crataegus
  ornamental species 104, 106–7
  fruit characters 241
  recent speciation 242
Crataegus laevigata
  presence in study area 45
  fruit ripening season 201
Crataegus monogyna **45–8**
  fruit ripening season 201
Crataegus oxyacantha 48
Crossbill
  eating rowan fruit 51
Crossbill, Parrot
  digestive system 217
Crow, Carrion, **165–6**
Crowberry
  eaten by birds 124, 137, 167
Cuckoo
  eating fruit 97, 167
Currant, mountain
  eaten by birds 12, 156
Currant, wild **84**

Curant, cultivated
  red fruit preferred to white 196
Cyphomendra betacea
  ultra-violet reflectance of fruit 196

Daphne laureola see spurge laurel
Daphne mezereum see mezereon
Darwin, Charles
  on fruit colours 194
Deer, Fallow
  eating lords and ladies 98–9
Dendrocopos major see Greater Spotted Woodpecker
Dewberry
  eaten by birds 100–1
Dispersers
  definition 11
Dogwood **75–6**
Drupe
  terminology 16, 189
Duke of Argyll's tea tree
  fruit eaten by Blackcap 151
Dunnock
  eating fruit 97

Elaeagnus
  fruit eaten by Starlings 162
Elder **65–8**
Elder, dwarf 68
Elder, red 67–8
Erithacus rubecula see Robin
Euonymus europaeus **81–3**
Euphonia
  as mistletoe fruit specialists 217
Euphonia gouldi
  rapid regurgitation of seeds 217

Falco tinnunculus
  attacking fruit-eating birds 228
Fig
  eaten by birds 155, 164
Fieldfare **133–7**
  defence of rose hips 42
  defence of haws 47
  digestive system 217
  fruit preferences 222–3
  exclusive fruit diet in winter 231
Ficedula hypoleuca
  eating fruit 101, 167
Flower-pecker
  as mistletoe fruit specialist 217

Flycatcher, Pied
  eating fruit 101, 167
Flycatcher, Spotted **157**
  eating guelder rose fruit 71
  eating currant 84
Fragaria vesca **100**
  seed size 191
Fragaria virginiana
  seed dispersal 100
Fringilla coelebes **180–1**
Fringilla montifringilla **180–1**
Fruit
  nutritive quality 14, 204–5
  seed burden 14
  relative yield 14
  terminology 15–16
  ripening seasons in England 18–21
  physical characters **189–93**
  colour and taste **194–8**
  ripening seasons in Europe **202–5**
  preferences by birds **219–26**
  defence of fruit supply **229–30**
Fruit-crow, Purple-throated
  method of taking fruit 211
Fruit-pigeon
  digestive system 217

Gallinula chloropus
  eating fruit 67, 166
Garrulus glandarius **164–5**
Gooseberry, wild **85**
Grape
  eaten by Song Thrush 124
  eaten by Magpie 165
  u/v reflectance 196
Greenfinch, **178–80**
  possible immunity to seed poisons 208
Grosbeak, Pine
  digestive system 217
Grouse, Hazel
  eating mezereon fruit 95
Guelder rose **70–1**
Gull, Black-headed
  eating cherries 56, 167
Gull, Common
  eating fruit 167
Gull, Herring
  eating *Vaccinium* berries 167

Haws **45–8**

Hawfinch
    eating various seeds 48, 56, 59, 71, 79, 83, 94
Hedera helix see ivy
Hippolais
    bill-shape and frugivory 216
Holly **22–28**
    in Cantabrian Mountains 141
    seed-defences 208
Honeyguide, Orange-rumped
    defence of honeycombs 229
Honeysuckle **72–74**

Ilex
    conservatism of fruit characters 190, 241
Ilex aquifolium see holly
Indicator xanthonotus
    defence of honeycombs 229
Iris, stinking **101**
Ivy **31–34**
    fruit eaten by Carrion Crow 166
    pulp:seed ratio 192

Jay **164–5**
Juniper **96–8**
    eaten fruit by Song Thrush 124
    fruit eaten by Mistle Thrush 132
Juniperus oxycedrus 97
Juniperus phoenicea 97–8

Kestrel
    attacking fruit-eating birds 228

Larus spp.
    eating fruit 56, 167
Lauraceae
    in Canary Islands 174
    seed defences 208
Laurus
    conservatism of fruit characters 241
Laurus nobilis
    fruit eaten by Blackcap 153
Ligustrum vulgare **77–9**
Locustella
    bill-shape and frugivory 216
Locustella naevia
    eating fruit 167
Lonicera **72–4**

Loranthus europaeus
    fruit eaten by Fieldfare 231
Lords and ladies **98–9**
    seed defences 209
Luscinia megarhynchos
    eating fruit 84, 167
Loxia pytyopsittacus
    digestive system 217
Lycium barbarum
    fruit eaten by Blackcap 151

Madder
    fruit eaten by Robin 146
Magpie **163–4**
Mahonia aquifolium
    fruit eaten by birds 105, 108
Malus sylvestris see crab-apple
Manakin
    taste-discrimination 198
    rapid regurgitation of seeds 217
Manakin, Black-and-white
    time spent in display 235
Mespil, snowy
    fruit eaten by birds 104
Mezereon **95**
    fruit eaten by Lesser Whitethroat 157
Miconia
    fruit seasons in tropics 199–200
Mimus polyglottos see Mockingbird, Northern
Mistletoe **35–40**
    fruit eaten by Fieldfare 137, 231
    seeds eaten by Blue Tit 185
Mockingbird, Northern
    defence of fruit supply 230
    fruit fed to young 234
Monticola solitarius
    eating spurge laurel fruit 94
Moorhen
    eating fruit 67, 166
Motacilla alba
    defence of food supply 229
Mouse, field
    eating hips 44, 45
    eating juniper seeds 97
    eating acorns 146
Muntjac
    eating lords and ladies 98–9
Muscicapa striata **157**
Myiophonus
    bill-shape and frugivory 216

Myrtus communis
  fruit eaten by Song Thrush 124

Nightingale
  eating fruit 84, 167
Nightshade, deadly **89**
  fruit eaten by Blackcap 152
  size of fruit 189
Nightshade, woody **86–8**

Oak, evergreen
  Robin feeding on acorn pulp 146
Oenanthe oenanthe
  eating fruit 167
Oilbird
  exclusive fruit diet of young 234
Olive (*Olea* spp.)
  fruit eaten by birds 119, 124, 153, 161, 184
Ornamental fruits
  exploited by thrushes 104–5
Ouzel, Ring **166**
  eating juniper fruit 98

Palm
  seed defences 208
Palm, date
  fruit eaten by Blackcap 152
Parakeet, Rose-ringed
  eating holly seeds 208 (footnote)
Parrot
  as seed-predators 208, 209
Partridge, Grey
  eating fruit 88
Parus caeruleus **184–5**
Parus major **183–4**
Parus montanus **186**
Parus palustris **185–6**
Pear, cultivated
  eaten by birds 102
Passer domesticus
  eating elderberries 182
Passer montanus
  eating elderberries 182
Phainopepla
  rapid regurgitation of seeds 217
Pheasant
  eating fruit 89, 95

Phoenicurus phoenicurus
  eating fruit 167
Phoenicurus ochruros
  eating fruit 167
Phoenix
  fruit eaten by Blackcap 152
Phylloscopus
  bill-shape and frugivory 216
Phylloscopus collybita
  eating fruit 167
Phylloscopus sibilatrix
  eating fruit 167
Phylloscupus trochilus
  eating fruit 167
Phytolacca decandra
  fruit eaten by Song Thrush 124
Pica pica **163–4**
Picus viridis
  eating dogwood fruit 75, 76, 166
Pigeons
  in Canary Islands 174
Pinicola enucleator
  digestive system 217
Pistacia lentiscus
  fruit eaten by birds 124, 146
Polygonatum multiflorum
  fruit eaten by Blackbird 101
Polyscias fulva
  fruit eaten by Blackcap 153
Predation
  effect on fruit foraging 228–9
Privet **77–9**
Procnias
  bill-shape and frugivory 216
Procnias averano see Bellbird
Prunus
  ornamental species 107
  pulp:seed ratios 192
  fruit ripening seasons 200
  seed defences 208
  recent speciation 242
Prunus avium see cherry, wild
Prunus cerasus 166
Prunus domestica 58
Prunus padus **56–7**, 107
Prunus spinosa **58–9**
Psittacula krameri
  eating holly seeds 209 (footnote)
Pteroglossus torquatus
  method of taking fruit 211
Pulp-predators 11, 175, **183–7**

Pyracantha **106**
    fruit eaten by birds 104, 106, 115
Pyrrhula pyrrhula **176–8**
Pyrus bourgeana
    fruit eaten by Jay 165

Quercus ilex
    Robin feeding on acorn pulp 146
Querula purpurata
    method of taking fruit 211
Quetzal
    specialist on lauraceous fruit 208

Raspberry, wild **63**
Redstart, Black
    eating fruit 68, 84, 167
Redstart, Common
    eating fruit 167
Redwing **138–41**
    method of taking fruit 213
    fruit preferences 222
Rhamnus catharticus **79–81**
Rhamnus frangula
    fruit eaten by Cuckoo 167
Ribes alpinum
    fruit eaten by birds 124, 156
Ribes rubrum **84**
Ribes uva-crispi **85**
Ring Ouzel **166**
    eating juniper fruit 98
Robin **142–6**
    method of taking fruit 213
    captive experiments 232–3
Rosa **44–5**
    fruit ripening seasons 200
    fruit characters 241
    recent speciation 242
Rosa arvensis **44**
Rosa canina see rose, dog
Rosa rubiginosa **44**
Rosa rubrifolia
    fruit eaten by birds 107
Rosa rugosa
    fruit eaten by Garden Warbler 155
Rosaceae
    vulnerability to seed-predators 208
Rose, dog **41–3**
    seed dispersal 44–5
Rose, field **44**

Rowan **49–51**
    effect on Fieldfare movement 137
    fruit eaten by migrating Redwings 141
Rubia peregrina
    fruit eaten by Robin 146
Rubus
    fruit ripening seasons 200
Rubus caesius
    fruit eaten by birds 100–1
Rubus fruticosus **60–2**
Rubus idaeus **63**
Rupicola peruviana
    feeding vertibrates to young 231

Sambucus ebulus
    fruit eaten by birds 68
Sambucus nigra **65–8**
Sambucus racemosa
    fruit eaten by birds 67–8, 155
Saxicola rubetra
    eating fruit 167
Seed
    terminology 16
    defences against predation 207–9
Seed-predator **206–9**
    definition 11
Sloe **58–9**
Smilax aspera
    seed number 193, 199
Snowberry
    fruit colour 104
    fruit eaten by Bullfinch 178
    fruit eaten by Greenfinch 180
Solanum
    fruit ripening seasons 201
Solanum dulcamara **86–8**
Solanum nigrum **88**
Solomon's seal
    fruit eaten by Blackbird 101
Sorbus
    fruit ripening seasons 201
    recent speciation 242
    white-fruited species 104
Sorbus aria **51–3**
Sorbus aucuparia see rowan
Sorbus torminalis **54**
Sparrow, House **182**
    eating elderberries 67
Sparrow, Tree **182**
    eating elderberries 182

Sparrowhawk
  attacking fruit-eating birds 228, 229
Spindle **81–3**
  seed defences 208
Spurge laurel **94**
Starling **158–62**
  method of taking fruit 213–14
  digestive system 217
  fruit preferences **223**
Strawberry, cultivated
  eaten by birds 102
Strawberry, wild
  eaten by birds 100
  seed size 191
Strawberry tree
  fruit eaten by Jay 164
Sturnus vulgaris see Starling
Sweet-briar **44**
Sylvia
  fruit-eating related to size 215
  fruit-eating related to bill-shape 215
  digestive system 217
Sylvia atricapilla see Blackcap
Sylvia borin **154–6**
Sylvia communis **156**
Sylvia curruca **156–7**
Sylvia undata
  degree of frugivory 216
Symphoricarpus rivularis see snowberry

Tamus communis **92–3**
Tanager
  taste discrimination 197–8
  method of taking fruit 211
Taxus baccata see yew
Thrush
  method of taking fruit 211, 212
  degree of frugivory related to size 216
Thrush, Blue Rock
  eating spurge laurel fruit 94
Thrush, Mistle **125–33**
  defence of holly 22
  defence of yew 29
  defence of mistletoe 36, 39
  fruit preferences 222
  defence of fruit supply 229
  physical characters related to fruit defence 230
  daily energy budget 236–7
Thrush, Song **120–4**
  taste preferences 196–7
  fruit preferences 221–2

Tit, Blue **184–5**
Tit, Great **183–4**
Tit, Long-tailed **187**
Tit, Marsh **185–6**
Tit, Willow **186**
Toucan
  method of taking fruit 211
Trogon, Slaty-tailed
  method of taking fruit 211
Turdus
  bill-shape and frugivory 216
Turdus iliacus see Redwing
Turdus merula see Blackbird
Turdus philomelos see Thrush, Song
Turdus pilaris see Fieldfare
Turdus torquatus see Ring Ouzel
Turdus viscivorus see Thrush, Mistle

Ultra-violet
  reflectance by fruits 196

Vaccinium
  fruit eaten by Herring Gull 167
Viburnum
  fruit ripening seasons 201
Viburnum lantana **68–9**
Viburnum opulus **70–1**
Viburnum tinus
  fruit eaten by Robin 146
Viscum album see mistletoe
Vitis vinifera
  ultra-violet reflectance of fruit 196

Wagtail, White
  defence of food supply 229
Warbler, Dartford
  degree of frugivory 216
Warbler, Garden **154–5**
  captive experiments 232–3
Warbler, Grasshopper
  eating fruit 167
Warbler, Reed
  eating fruit 167
Warbler, Sedge
  eating fruit 167
Warbler, Willow
  eating fruit 84, 99, 154, 167
Warbler, Wood
  eating fruit 68, 84, 167

Waxwing **168–70**
  digestive system 217
  daily energy budget 237
Wayfaring tree **68–9**
Wheatear
  eating fruit 167
Whinchat
  eating fruit 101, 167
White, Gilbert
  observation on fruit colour 194, 196
Whitebeam **51–3**
Whitethroat, Common **156**
Whitethroat, Lesser **156–7**
Wild service tree **54**
Wild strawberry **100**

Woodpecker, Greater Spotted
  eating rowan fruit 51
  eating cherries 166
Woodpecker, Green
  eating dogwood fruit 76, 166
Woodpigeon **171–4**
  immunity to ivy poison 208

Yew **28–31**
  conservatism of fruit characters 189, 240
  seed defences 209

Zoothera
  bill-shape and frugivory 216